Published for
OXFORD INTERNATIONAL AQA EXAMINATIONS

International AS Level
MATHEMATICS

Sue Chandler
Janet Crawshaw
Joan Chambers

Great Clarendon Street, Oxford, OX2 6DP, United Kingdom

Oxford University Press is a department of the University of
Oxford. It furthers the University's objective of excellence in
research, scholarship, and education by publishing worldwide.
Oxford is a registered trade mark of Oxford University Press in
the UK and in certain other countries

British Library Cataloguing in Publication Data
Data available

978-0-19-837596-8

17

Paper used in the production of this book is a natural,
recyclable product made from wood grown in sustainable
forests. The manufacturing process conforms to the
environmental regulations of the country of origin.

Printed and bound by CPI Group (UK) Ltd, Croydon, CR0 4YY

Acknowledgements
The publishers would like to thank the following for
permissions to use their photographs:

Cover: Colin Anderson/Getty Images.

Header: Shutterstock.

Although we have made every effort to trace and contact all
copyright holders before publication this has not been possible
in all cases. If notified, the publisher will rectify any errors or
omissions at the earliest opportunity.

Links to third party websites are provided by Oxford in
good faith and for information only. Oxford disclaims any
responsibility for the materials contained in any third party
website referenced in this work.

AQA material is reproduced by permission of AQA.

Contents

About this book v

1 Expanding Brackets, Surds and Indices

1.1 Algebraic expressions 2
1.2 Expansion of Two Brackets 3
1.3 Square roots and other roots 6
1.4 Surds 7
1.5 Indices 10
Summary 14
Review 14
Assessment 15

2 Quadratic Polynomials and Equations

2.1 Quadratic polynomials 16
2.2 Quadratic equations 19
2.3 Solution by completing the square 22
2.4 The formula for solving a quadratic equation 23
2.5 Properties of the roots of a quadratic equation 24
2.6 Simultaneous equations 27
Summary 29
Review 29
Assessment 30

3 Algebraic Division

3.1 Division of a polynomial by $x - a$ 32
3.2 The remainder theorem and the factor theorem 33
3.3 The factors of $a^3 - b^3$ and $a^3 + b^3$ 35
Summary 36
Review 37
Assessment 37

4 Functions and Graphs

4.1 Functions 40
4.2 Graphical interpretation of equations 43
4.3 Inequalities 46
4.4 Transformations of graphs 48
Summary 52
Review 52
Assessment 53

5 Coordinate Geometry

5.1 Lines joining two points 56
5.2 Gradient 59
5.3 The equation of a straight line 61
5.4 Intersection 66
Summary 67
Review 68
Assessment 68

6 Differentiation

6.1 Chords, tangents, normals and gradients 70
6.2 Differentiation 72
6.3 Gradients of tangents and normals 75
6.4 Increasing and decreasing functions 78
6.5 Stationary points 79
6.6 Maximum and minimum points 80
Summary 85
Review 86
Assessment 87

7 Integration

7.1 Indefinite integration 90
7.2 Using integration to find an area 92
7.3 The trapezium rule 97
Summary 99
Review 99
Assessment 100

8 Sequences and Series

8.1 Defining a sequence 102
8.2 Series 105
8.3 Arithmetic series 107
8.4 Geometric series 110
8.5 The binomial theorem 115
Summary 120
Review 120
Assessment 121

9 Coordinate Geometry and Circles

9.1 The equation of a circle 124
9.2 Geometric properties of circles 127
9.3 Tangents to circles 128
Summary 131
Review 131
Assessment 132

10 Trigonometry

10.1 Trigonometric ratios of acute angles 134
10.2 The sine rule and cosine rule 137
10.3 The area of a triangle 144
Summary 145
Review 145
Assessment 146

11 Trigonometric Functions and Equations

11.1 Angle units 148
11.2 The length of an arc 150
11.3 The area of a sector 152
11.4 The trigonometric functions 154
11.5 Solving trigonometric equations 159
Summary 162
Review 163
Assessment 163

12 Exponentials and Logarithms

12.1 Exponential functions 166
12.2 Logarithms 168
12.3 The laws of logarithms 170
12.4 Equations containing logarithms or x
    as a power 172
Summary 173
Review 173
Assessment 174

13 Probability

13.1 Introduction to probability 176
13.2 Combined events 182
13.3 Conditional events 191
13.4 Tree diagrams 200
13.5 Further applications 209
Summary 212
Review 214
Assessment 217

14 Discrete Random Variables

14.1 Discrete variables 220
14.2 Discrete random variables 229
14.3 E(X), the expectation of X 235
14.4 The variance and standard deviation of X ... 243
14.5 Sum or difference of two independent random
    variables 249
14.6 Further applications 255
Summary 258
Review 260
Assessment 263

15 Bernoulli and Binomial Distributions

15.1 The Bernoulli distribution 266
15.2 The binomial distribution 269
15.3 The cumulative binomial distribution
    function 276
15.4 Mean, variance and standard deviation of a
    binomial distribution 283
15.5 Further applications 287
Summary 290
Review 291
Assessment 293

16 Displacement, Speed, Velocity and Acceleration

16.1 Displacement, speed, velocity and
    acceleration 296
16.2 Displacement-time and velocity-time graphs ... 300
Summary 308
Review 309
Assessment 310

17 Motion in a Straight Line

17.1 Equations of motion with constant
    acceleration 312
17.2 Free fall motion under gravity 317
17.3 Motion in a straight line with variable
    acceleration 321
Summary 324
Review 325
Assessment 326

18 Forces and Newton's Laws

18.1 Types of force 328
18.2 Newton's first law of motion 331
18.3 Connected particles 340
18.4 Dynamic friction 343
Summary 346
Review 347
Assessment 347

19 Momentum and Impulse

19.1 Momentum and impulse 350
19.2 Conservation of linear momentum 355
Summary 358
Review 358
Assessment 359

Glossary 362
Answers 366
Index 386

About this book

This book has been specially created for the Oxford AQA International AS Level Mathematics examination (9660).

It has been written by an experienced team of teachers, consultants and examiners and is designed to help you obtain the best possible grade in your maths qualification.

In each chapter the lessons are organised in a logical order to help you to progress through each topic. At the start of each chapter you can see an Introduction, to show you how you will use the knowledge in this chapter, a Recap of what prior knowledge you will need to recall and a clear list of the Objectives that you will fulfil by the end of the chapter.

The Note boxes give you help and support as you work through the examples and exercises.

Clear, worked examples show you how to tackle each question and the steps needed to reach the answer.

Key points are in bold and the chapter colour to make it clear that this information is important.

Exercises allow you to apply the skills that you have learned, and give the opportunity to practice your reasoning and problem solving abilities.

At the end of a chapter you will find a summary of what you have learned, together with a review section that allows you to test your fluency in the basic skills. Finally there is an Assessment section where you can practise exam-style questions.

At the end of the book you will find a comprehensive glossary of key phrases and terms and a full set of answers to all of the exercises.

We wish you well with your studies and hope that you enjoy this course and achieve exam success.

1 Expanding Brackets, Surds and Indices

Introduction

Working with algebraic expressions is needed in any mathematics course beyond GCSE. This chapter gives the facts and the practice necessary for you to develop these skills.

Recap

You need to remember...

▶ An algebraic expression does not contain an equals sign but an equation does contain an equals sign.
  For example, $3x^2 - 4$ is an expression, $2x + 6y = 7$ is an equation.
▶ If a string of numbers and letters are multiplied, the multiplication can be done in any order. For example,
  $2p \times 3q = 2 \times p \times 3 \times q = 2 \times 3 \times p \times q = 6pq$
▶ $x^2 = x \times x$, $x^5 = x \times x \times x \times x \times x$, etc.

Objectives

By the end of this chapter, you should know how to...

▶ Identify like and unlike terms.
▶ Expand brackets.
▶ Explain the meaning of a surd.
▶ Simplify an expression containing surds.
▶ Work with numbers in index form.

1.1 Algebraic expressions

The **terms** in an algebraic expression are the parts separated by a plus or minus sign.

Like terms contain the same letters to the same powers; like terms can be added or subtracted.

For example, $2ab$ and $5ab$ are like terms and can be added,

so $\qquad\qquad 2ab + 5ab = 7ab$

Unlike terms contain different letters; they cannot be added or subtracted.

For example, ab and ac are unlike terms because they contain different letters. Also x^2 and x^3 cannot be added because they are to different powers.

Example 1

Question

Simplify $5x - 3(4 - x)$

Answer

$5x - 3(4 - x) = 5x - 12 + 3x$
$\qquad\qquad\qquad = 8x - 12$

Note

Remember that $-3(4 - x)$ means 'multiply every term inside the bracket by minus 3. Remember that $(-3) \times (-x) = +3x$.

Coefficients

We can identify a term in an expression by using the letter, or combination of letters, involved.

For example
$2x^2$ is 'the term in x^2,

$3xy$ is 'the term in xy'.

The number (including its sign) in front of the letters is called the **coefficient**.

For example in the term $2x^2$, 2 is the coefficient of x^2

in the term $3xy$, 3 is the coefficient of xy.

If no number is written in front of a term, the coefficient is 1 or −1, depending on the sign of the term.

For example, in the expression $x^3 + 5x^2y - y^3 + 2$

the coefficient of x^3 is 1

the coefficient of x^2y is 5

the coefficient of y^3 is − 1.

The term 2 has no variable. This is called a constant term.

Exercise 1

Simplify.

1 $2x^2 - 4x + x^2$

2 $5a - 4(a + 3)$

3 $2y - y(x - y)$

4 $8pq - 9p^2 - 3pq$

5 $4xy - y(x - y)$

6 $x^3 - 2x^2 + x^2 - 4x + 5x + 7$

7 $t^2 - 4t + 3 - 2t^2 + 5t + 2$

8 $2(a^2 - b) - a(a + b)$

9 $3 - (x - 4)$

10 $5x - 2 - (x + 7)$

11 $3x(x + 2) + 4(3x - 5)$

12 $a(b - c) - c(a - b)$

> **Note**
>
> $-(x - 4)$ means $-1(x - 4)$.

13 $2ct(3 - t) + 5t(c - 11t)$

14 $x^2(x + 7) - 3x^3 + x(x^2 - 7)$

15 $(3y^2 + 4y - 2) - (7y^2 - 20y + 8)$

16 Write down the coefficient of x in $x^2 - 7x + 4$.

17 What is the coefficient of xy^2 in the expression $y^3 + 2xy^2 - 7xy$?

18 For the expression $x^3 - 3x + 7$, write down the coefficient of

    a x^3 **b** x^2 **c** x

1.2 Expansion of Two Brackets

Expanding brackets means multiplying them out to remove the brackets.

To expand $(2x + 4)(x - 3)$ each term in the first bracket is multiplied by each term in the second bracket.

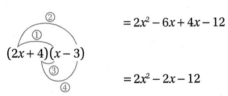

$= 2x^2 - 6x + 4x - 12$

$= 2x^2 - 2x - 12$

Exercise 2

Expand and simplify.

1. $(x+2)(x+4)$
2. $(x+5)(x+3)$
3. $(a+6)(a+7)$
4. $(t+8)(t+7)$
5. $(s+6)(s+11)$
6. $(2x+1)(x+5)$
7. $(5y+3)(y+5)$
8. $(2a+3)(3a+4)$
9. $(7t+6)(5t+8)$
10. $(11s+3)(9s+2)$
11. $(x-3)(x-2)$
12. $(y-4)(y-1)$
13. $(a-3)(a-8)$
14. $(b-8)(b-9)$
15. $(p-3)(p-12)$
16. $(2y-3)(y-5)$
17. $(x-4)(3x-1)$
18. $(2r-7)(3r-2)$
19. $(4x-3)(5x-1)$
20. $(2a-b)(3a-2b)$
21. $(x-3)(x+2)$
22. $(a-7)(a+8)$
23. $(y+9)(y-7)$
24. $(s-5)(s+6)$
25. $(q-5)(q+13)$
26. $(2t-5)(t+4)$
27. $(x+3)(4x-1)$
28. $(2q+3)(3q-5)$
29. $(x+y)(x-2y)$
30. $(s+2t)(2s-3t)$

Difference of two squares

The expansion of $(x-4)(x+4)$ is a special case.

$$(x-4)(x+4) = x^2 - 4x + 4x - 16$$
$$= x^2 - 16$$

Any expansion of the form $(x+b)(x-b) = x^2 - b^2$ is known as the difference of two squares.

Squares

$(2x+3)^2$ means $(2x+3)(2x+3)$

$\therefore \qquad (2x+3)^2 = (2x+3)(2x+3)$
$$= (2x)^2 + (2)(2x)(3) + (3)^2$$
$$= 4x^2 + 12x + 9$$

In general, $\quad (ax+b)^2 = a^2x^2 + (2)(ax)(b) + b^2$
$$= a^2x^2 + 2abx + b^2$$

and $\qquad (ax-b)^2 = a^2x^2 - 2abx + b^2$

Exercise 3

Expand and simplify.

1. $(x-2)(x+2)$
2. $(5+x)(5-x)$
3. $(x+3)(x-3)$
4. $(2x-1)(2x+1)$
5. $(x-8)(x+8)$
6. $(x-a)(x+a)$
7. $(x-1)(x+1)$
8. $(3b+4)(3b-4)$
9. $(2y-3)(2y+3)$
10. $(ab+6)(ab-6)$
11. $(5x+1)(5x-1)$
12. $(xy+4)(xy-4)$

> **Note**
>
> Questions 1 to 6 show that the expansion of $(ax+b)(ax-b)$ can be written down directly, so $(ax+b)(ax-b) = a^2x^2 - b^2$. Use this result to expand the brackets in Questions 7 to 12.

Expand.

13. $(x+4)^2$
14. $(x+2)^2$
15. $(2x+1)^2$
16. $(3x+5)^2$
17. $(2x+7)^2$
18. $(x-1)^2$
19. $(x-3)^2$
20. $(2x-1)^2$
21. $(4x-3)^2$
22. $(5x-2)^2$
23. $(3t-7)^2$
24. $(x+y)^2$
25. $(2q+9)^2$
26. $(3q-11)^2$
27. $(2x-5y)^2$
28. Expand and simplify $(x-2)^2(3x-4)$. Write down the coefficients of x^2 and x.

Important expansions

These general results should be memorised.

$(ax + b)^2 = a^2x^2 + 2abx + b^2$

$(ax - b)^2 = a^2x^2 - 2abx + b^2$

$(ax + b)(ax - b) = a^2x^2 - b^2$

The next exercise has different expansions including some given above.

Example 2

Question

Expand $(4p + 5)(3 - 2p)$

Answer

$$\begin{aligned} (4p + 5)(3 - 2p) &= (5 + 4p)(3 - 2p) \\ &= 15 - 10p + 12p - 8p^2 \\ &= 15 + 2p - 8p^2 \end{aligned}$$

Harder expansions

Expanding expressions such as $(x - 2)(x^2 - x + 5)$ should be done systematically.

First multiply each term of the second bracket by x, writing down the separate results as they are found. Then multiply each term of the second bracket by -2. Do not collect like terms at this stage.

$$(x - 2)(x^2 - x + 5)$$
$$= x^3 - x^2 + 5x - 2x^2 + 2x - 10$$

Now collect like terms

$$= x^3 - 3x^2 + 7x - 10$$

Example 3

Question

Expand $(x + 2)(2x - 1)(x + 4)$

Answer

$$\begin{aligned} (x + 2)(2x - 1)(x + 4) &= (x + 2)(2x^2 + 7x - 4) \\ &= 2x^3 + 7x^2 - 4x + 4x^2 + 14x - 8 \\ &= 2x^3 + 11x^2 + 10x - 8 \end{aligned}$$

> **Note**
>
> First expand the last two brackets.

Exercise 4

Expand.

1. $(2x - 3)(4 - x)$
2. $(x - 7)(x + 7)$
3. $(6 - x)(1 - 4x)$
4. $(7p + 2)(2p - 1)$
5. $(3p - 1)^2$
6. $(5t + 2)(3t - 1)$
7. $(4 - p)^2$
8. $(4t - 1)(3 - 2t)$
9. $(x + 2y)^2$
10. $(4x - 3)(4x + 3)$
11. $(3x + 7)^2$
12. $(R + 3)(5 - 2R)$
13. $(a - 3b)^2$
14. $(2x - 5)^2$
15. $(7a + 2b)(7a - 2b)$
16. $(3a + 5b)^2$

Expand and simplify

17 $(x-2)(x^2+x+1)$

18 $(3x-2)(x^2-x-1)$

19 $(2x-1)(2x^2-3x+5)$

20 $(x-1)(x^2-x-1)$

21 $(2x+3)(x^2-6x-3)$

22 $(x+1)(x+2)(x+3)$

23 $(x+4)(x-1)(x+1)$

24 $(x-2)(x-3)(x+1)$

25 $(x+1)(2x+1)(x+2)$

26 $(x+2)(x+1)^2$

27 $(2x-1)^2(x+2)$

28 $(3x-1)^3$

29 $(4x+3)(x+1)(x-4)$

30 $(x-1)(2x-1)(2x+1)$

31 $(2x+1)(x+2)(3x-1)$

32 $(x+1)^3$

33 $(x-2)(x+2)(x+1)$

34 $(x+3)(2x+3)(x-1)$

35 $(3x-2)(2x+5)(4x-1)$

36 $2(x-7)(2x+3)(x-5)$

37 $(x+y)^3$

38 $(x+y)^4$

39 $(x^2-5)^3$

40 $(2-3x^2)^3$

41 Find the coefficients of x^3 and x^2 in the expansion of $(x-4)(2x+3)(3x-1)$.

42 State the coefficient of x^6 in the expansion of $(2x^3-3)^3$.

> **Note**
>
> The result from question 37 should be memorized:
> $(a+b)^3 = a^3 + 3a^2b + 3ab^2 + b^3$.

> **Note**
>
> For question 42 use the general result of the expansion of $(a+b)^3$ and replace a by $2x^3$ and b by -3.

1.3 Square roots and other roots

When a number is given as the product of two equal factors, that factor is called the **square root** of the number, for example

$4 = 2 \times 2 \Rightarrow 2$ is the square root of 4.

This is written $2 = \sqrt{4}$.

-2 is also a square root of 4, as $4 = -2 \times -2$ but $\sqrt{4} \neq -2$.

The symbol $\sqrt{}$ is used *only for the positive square root.*

So, although $x^2 = 4 \Rightarrow x = \pm 2$, the only value of $\sqrt{4}$ is 2. The negative square root of 4 is written $-\sqrt{4}$.

When both square roots are wanted, we write $\pm\sqrt{4}$.

> **Note**
>
> The symbol $\Rightarrow$ means gives or implies.

Cube roots

When a number is given as the product of three equal factors, that factor is called the **cube root** of the number.

For example $27 = 3 \times 3 \times 3$ so 3 is the cube root of 27.

This is written $\sqrt[3]{27} = 3$.

Other roots

The notation used for square and cube roots can be extended to represent fourth roots, fifth roots, and so on.

For example $16 = 2 \times 2 \times 2 \times 2 \Rightarrow \sqrt[4]{16} = 2$

and $243 = 3 \times 3 \times 3 \times 3 \times 3 \Rightarrow \sqrt[5]{243} = 3$

In general, if a number, n, can be written as the product of p equal factors then each factor is called the pth root of n and is written $\sqrt[p]{n}$.

1.4 Surds

A number which is either an integer, or a fraction whose numerator and denominator are both integers, is called a **rational number**.

The square roots of some numbers are rational. For example

$$\sqrt{9} = 3, \sqrt{25} = 5, \sqrt{\frac{4}{49}} = \frac{2}{7}$$

This is not true of all square roots. For example $\sqrt{2}$, $\sqrt{5}$ and $\sqrt{11}$ are not rational numbers. Such square roots can be given to as many decimal places as are needed, for example

$$\sqrt{3} = 1.73 \qquad \text{correct to 2 d.p.}$$
$$\sqrt{3} = 1.73205 \qquad \text{correct to 5 d.p.}$$

but they can not be expressed exactly as a decimal. They are called **irrational numbers**, and cannot be written as $\frac{a}{b}$ where a and b are integers.

The only way to give an exact answer for an irrational number is to leave them in the form $\sqrt{2}, \sqrt{7}$ and so on.

In this form they are called **surds**. At this level of mathematics answers should be given exactly unless an approximate answer is asked for; for example, give your answer correct to 3 significant figures.

Simplifying surds

Consider $\sqrt{18}$.

One of the factors of 18 is 9, and 9 has an exact square root,

so $\quad \sqrt{18} = \sqrt{(9 \times 2)} = \sqrt{9} \times \sqrt{2}$

But $\sqrt{9} = 3$, therefore $3\sqrt{2}$ is the simplest possible surd form for $\sqrt{18}$

Similarly $\sqrt{\dfrac{2}{25}} = \sqrt{\dfrac{2}{5 \times 5}} = \dfrac{\sqrt{2}}{5}$.

Multiplying surds

Consider $(4 - \sqrt{5})(3 + \sqrt{2})$

The multiplication is carried out in the same way and order as when multiplying two linear brackets.

so $(4 - \sqrt{5})(3 + \sqrt{2}) = (4)(3) + (4)(\sqrt{2}) - (3)(\sqrt{5}) - (\sqrt{5})(\sqrt{2})$
$$= 12 + 4\sqrt{2} - 3\sqrt{5} - (\sqrt{5})(\sqrt{2})$$
$$= 12 + 4\sqrt{2} - 3\sqrt{5} - \sqrt{10}$$

Example 4

Question

Expand and simplify $(2 + 2\sqrt{7})(5 - \sqrt{7})$.

Answer

$(2 + 2\sqrt{7})(5 - \sqrt{7}) \quad = (2)(5) - (2)(\sqrt{7}) + (5)(2\sqrt{7}) - (2\sqrt{7})(\sqrt{7})$
$$= 10 - 2\sqrt{7} + 10\sqrt{7} - 14$$

Collect like terms $\quad = 8\sqrt{7} - 4$

Example 5

Expand and simplify $(4-\sqrt{3})(4+\sqrt{3})$.

$$(4-\sqrt{3})(4+\sqrt{3})=4^2-(\sqrt{3})^2$$
$$=16-3=13$$

This example is special because the result is a rational number.

This is because the two given brackets are of the form $(x-a)(x+a)$ which when expanded give x^2-a^2.

The product of any two brackets of the type $(p-\sqrt{q})\,(p+\sqrt{q})$ is
$p^2+(\sqrt{q})^2=p^2-q$, which is always rational.

Exercise 5

Express in terms of the simplest possible surd.

1 $\sqrt{12}$ **2** $\sqrt{32}$ **3** $\sqrt{27}$

4 $\sqrt{50}$ **5** $\sqrt{200}$ **6** $\sqrt{72}$

7 $\sqrt{162}$ **8** $\sqrt{288}$ **9** $\sqrt{75}$

10 $\sqrt{48}$ **11** $\sqrt{500}$ **12** $\sqrt{20}$

Expand and simplify where possible.

13 $\sqrt{3}(2-\sqrt{3})$ **14** $\sqrt{2}(5+4\sqrt{2})$

15 $\sqrt{5}(2+\sqrt{75})$ **16** $\sqrt{2}(\sqrt{32}-\sqrt{8})$

17 $(\sqrt{3}+1)(\sqrt{2}-1)$ **18** $(\sqrt{3}+2)(\sqrt{3}+5)$

19 $(\sqrt{5}-1)(\sqrt{5}+1)$ **20** $(2\sqrt{2}-1)(\sqrt{2}-1)$

21 $(\sqrt{5}-3)(2\sqrt{5}-4)$ **22** $(4+\sqrt{7})(4-\sqrt{7})$

23 $(\sqrt{6}-2)^2$ **24** $(2+3\sqrt{3})^2$

Multiply by a bracket which will make that product rational.

25 $(4-\sqrt{5})$ **26** $(\sqrt{11}+3)$

27 $(2\sqrt{3}-4)$ **28** $(\sqrt{6}-\sqrt{5})$

29 $(3-2\sqrt{3})$ **30** $(2\sqrt{5}-\sqrt{2})$

Rationalising a denominator

A fraction whose denominator contains a surd is more awkward to deal with than one where a surd occurs only in the numerator.

There is a technique for transferring the surd expression from the denominator to the numerator; it is called *rationalising the denominator* (that is making the denominator into a rational number).

Example 6

Question

Rationalise the denominator of $\dfrac{2}{\sqrt{3}}$

Answer

$$\frac{2}{\sqrt{3}}=\frac{2\sqrt{3}}{(\sqrt{3})(\sqrt{3})}=\frac{2\sqrt{3}}{3}$$

> **Note**
>
> The square root in the denominator can be removed if we multiply it by another $\sqrt{3}$. We must also multiply the numerator by $\sqrt{3}$, so the value of the fraction is not changed.

Example 7

Question

Rationalise the denominator and simplify $\dfrac{3\sqrt{2}}{5-\sqrt{2}}$.

Answer

$$\frac{3\sqrt{2}}{5-\sqrt{2}}=\frac{3\sqrt{2}(5+\sqrt{2})}{(5-\sqrt{2})(5+\sqrt{2})}$$

$$=\frac{15\sqrt{2}+3(\sqrt{2})(\sqrt{2})}{25-(\sqrt{2})(\sqrt{2})}$$

$$=\frac{15\sqrt{2}+6}{23}$$

> **Note**
>
> A product of the type $(a-\sqrt{b})(a+\sqrt{b})$ is wholly rational. In this question multiply numerator and denominator by $5+\sqrt{2}$.

Example 8

Question

Simplify $\dfrac{3\sqrt{2}+\sqrt{3}}{2\sqrt{3}+\sqrt{2}}$.

Answer

First rationalise the denominator

$$\frac{3\sqrt{2}+\sqrt{3}}{2\sqrt{3}+\sqrt{2}}=\frac{\left(3\sqrt{2}+\sqrt{3}\right)\left(2\sqrt{3}-\sqrt{2}\right)}{\left(2\sqrt{3}+\sqrt{2}\right)\left(2\sqrt{3}-\sqrt{2}\right)}=\frac{\left(3\sqrt{2}+\sqrt{3}\right)\left(2\sqrt{3}-\sqrt{2}\right)}{12-2}$$

then expand the brackets

$$=\frac{6\sqrt{6}-6+6-\sqrt{6}}{10}=\frac{5\sqrt{6}}{10}=\frac{\sqrt{6}}{2}$$

Exercise 6

Rationalise the denominator and simplify where possible.

1. $\dfrac{3}{\sqrt{2}}$

2. $\dfrac{1}{\sqrt{7}}$

3. $\dfrac{2}{\sqrt{11}}$

4. $\dfrac{3\sqrt{2}}{\sqrt{5}}$

5. $\dfrac{1}{\sqrt{27}}$

6. $\dfrac{\sqrt{5}}{\sqrt{10}}$

7. $\dfrac{1}{\sqrt{2}-1}$

8. $\dfrac{3\sqrt{2}}{5+\sqrt{2}}$

9. $\dfrac{2}{2\sqrt{3}-3}$

(10) $\dfrac{5}{2-\sqrt{5}}$ (11) $\dfrac{1}{\sqrt{7}-\sqrt{3}}$ (12) $\dfrac{4\sqrt{3}}{2\sqrt{3}-3}$

(13) $\dfrac{3-\sqrt{5}}{\sqrt{5}+1}$ (14) $\dfrac{2\sqrt{3}-1}{4-\sqrt{3}}$ (15) $\dfrac{\sqrt{5}-1}{\sqrt{5}-2}$

(16) $\dfrac{3}{\sqrt{3}-\sqrt{2}}$ (17) $\dfrac{3\sqrt{5}}{2\sqrt{5}+1}$ (18) $\dfrac{\sqrt{2}+1}{\sqrt{2}-1}$

(19) $\dfrac{2\sqrt{7}}{\sqrt{7}+2}$ (20) $\dfrac{\sqrt{5}-1}{3-\sqrt{5}}$ (21) $\dfrac{1}{\sqrt{11}-\sqrt{7}}$

(22) $\dfrac{4-\sqrt{3}}{3-\sqrt{3}}$ (23) $\dfrac{1-3\sqrt{2}}{3\sqrt{2}+2}$ (24) $\dfrac{1}{3\sqrt{2}-2\sqrt{3}}$

(25) $\dfrac{\sqrt{3}}{\sqrt{2}(\sqrt{6}-\sqrt{3})}$ (26) $\dfrac{1}{\sqrt{3}(\sqrt{21}+\sqrt{7})}$ (27) $\dfrac{\sqrt{2}}{\sqrt{3}(\sqrt{5}-\sqrt{2})}$

(28) $\dfrac{2\sqrt{3}+\sqrt{2}}{3\sqrt{2}+\sqrt{3}}$ (29) $\dfrac{2\sqrt{5}+\sqrt{2}}{5\sqrt{2}+\sqrt{5}}$ (30) $\dfrac{3\sqrt{6}+\sqrt{3}}{6\sqrt{3}+\sqrt{6}}$

1.5 Indices

Base and index

In an expression such as 3^4, the **base** is 3 and the 4 is the **power** or **index** (the plural is **indices**).

Working with indices involves using properties which apply to any base, so these laws are given in terms of a general base a where a stands for any number.

Law 1

Because a^3 means $a \times a \times a$ and a^2 means $a \times a$ it follows that
$$a^3 \times a^2 = (a \times a \times a) \times (a \times a) = a^5$$

so $a^3 \times a^2 = a^{3+2}$

Examples with different powers all show the general law that

$\qquad a^p \times a^q = a^{p+q}$

Law 2

For division
$$a^7 \div a^4 = \frac{\cancel{a} \times \cancel{a} \times \cancel{a} \times \cancel{a} \times a \times a \times a}{\cancel{a} \times \cancel{a} \times \cancel{a} \times \cancel{a}} = a^3$$

so $\quad a^7 \div a^4 = a^{7-4}$

This is one example of the general law that

$\qquad a^p \div a^q = a^{p-q}$

When this law is applied to some fractions, interesting cases arise.

Consider $a^3 \div a^5$.
$$\frac{a^3}{a^5} = \frac{\cancel{a} \times \cancel{a} \times \cancel{a}}{\cancel{a} \times \cancel{a} \times \cancel{a} \times a \times a} = \frac{1}{a^2}$$

But from Law 2 we have $a^3 \div a^5 = a^{3-5} = a^{-2}$.

Therefore a^{-2} means $\dfrac{1}{a^2}$.

In general

$$a^{-p} = \frac{1}{a^p}$$

so

a^{-p} means the reciprocal of a^p

Now look at $a^4 \div a^4$.

$$\frac{a^4}{a^4} = \frac{\not a \times \not a \times \not a \times \not a}{\not a \times \not a \times \not a \times \not a} = 1$$

From Law 2, $\dfrac{a^4}{a^4} = a^{4-4} = a^0$.

Therefore $\boldsymbol{a^0 = 1}$

so

any base to the power zero is equal to 1.

Law 3

$$\begin{aligned}
(a^2)^3 \quad &= (a \times a)^3 \\
&= (a \times a) \times (a \times a) \times (a \times a) \\
&= a^6
\end{aligned}$$

so $\quad (a^2)^3 = a^{2 \times 3}$

In general

$$(a^p)^q = a^{pq}$$

Law 4

This law explains the meaning of a fractional index. From the first law

$$a^{\frac{1}{2}} \times a^{\frac{1}{2}} = a^{\frac{1}{2} + \frac{1}{2}} = a^1 = a$$

so $\qquad a = a^{\frac{1}{2}} \times a^{\frac{1}{2}}$

But $\qquad a = \sqrt{a} \times \sqrt{a}$

Therefore $a^{\frac{1}{2}}$ means $\sqrt{a}$, which is the positive square root of a.

Also $\qquad a^{\frac{1}{3}} \times a^{\frac{1}{3}} \times a^{\frac{1}{3}} = a^{\frac{1}{3} + \frac{1}{3} + \frac{1}{3}} = a^1 = a$

But $\qquad \sqrt[3]{a} \times \sqrt[3]{a} \times \sqrt[3]{a} = a$

Therefore $a^{\frac{1}{3}}$ means $\sqrt[3]{a}$, which is the cube root of a.

In general

$$a^{\frac{1}{p}} = \sqrt[p]{a}\,, \text{ that is the } p\text{th root of } a.$$

For a more general fraction index, $\dfrac{p}{q}$, the third law shows that

$$a^{\frac{p}{q}} = (a^p)^{\frac{1}{q}} \text{ or } (a^{\frac{1}{q}})^p$$

$$= \sqrt[q]{(a)^p} \text{ or } \left(\sqrt[q]{a}\right)^p$$

For example

$$a^{\frac{3}{4}} = (a^3)^{\frac{1}{4}} \text{ or } (a^{\frac{1}{4}})^3$$
$$= \sqrt[4]{a^3} \text{ or } (\sqrt[4]{a})^3$$

so $a^{\frac{3}{4}}$ means either the fourth root of a^3 or the cube of the fourth root of a.

All the laws can be applied to simplify a wide range of expressions containing indices when the terms all have the same base.

Example 9

Question

Simplify

a $\dfrac{2^3 \times 2^7}{4^3}$
b $(x^2)^7 \times x^{-3}$
c $\sqrt[3]{a^4 b^5} \times \dfrac{b^{\frac{1}{3}}}{a}$

Answer

a $\dfrac{2^3 \times 2^7}{4^3} = \dfrac{2^3 \times 2^7}{(2^2)^3} = \dfrac{2^{3+7}}{2^{2\times 3}} = \dfrac{2^{10}}{2^6} = 2^4$

> **Note**
>
> First express all the terms to base 2.

b $(x^2)^7 \times x^{-3} = x^{2 \times 7} \times \dfrac{1}{x^3} = x^{14} \times \dfrac{1}{x^3} = x^{11}$

c $\sqrt[3]{a^4 b^5} \times \dfrac{b^{\frac{1}{3}}}{a} = (a^{\frac{4}{3}})(b^{\frac{5}{3}})(b^{\frac{1}{3}})(a^{-1}) = (a^{\frac{4}{3}-1})(b^{\frac{5}{3}+\frac{1}{3}}) = a^{\frac{1}{3}} b^2$

Example 10

Question

Evaluate

a $(64)^{-\frac{1}{3}}$
b $\left(\dfrac{25}{9}\right)^{-\frac{3}{2}}$

Answer

a $(64)^{-\frac{1}{3}} = \dfrac{1}{(64)^{\frac{1}{3}}}$

$= \dfrac{1}{\sqrt[3]{64}} = \dfrac{1}{4}$

> **Note**
>
> $\left(\dfrac{9}{25}\right)^{\frac{3}{2}}$ could have been written as $\sqrt{\left(\dfrac{9}{25}\right)^3}$ but this involves much bigger numbers.

b $\left(\dfrac{25}{9}\right)^{-\frac{3}{2}} = \left(\dfrac{9}{25}\right)^{\frac{3}{2}} = \left(\sqrt{\dfrac{9}{25}}\right)^3 = \left(\dfrac{3}{5}\right)^3 = \dfrac{27}{125}$

Example 11

Question

State the value of p given that

a $125 = 5^p$
b $\left(\dfrac{1}{7}\right)^p = 1$
c $\dfrac{4}{9} = \left(\dfrac{3}{2}\right)^p$

Answer

a $125 = 5^3$

Therefore $p = 3$.

b Any number to the power zero is 1, so $p = 0$.

c $\dfrac{4}{9} = \left(\dfrac{2}{3}\right)^2 = \left(\dfrac{3}{2}\right)^{-2}$ so $p = -2$.

Exercise 7

Simplify.

1 $\dfrac{2^4}{2^2 \times 4^3}$

2 $4^{\frac{1}{2}} \times 2^{-3}$

3 $(3^3)^{\frac{1}{2}} \times 9^{\frac{1}{4}}$

4 $\dfrac{x^{\frac{1}{3}} \times x^{\frac{4}{3}}}{x^{-\frac{1}{3}}}$

5 $\dfrac{p^{\frac{1}{2}} \times p^{-\frac{3}{4}}}{p^{-\frac{1}{4}}}$

6 $(\sqrt{t})^3 \times (\sqrt{t})^5$

7 $(y^2)^{\frac{3}{2}} \times y^{-3}$

8 $(16)^{\frac{5}{4}} \div 8^{\frac{4}{3}}$

9 $\dfrac{y^{\frac{1}{2}}}{y^{-\frac{3}{4}}} \times \sqrt{y^{\frac{1}{2}}}$

10 $x^2 \times x^{\frac{5}{2}} \div x^{-\frac{1}{2}}$

11 $\dfrac{y^{\frac{1}{6}} \times y^{-\frac{2}{3}}}{y^{\frac{1}{4}}}$

12 $\left(p^{\frac{1}{3}}\right)^2 \times (p^2)^{\frac{1}{3}} \div \sqrt[3]{p}$

Evaluate.

13 $\left(\dfrac{1}{3}\right)^{-1}$

14 $\left(\dfrac{1}{4}\right)^{\frac{5}{2}}$

15 $(8)^{-\frac{1}{3}}$

16 $\dfrac{1}{(16)^{-\frac{1}{4}}}$

17 $\left(\dfrac{1}{9}\right)^{-\frac{3}{2}}$

18 $\left(\dfrac{27}{8}\right)^{\frac{2}{3}}$

19 $\left(\dfrac{100}{9}\right)^0$

20 $\dfrac{1}{4^{-2}}$

21 $(0.64)^{-\frac{1}{2}}$

22 $\left(-\dfrac{1}{5}\right)^{-1}$

23 $(121)^{\frac{3}{2}}$

24 $\left(\dfrac{125}{27}\right)^{-\frac{1}{3}}$

25 $18^{\frac{1}{2}} \times 2^{\frac{1}{2}}$

26 $3^{-3} \times 2^0 \times 4^2$

27 $\dfrac{8^{\frac{1}{2}} \times 32^{\frac{1}{2}}}{(16)^{\frac{1}{4}}}$

28 $5^{\frac{1}{3}} \times 25^0 \times 25^{\frac{1}{3}}$

29 $27^{\frac{1}{4}} \times 3^{\frac{1}{4}} \times (\sqrt{3})^{-2}$

30 $\dfrac{9^{\frac{1}{3}} \times 27^{-\frac{1}{2}}}{3^{-\frac{1}{6}} \times 3^{-\frac{2}{3}}}$

State the value of p given that

31 $121 = 11^p$

32 $\dfrac{1}{7} = 7^p$

33 $1 = \left(\dfrac{1}{2}\right)^p$

34 $\dfrac{1}{5} = 5^p$

35 $\dfrac{1}{32} = 2^p$

36 $\dfrac{2}{5} = \left(\dfrac{25}{4}\right)^p$

37 $8^{-\frac{1}{3}} \times 4^3 = 2^p$

38 $3^2 \times 4^p = \dfrac{9}{2}$

39 $x^p \times x^{-2} = x$

40 $(x^2)^p \times x^{-4} = x^{-1}$

Summary

Expansions

$(ax+b)^2 = a^2x^2 + 2abx + b^2$

$(ax-b)^2 = a^2x^2 - 2abx + b^2$

$(ax+b)(ax-b) = a^2x^2 - b^2$

$(a+b)^3 = a^3 + 3a^2b + 3ab^2 + b^3$.

Surds

The denominator of $\dfrac{a}{\sqrt{b}}$ can be rationalized by multiplying numerator and denominator by $\sqrt{b}$.

The denominator of $\dfrac{a}{b+\sqrt{c}}$ can be rationalized by multiplying numerator and denominator by $b-\sqrt{c}$.

Indices

$a^n \times a^m = a^{n+m}$

$a^n \div a^m = a^{n-m}$

$(a^n)^m = a^{nm}$

$\sqrt[n]{a} = a^{\frac{1}{n}}$

$a^0 = 1$

Review

1. Simplify $a^2(3-a) - (a-a^3)$.

2. Expand and simplify $(2x-7)(x+5)$.

3. Expand and simplify $(4x-3)^2$.

4. Expand and simplify $(2x-3)(2x+3)$.

5. Expand and simplify $(3-5x)(2x+1)$.

6. Expand and simplify $(2x-3y)(x+y)^2$.

7. State the coefficient of a^2b in the expansion of $(3a-2b)^3$.

8. Express $\sqrt{150}$ in terms of the simplest possible surd.

9. Expand $\left(4-3\sqrt{3}\right)^2$ and simplify if possible.

For Questions 10, 11 and 12 state the letter that gives the correct answer.

10. $\dfrac{1-\sqrt{2}}{1+\sqrt{2}}$ is equal to

   a 1   **b** −1   **c** $3-2\sqrt{2}$ **d** $2\sqrt{2}-3$

11. $\dfrac{p^{-\frac{1}{2}} \times p^{\frac{3}{4}}}{p^{-\frac{1}{4}}}$ simplifies to

   a $p^{\frac{1}{2}}$ **b** $p^{-\frac{1}{2}}$ **c** $p^{\frac{3}{4}}$ **d** p

12 $\dfrac{5^{\frac{1}{4}} \times 5 \times 5^{\frac{1}{6}}}{\sqrt{5}} = 5^p$. The value of p is

 a $-\dfrac{1}{12}$ **b** $\dfrac{11}{12}$ **c** $1\dfrac{11}{12}$

Assessment

1 **a** Given that $\dfrac{1}{27} = 3^r$ state the value of r.

 b Given that $\sqrt{3} = 3^r$ state the value of r.

2 **a** The expression $(x-3)(x^2-5x+6)$ can be written in the form
  $x^3 + px^2 + qx - 18$. Show that $p = -8$ and find the value of q.

 b The expression $(x^2+6)^3$ can be written in the form $x^6 + px^4 + qx^2 + 216$.
  Find the values of p and q.

3 **a** Show that $\sqrt{72} = p\sqrt{2}$ giving the value of p.

 b Show that $\dfrac{\sqrt{8}+\sqrt{18}}{\sqrt{2}} = n$ where n is an integer. State the value of n.

 c Show that $\dfrac{2\sqrt{2}-1}{2-\sqrt{2}}$ can be expressed in the form $p + q\sqrt{2}$ where p and q
  are rational numbers. State the values of p and q.

4 **a** Express $\dfrac{2^{\frac{1}{2}} \times 2^{-\frac{1}{4}}}{2^{\frac{3}{4}}}$ as 2^r. State the value of r.

 b Express $\left(\dfrac{36x^2}{16}\right)^{\frac{1}{2}}$ in the form ax^b giving the values of a and b.

5 **a** Expand $(x-1)^2(2x+3)$.

 b Find the coefficients of x^2 and x in the expansion of $(x-2)(2x+3)(x+2)$.

6 **a** **i** Simplify $\left(3\sqrt{2}\right)^2$

  ii Show that $\left(3\sqrt{2}-2\right)^2 + \left(3+\sqrt{2}\right)^2$ is an integer and find its value.

 b Express $\dfrac{4\sqrt{5}-7\sqrt{2}}{2\sqrt{5}+\sqrt{2}}$ in the form $m - \sqrt{n}$, where m and n are integers.

AQA MPC1 January 2012

7 A rectangle has length $\left(9+5\sqrt{3}\right)$ cm and area $\left(15+7\sqrt{3}\right)$ cm².

 Find the width of the rectangle, giving your answer in the form $\left(m+n\sqrt{3}\right)$ cm,
 where m and n are integers.

AQA MPC1 June 2014

2 Quadratic Polynomials and Equations

Introduction

Quadratic polynomials and equations appear in many topics in A Level Mathematics. This chapter gives the facts and practice needed to work with them.

Recap

You need to remember how to...
- Expand two brackets such as $(2x-3)(3x+6)$.
- Simplify expressions containing surds.
- Solve a linear equation, for example $2x-5=10$.

Objectives

By the end of this chapter, you should know how to...
- Factorise quadratic expressions.
- Solve quadratic equations.
- Solve one linear equation and one quadratic equation simultaneously.
- Find the nature of the roots of a quadratic equation.

2.1 Quadratic polynomials

A **polynomial** is a collection of terms containing powers of x which are positive integers. For example,

$$2x^3 - 5x + 1, \, x^2 + 8, \, 5x - 7$$

The highest power in a quadratic polynomial is 2, so $x^2 + 8$ is a quadratic polynomial.

The general form of a quadratic polynomial is
$ax^2 + bx + c$
where a, b and c are constants (that is fixed numbers) and $a \neq 0$.

When the highest power of x is 1, the polynomial is called a linear polynomial (also called a linear expression).

For example $5x - 7$ is a linear expression.

Factorising quadratic polynomials

The product of two linear brackets is quadratic. In some cases we can reverse this process. For example, given a quadratic such as $x^2 - 5x + 6$, we can try to find two linear expressions in x whose product is $x^2 - 5x + 6$.

Expressing a quadratic as the product of two linear brackets is called factorising the quadratic.

To be able to do this we need to understand the pattern between what is inside the brackets and the resulting quadratic.

> **Note**
>
> A quadratic polynomial is also called a quadratic expression or just a quadratic.

Look at these examples

$$(2x+1)(x+5) = 2x^2 + 11x + 5 \qquad [1]$$
$$(3x-2)(x-4) = 3x^2 - 14x + 8 \qquad [2]$$
$$(x-5)(4x+2) = 4x^2 - 18x - 10 \qquad [3]$$

Notice that for the quadratic in each example

the coefficient of x^2 is the product of the coefficients of x in the two brackets,

the constant is the product of the numbers in the two brackets,

the coefficient of x is found by adding the coefficients formed by multiplying the x term in one bracket by the constant term in the other bracket.

Also notice that there is a relationship between the signs.

Positive signs throughout the quadratic come from positive signs in both brackets, as in [1].

A positive constant term and a negative coefficient of x in the quadratic come from a negative sign in each bracket, as in [2].

A negative constant term in the quadratic comes from a negative sign in one bracket and a positive sign in the other, as in [3].

Example 1

Question

Factorise $x^2 - 5x + 6$.

Answer

The x term in each bracket is x as x^2 can only be $x \times x$.

The sign in each bracket is $-$, so $x^2 - 5x + 6 = (x- \quad)(x- \quad)$.

The numerical terms in the brackets could be 6 and 1 or 2 and 3.

Checking the middle term tells us that the numbers must be 2 and 3.

$x^2 - 5x + 6 = (x-2)(x-3)$

> **Note**
>
> Mentally expanding the brackets checks that they are correct.

Example 2

Question

Factorise $x^2 - 3x - 10$.

Answer

The x term in each bracket is x as x^2 can only be $x \times x$. The sign in one bracket is positive and the other is negative so $x^2 - 3x - 10 = (x- \quad)(x+ \quad)$.

The numbers could be 10 and 1 or 5 and 2.

Checking the middle term shows that they are 5 and 2.

$x^2 - 3x - 10 = (x-5)(x+2)$

> **Note**
>
> Mentally expanding the brackets checks that they are correct.

Exercise 1

Factorise.

1 $x^2 + 8x + 15$ **2** $x^2 + 11x + 28$ **3** $x^2 + 7x + 6$

4 $x^2 + 7x + 12$ **5** $x^2 - 10x + 9$ **6** $x^2 - 6x + 9$

7 $x^2 + 8x + 12$ **8** $x^2 - 9x + 8$ **9** $x^2 + 5x - 14$

10 $x^2 + x - 12$ **11** $x^2 - 4x - 5$ **12** $x^2 - 10x - 24$

13 $x^2 + 9x + 14$ **14** $x^2 - 2x + 1$ **15** $x^2 - 9$

16 $x^2 + 5x - 24$ **17** $x^2 + 4x + 4$ **18** $x^2 - 1$

19 $x^2 - 3x - 18$ **20** $x^2 + 10x + 25$ **21** $x^2 - 16$

22 $4 + 5x + x^2$ **23** $2x^2 - 3x + 1$ **24** $3x^2 + 4x + 1$

25 $9x^2 - 6x + 1$ **26** $6x^2 - x - 1$ **27** $9 + 6x + x^2$

28 $4x^2 - 9$ **29** $x^2 + 2ax + a^2$ **30** $x^2y^2 - 2xy + 1$

Harder factorising

When there is more than one possible combination of terms for the brackets, recognising patterns can help to reduce the possibilities.

For example, when the coefficient of x^2 is 1, if the coefficient of x in the quadratic is odd, then there must be an even number in one bracket and an odd number in the other.

Example 3

Factorise $12 - x - 6x^2$.

The x terms in the brackets could be $6x$ and x, or $3x$ and $2x$, one positive and the other negative.

The number terms could be 12 and 1 or 3 and 4 (not 6 and 2 because the coefficient of x in the quadratic is odd).

Now try different combinations to find the correct one.

$12 - x - 6x^2 = (3 + 2x)(4 - 3x)$

Common factors

A **common factor** is a number or letter that is a factor of each term in an expression.

When the terms in a quadratic equation have a common factor it should be factorised first.

For example $4x^2 + 8x + 4 = 4(x^2 + 2x + 1)$

The quadratic inside the bracket now has smaller coefficients and can be factorised more easily

$$4x^2 + 8x + 4 = 4(x + 1)(x + 1)$$
$$= 4(x + 1)^2$$

Not all quadratics factorise.

Look at $3x^2 - x + 5$.

The options we can try are $(3x-5)(x-1)$ [1]

and $(3x-1)(x-5)$ [2]

From [1], $(3x-5)(x-1) = 3x^2 - 8x + 5$.

From [2], $(3x-1)(x-5) = 3x^2 - 16x + 5$.

As neither of the these pairs of brackets expand to give $3x^2 - x + 5$, this shows that $3x^2 - x + 5$ has no factors of the form $ax + b$ where a and b are integers.

Example 4

Question

Factorise $2x^2 - 8x + 16$.

Answer

$2x^2 - 8x + 16 = 2(x^2 - 4x + 8)$

The possible brackets are $(x-1)(x-8)$ and $(x-2)(x-4)$. Neither pair expands to $x^2 - 4x + 8$, so there are no further factors.

Exercise 2

Factorise.

1. $6x^2 + x - 12$
2. $4x^2 - 11x + 6$
3. $4x^2 + 3x - 1$
4. $3x^2 - 17x + 10$
5. $4x^2 - 12x + 9$
6. $3 - 5x - 2x^2$
7. $25x^2 - 16$
8. $3 - 2x - x^2$
9. $5x^2 - 61x + 12$
10. $9x^2 + 30x + 25$
11. $3 + 2x - x^2$
12. $12 + 7x - 12x^2$
13. $1 - x^2$
14. $9x^2 + 12x + 4$
15. $x^2 + 2xy + y^2$
16. $1 - 4x^2$
17. $4x^2 - 4xy + y^2$
18. $9 - 4x^2$
19. $36 + 12x + x^2$
20. $40x^2 - 17x - 12$
21. $7x^2 - 5x - 150$
22. $36 - 25x^2$
23. $x^2 - y^2$
24. $81x^2 - 36xy + 4y^2$
25. $49 - 84x + 36x^2$
26. $25x^2 - 4y^2$
27. $36x^2 + 60xy + 25y^2$
28. $4x^2 - 4xy - 3y^2$
29. $6x^2 + 11xy + 4y^2$
30. $49p^2q^2 - 28pq + 4$

Factorise where possible.

31. $x^2 + x + 1$
32. $2x^2 + 4x + 2$
33. $x^2 + 3x + 2$
34. $3x^2 + 12x - 15$
35. $x^2 + 4$
36. $x^2 - 4x - 6$
37. $x^2 + 3x + 1$
38. $2x^2 - 8x + 8$
39. $3x^2 - 3x - 6$
40. $2x^2 - 6x + 8$
41. $3x^2 - 6x - 24$
42. $x^2 - 4x - 12$
43. $x^2 + 1$
44. $4x^2 - 100$
45. $5x^2 - 25$
46. $7x^2 + x + 4$
47. $10x^2 - 39x - 36$
48. $x^2 + xy + y^2$

2.2 Quadratic equations

When a quadratic expression has a particular value this is a quadratic equation,

for example $2x^2 - 5x + 1 = 0$

Using a, b and c to stand for any numbers, any quadratic equation can be written in the general form $ax^2 + bx + c = 0$ where $a \neq 0$.

Solution by factorising

The quadratic expression on the left-hand side of the equation $x^2 - 3x + 2 = 0$ factorises.

So $x^2 - 3x + 2 = (x - 2)(x - 1)$

Therefore the equation becomes

$$(x - 2)(x - 1) = 0 \qquad\qquad\qquad [1]$$

> If the product of two quantities is zero then one, or both, of those quantities must be zero.

Applying this fact to equation [1] gives

$x - 2 = 0$ or $x - 1 = 0$

so $\qquad x = 2$ or $x = 1$

This is the solution of the given equation.

The values 2 and 1 are called the **roots** of the equation.

This method of solution can be used for any quadratic equation in which the quadratic expression factorises.

Example 5

Question

Find the roots of the equation

$$x^2 + 6x - 7 = 0$$

Answer

$$x^2 + 6x - 7 = 0$$
$\Rightarrow \qquad\qquad (x - 1)(x + 7) = 0$
$\Rightarrow \qquad x - 1 = 0 \quad \text{or} \quad x + 7 = 0$
$\Rightarrow \qquad\qquad x = 1 \quad \text{or} \quad x = -7$

The roots of the equation are 1 and −7.

Rearranging the equation

The terms in a quadratic equation are not always in the order $ax^2 + bx + c = 0$. When they are in a different order they should be rearranged.

For example $\qquad x^2 - x = 4$ becomes $x^2 - x - 4 = 0$

$\qquad\qquad\qquad\quad 3x^2 - 1 = 2x$ becomes $3x^2 - 2x - 1 = 0$

$\qquad\qquad\qquad\quad x(x - 1) = 2$ becomes $x^2 - x = 2 \quad \Rightarrow \quad x^2 - x - 2 = 0$

Collect the terms on the side where the x^2 term is positive, for example

$\qquad 2 - x^2 = 5x$ becomes $0 = x^2 + 5x - 2$

so $\quad x^2 + 5x - 2 = 0$

Example 6

Question

Solve the equation $4x - x^2 = 3$.

Answer

$4x - x^2 = 3$
$\Rightarrow \quad 0 = x^2 - 4x + 3$
$\Rightarrow \quad x^2 - 4x + 3 = 0 \quad \Rightarrow \quad (x-3)(x-1) = 0$
$\Rightarrow \quad x - 3 = 0 \text{ or } x - 1 = 0$
$\Rightarrow \quad x = 3 \text{ or } x = 1$

Losing a solution

Quadratic equations sometimes have a common factor containing the unknown quantity. Do not divide by the common factor because this loses part of the solution. The example below shows this.

Correct solution

$x^2 - 5x = 0$
$x(x - 5) = 0$
$x = 0 \text{ or } x - 5 = 0$
$\Rightarrow \quad x = 0 \text{ or } x = 5$

Faulty solution

$x^2 - 5x = 0$
$x - 5 = 0 \text{ (Dividing by } x.)$
$x = 5$
The solution $x = 0$ has been lost.

Dividing an equation by a numerical common factor is correct and sensible, but dividing by a common factor containing the unknown quantity means losing a solution.

Exercise 3

Solve the equations.

1. $x^2 + 5x + 6 = 0$
2. $x^2 + x - 6 = 0$
3. $x^2 - x - 6 = 0$
4. $x^2 + 6x + 8 = 0$
5. $x^2 - 4x + 3 = 0$
6. $x^2 + 2x - 3 = 0$
7. $2x^2 + 3x + 1 = 0$
8. $4x^2 - 9x + 2 = 0$
9. $x^2 + 4x - 5 = 0$
10. $x^2 + x - 72 = 0$

Find the roots of these equations.

11. $x^2 - 2x - 3 = 0$
12. $x^2 - 5x + 4 = 0$
13. $x^2 - 6x + 5 = 0$
14. $x^2 + 3x - 10 = 0$
15. $x^2 - 5x - 14 = 0$
16. $x^2 - 9x + 14 = 0$

Solve each equation, making sure that you give all the roots.

17. $x^2 + 10 - 7x = 0$
18. $15 - x^2 - 2x = 0$
19. $x^2 - 3x = 4$
20. $12 - 7x + x^2 = 0$
21. $2x - 1 + 3x^2 = 0$
22. $x(x + 7) + 6 = 0$
23. $2x^2 - 4x = 0$
24. $x(4x + 5) = -1$
25. $2 - x = 3x^2$
26. $6x^2 + 3x = 0$
27. $x^2 + 6x = 0$
28. $x^2 = 10x$
29. $x(4x + 1) = 3x$
30. $20 + x(1 - x) = 0$
31. $x(3x - 2) = 8$
32. $x^2 - x(2x - 1) + 2 = 0$
33. $x(x + 1) = 2x$
34. $4 + x^2 = 2(x + 2)$
35. $x(x - 2) = 3$
36. $1 - x^2 = x(1 + x)$

2.3 Solution by completing the square

When there are no obvious factors, another method involves adding a constant to the x^2 term and the x term, to make a perfect square. This technique is called **completing the square**.

For example, adding 1 to $x^2 - 2x$ gives $x^2 - 2x + 1$

$$x^2 - 2x + 1 = (x - 1)^2 \text{ which is a perfect square.}$$

Adding the number 1 was not a guess, it was found by using the fact that

$$x^2 + 2ax + a^2 = (x + a)^2$$

You saw this relationship in Chapter 1.1.

This shows that the number to be added is always (half the coefficient of x)^2.

So $x^2 + 6x$ needs 3^2 to be added to make a perfect square,

giving $x^2 + 6x + 9 = (x + 3)^2$

To complete the square when the coefficient of x^2 is not 1, first take out the coefficient of x^2 as a factor.

For example $2x^2 + x = 2(x^2 + \dfrac{1}{2}x)$

Then add $\left(\dfrac{1}{2} \times \dfrac{1}{2}\right)^2$ inside the bracket, giving

$$2\left(x^2 + \dfrac{1}{2}x + \dfrac{1}{16}\right) = 2\left(x + \dfrac{1}{4}\right)^2$$

Be careful when the coefficient of x^2 is negative.

For example $\quad -x^2 + 4x = -(x^2 - 4x)$

Then $\quad -(x^2 - 4x + 4) = -(x - 2)^2$

$\Rightarrow \quad -x^2 + 4x - 4 = -(x - 2)^2$

Example 7

Solve the equation $x^2 - 4x - 2 = 0$. Give the solution in surd form.

No factors can be found so isolate the two terms with x in them.

$$x^2 - 4x - 2 = 0 \Rightarrow x^2 - 4x = 2$$

Now complete the square on the left-hand side and add the same number to the right-hand side.

Add $\left\{\dfrac{1}{2} \times (-4)\right\}^2$ to *both* sides, giving $x^2 - 4x + 4 = 2 + 4$

$\Rightarrow \quad (x - 2)^2 = 6$

$\Rightarrow \quad x - 2 = \pm\sqrt{6}$

$\Rightarrow \quad x = 2 + \sqrt{6} \ \text{ or } \ x = 2 - \sqrt{6}$

Example 8

Question

Find in surd form the roots of the equation $2x^2 - 3x - 3 = 0$.

Answer

$2x^2 - 3x - 3 = 0$

$2\left(x^2 - \dfrac{3}{2}x\right) = 3 \qquad \Rightarrow \quad x^2 - \dfrac{3}{2}x = \dfrac{3}{2}$

$x^2 - \dfrac{3}{2}x + \dfrac{9}{16} = \dfrac{3}{2} + \dfrac{9}{16} \qquad \Rightarrow \quad \left(x - \dfrac{3}{4}\right)^2 = \dfrac{33}{16}$

$\therefore \quad \left(x - \dfrac{3}{4}\right) = \pm\sqrt{\dfrac{33}{16}} = \pm\dfrac{1}{4}\sqrt{33} \quad \Rightarrow \quad x = \dfrac{3}{4} \pm \dfrac{1}{4}\sqrt{33}$

The roots of the equations are $\dfrac{1}{4}(3+\sqrt{33})$ and $\dfrac{1}{4}(3 - \sqrt{33})$.

Exercise 4

Add a number to each expression so that the result is a perfect square.

1. $x^2 - 4x$
2. $x^2 + 2x$
3. $x^2 - 6x$
4. $x^2 + 10x$
5. $2x^2 - 4x$
6. $x^2 + 5x$
7. $3x^2 - 48x$
8. $x^2 + 18x$
9. $2x^2 - 40x$
10. $x^2 + x$
11. $3x^2 - 2x$
12. $2x^2 + 3x$

Solve the equations by completing the square, giving the solutions in surd form.

13. $x^2 + 8x = 1$
14. $x^2 - 2x - 2 = 0$
15. $x^2 + x - 1 = 0$
16. $2x^2 + 2x = 1$
17. $x^2 + 3x + 1 = 0$
18. $2x^2 - x - 2 = 0$
19. $x^2 + 4x = 2$
20. $3x^2 + x - 1 = 0$
21. $2x^2 + 4x = 7$
22. $x^2 - x = 3$
23. $4x^2 + x - 1 = 0$
24. $2x^2 - 3x - 4 = 0$

2.4 The formula for solving a quadratic equation

When completing the square is applied to a general quadratic equation, a formula is derived which can be used to solve any quadratic equation.

The general quadratic equation is

$$ax^2 + bx + c = 0 \qquad a \neq 0$$

Using the method of completing the square for this equation gives

$$ax^2 + bx = -c$$

so $\quad a\left(x^2 + \dfrac{b}{a}x\right) = -c \implies x^2 + \dfrac{b}{a}x = -\dfrac{c}{a}$

$\therefore \quad x^2 + \dfrac{b}{a}x + \left(\dfrac{b}{2a}\right)^2 = \left(\dfrac{b}{2a}\right)^2 - \dfrac{c}{a}$

$\therefore \quad \left(x + \dfrac{b}{2a}\right)^2 = \dfrac{b^2}{4a^2} - \dfrac{c}{a} = \dfrac{b^2 - 4ac}{4a^2}$

$\implies \quad x + \dfrac{b}{2a} = \pm\sqrt{\dfrac{b^2 - 4ac}{4a^2}}$

$\implies \quad x = -\dfrac{b}{2a} \pm \dfrac{\sqrt{b^2 - 4ac}}{2a}$

$\implies \quad x = \dfrac{-b \pm \sqrt{b^2 - 4ac}}{2a}$

> **Note**
>
> You need to learn this formula.

Example 9

Question

Find, by using the formula, the roots of the equation $2x^2 - 7x - 1 = 0$ giving them correct to 3 decimal places.

Answer

$$2x^2 - 7x - 1 = 0$$

Comparing with $\qquad ax^2 + bx + c = 0$ gives $a = 2$, $b = -7$, $c = -1$

Using the formula gives $\quad x = \dfrac{-b \pm \sqrt{b^2 - 4ac}}{2a} = \dfrac{7 \pm \sqrt{49 - 4(2)(-1)}}{4}$

Therefore, in surd form, $\quad x = \dfrac{7 \pm \sqrt{57}}{4}$

The roots are 3.637 and -0.137 correct to 3 decimal places.

Exercise 5

Solve the equations by using the formula. Give the solutions in surd form.

1. $x^2 + 4x + 2 = 0$
2. $2x^2 - x - 2 = 0$
3. $x^2 + 5x + 1 = 0$
4. $2x^2 - x - 4 = 0$
5. $x^2 + 1 = 4x$
6. $2x^2 - x = 5$
7. $1 + x - 3x^2 = 0$
8. $3x^2 = 1 - x$

Find, correct to 3 decimal places, the roots of these equations.

9. $5x^2 + 9x + 2 = 0$
10. $2x^2 - 7x + 4 = 0$
11. $4x^2 - 7x - 1 = 0$
12. $3x = 5 - 4x^2$
13. $4x^2 + 3x = 5$
14. $1 = 5x - 5x^2$
15. $8x - x^2 = 1$
16. $x^2 - 3x = 1$

2.5 Properties of the roots of a quadratic equation

Some facts about the roots of a quadratic equation can be found by looking at the formula used for solving a quadratic equation. First write it in the form

$$x = -\dfrac{b}{2a} \pm \dfrac{\sqrt{b^2 - 4ac}}{2a}$$

The separate roots are $-\dfrac{b}{2a}+\dfrac{\sqrt{b^2-4ac}}{2a}$ and $-\dfrac{b}{2a}-\dfrac{\sqrt{b^2-4ac}}{2a}$

When the roots are added, the terms containing the square root disappear giving

$$\text{sum of roots}=-\dfrac{b}{a}$$

This fact is useful as a check on the accuracy of roots that have been calculated.

The discriminant and the nature of the roots

In the formula there are two terms. The first, $-\dfrac{b}{2a}$, can always be found for any values of a and b.

There are three different types of value of the second term $\dfrac{\sqrt{b^2-4ac}}{2a}$.

1 When b^2-4ac is positive, its square root can be found and, whether it is a whole number, a fraction or a surd, it is a type of number we know – it is called a **real number**.

The two square roots, $\pm\sqrt{b^2-4ac}$, have different values giving two different real values of x.

So the equation has *two different real roots*.

2 If b^2-4ac is zero then its square root also is zero and $x=-\dfrac{b}{2a}-\dfrac{\sqrt{b^2-4ac}}{2a}$ gives

$$x=-\dfrac{b}{2a}+0 \text{ and } x=-\dfrac{b}{2a}-0$$

so there is just one value of x that satisfies the equation.

For example, using the formula to solve $x^2-2x+1=0$

gives $x=-\dfrac{(-2)}{2}\pm0$, so $x=1$ or $x=1$.

Solution by factorising shows that there are two equal roots,

$$(x-1)(x-1)=0 \Rightarrow x=1 \text{ or } x=1$$

This type of equation has *equal roots* (also called a *repeated root*).

3 If b^2-4ac is negative its square root cannot be found because there is no real number whose square is negative. Therefore the equation has *no real roots*.

These facts show that the roots of a quadratic equation can be

either   real and different

or   real and equal

or   not real

and that it is the value of b^2-4ac which determines the nature of the roots.

$b^2 - 4ac$ is called the discriminant

Condition	Nature of Roots
$b^2 - 4ac > 0$	Real and different
$b^2 - 4ac = 0$	Real and equal
$b^2 - 4ac < 0$	Not real

Sometimes it matters only that the roots are real, in which case the first two conditions are combined to give

If $b^2 - 4ac \geq 0$, the roots are real.

Example 10

Question

Determine the nature of the roots of the equation $x^2 - 6x + 1 = 0$.

Answer

$x^2 - 6x + 1 = 0$
$a = 1, b = -6, c = 1$
$b^2 - 4ac = (-6)^2 - 4(1)(1) = 32$
$b^2 - 4ac > 0$ so the roots are real and different.

Example 11

Question

If the roots of the equation $2x^2 - px + 8 = 0$ are equal, find the value of p.

Answer

$2x^2 - px + 8 = 0$
$a = 2, b = -p, c = 8$
The roots are equal so $b^2 - 4ac = 0$,

therefore $\qquad (-p)^2 - 4(2)(8) = 0$
$\Rightarrow \qquad\qquad p^2 - 64 = 0 \Rightarrow p^2 = 64 \quad$ so $\quad p = \pm 8$

Example 12

Question

Prove that the equation $(k - 2)x^2 + 2x - k = 0$ has real roots for any value of k.

Answer

$(k - 2)x^2 + 2x - k = 0$
$a = k - 2, b = 2, c = -k$
$b^2 - 4ac = 4 - 4(k - 2)(-k) = 4 + 4k^2 - 8k$
$\qquad\qquad = 4k^2 - 8k + 4 = 4(k^2 - 2k + 1) = 4(k - 1)^2$

$(k - 1)^2$ cannot be negative whatever the value of k, so $b^2 - 4ac$ cannot be negative.
Therefore the roots are real for any value of k.

Exercise 6

Without solving the equation, write down the sum of its roots.

1 $x^2 - 4x - 7 = 0$ **2** $3x^2 + 5x + 1 = 0$ **3** $2 + x - x^2 = 0$

4 $x^2 + 3x + 1 = 0$ **5** $3x^2 - 4x - 2 = 0$ **6** $7 + 2x - 5x^2 = 0$

Without solving the equation, find the nature of its roots.

7 $x^2 - 6x + 4 = 0$ **8** $3x^2 + 4x + 2 = 0$

9 $2x^2 - 5x + 3 = 0$ **10** $x^2 - 6x + 9 = 0$

11 $4x^2 - 12x - 9 = 0$ **12** $4x^2 + 12x + 9 = 0$

13 $x^2 + 4x - 8 = 0$ **14** $x^2 + ax + a^2 = 0$

15 $x^2 - ax - a^2 = 0$ **16** $x^2 + 2ax + a^2 = 0$

17 If the roots of $3x^2 + kx + 12 = 0$ are equal, find k.

18 If $x^2 - 3x + a = 0$ has equal roots, find a.

19 The roots of $x^2 + px + (p-1) = 0$ are equal. Find p.

20 Prove that the roots of the equation $kx^2 + (2k+4)x + 8 = 0$ are real for all values of k.

21 Show that the equation $ax^2 + (a+b)x + b = 0$ has real roots for all values of a and b.

22 Find the relationship between p and q if the roots of the equation $px^2 + qx + 1 = 0$ are equal.

2.6 Simultaneous equations

When only one unknown quantity has to be found, only one equation is needed for a solution.

When two unknown quantities have to be found, two equations are needed. The two equations can be used to eliminate one of the unknowns. This gives one equation containing just one unknown which can be found. The other unknown can then also be found.

The values found then satisfy both equations, so the equations are satisfied simultaneously.

For example, to solve the equations $y + x = 3$ [1]

and $y + 2x = 5$ [2]

we can eliminate y by subtracting [1] from [2] so

     $[2] - [1] \Rightarrow x = 2$, then from [1], $y + 2 = 3$ so $y = 1$

or we can use [1] to find y in terms of x and then substitute this expression in [2]

     $[1] \Rightarrow y = 3 - x$ so [2] becomes $3 - x + 2x = 5 \Rightarrow x = 2$ and again from [1], $y = 1$

Solution of one linear and one quadratic equation

A way to eliminate an unknown quantity from two equations is by substitution. From the linear equation, one unknown can be expressed in terms of the other. This can then be substituted in the quadratic equation.

Example 13

Solve the equations
$x - y = 2$
$2x^2 - 3y^2 = 15$

$x - y = 2$ [1]
$2x^2 - 3y^2 = 15$ [2]

Equation [1] is linear so use it for the substitution

$\quad x = y + 2$

Substituting $y + 2$ for x in [2] gives

$\quad 2(y + 2)^2 - 3y^2 = 15$

$\Rightarrow\quad 2(y^2 + 4y + 4) - 3y^2 = 15$

$\Rightarrow\quad 2y^2 + 8y + 8 - 3y^2 = 15$

Collecting terms on the side where y^2 is positive gives

$\quad 0 = y^2 - 8y + 7$

$\Rightarrow\quad 0 = (y - 7)(y - 1), \qquad$ so $\qquad y = 7$ or $y = 1$

Now use $x = y + 2$ to find corresponding values of x.

x	3	9
y	1	7

Therefore $x = 9$ and $y = 7$ or $x = 3$ and $y = 1$.

> **Note**
>
> The values of x and y must be given in *corresponding pairs*. It is incorrect to write the answer as $y = 7$ or $y = 1$ and $x = 9$ or $x = 3$ because
> $y = 7$ with $x = 3$
> $y = 1$ with $x = 9$
> are *not* solutions.

Exercise 7

Solve the following pairs of equations.

1 $x^2 + y^2 = 5$
     $y - x = 1$

2 $y^2 - x^2 = 8$
     $x + y = 2$

3 $3x^2 - y^2 = 3$
     $2x - y = 1$

4 $y = 4x^2$
     $y + 2x = 2$

5 $y^2 + xy = 3$
     $2x + y = 1$

6 $x^2 - xy = 14$
     $y = 3 - x$

7 $xy = 2$
     $x + y - 3 = 0$

8 $2x - y = 2$
     $x^2 - y = 5$

9 $y - x = 4$
     $y^2 - 5x^2 = 20$

10 $x + y^2 = 10$
      $x - 2y = 2$

11 $4x + y = 1$
      $4x^2 + y = 0$

12 $3xy - x = 0$
      $x + 3y = 2$

13 $x^2 + 4y^2 = 2$
      $2y + x + 2 = 0$

14 $x + 3y = 0$
      $2x + 3xy = 1$

15 $3x - 4y = 1$
      $6xy = 1$

16 $x^2 + 4y^2 = 2$
      $x + 2y = 2$

17 $xy = 9$
$x - 2y = 3$

18 $4x + y = 2$
$4x + y^2 = 8$

19 $1 + 3xy = 0$
$x + 6y = 1$

20 $x^2 - xy = 0$
$x + y = 1$

21 $xy + y^2 = 2$
$2x + y = 3$

22 $xy + x = -3$
$2x + 5y = 8$

Summary

Methods for solving quadratic equations

1 Collect the terms in the order $ax^2 + bx + c = 0$, then factorise the left-hand side if possible.

2 Arrange in the form $ax^2 + bx = -c$, then complete the square on the left-hand side, adding $\left(\dfrac{b}{2}\right)^2$ to *both* sides.

3 Use the formula $x = \dfrac{-b \pm \sqrt{b^2 - 4ac}}{2a}$

Roots that are not rational should be given in surd form unless an approximate form (such as correct to 3 significant figures) is asked for.

Properties of roots

The nature of the roots depends on the value of the discriminant, that is, on the value of $b^2 - 4ac$.

$b^2 - 4ac > 0 \Rightarrow$ real different roots

$b^2 - 4ac = 0 \Rightarrow$ real equal roots

$b^2 - 4ac \geq 0 \Rightarrow$ real roots

$b^2 - 4ac < 0 \Rightarrow$ no real roots

Review

1 Factorise $3x^2 - 9x + 6$.

2 Factorise $4x^2 - 36$.

3 Solve the equation $x^2 - 5x - 6 = 0$.

4 Solve the equation $x^2 - 6x - 5 = 0$.

5 Solve the equation $2x^2 + 3x = 1$.

6 Solve the equation $5 - 3x^2 = 4x$.

In Questions 7 and 8 solve the pair of equations simultaneously. (Choose your substitution carefully to keep the amount of squaring to a minimum.)

7 $2x^2 - y^2 = 7$
$x + y = 9$

8 $2x = y - 1$
$x^2 - 3y + 11 = 0$

9 Use the formula to solve the equation $3x^2 - 17x + 10 = 0$.

    a Are the roots of the equation rational or irrational?

    b What does your answer to part **a** show about the left-hand side of the equation?

10 Determine the nature of the roots of the equations

    a $x^2 + 3x + 7 = 0$

    b $3x^2 - x - 5 = 0$

    c $ax^2 + 2ax + a = 0$

    d $2 + 9x - x^2 = 0$

11 Find the values of p for which the equation $px^2 + 4x + (p - 3) = 0$ has equal roots.

Assessment

1 **a** The equation $x^2 - (p-2)x + (p-2) = 0$ has equal roots. Show that p satisfies the equation $p^2 - 8p + 12 = 0$.

    b Solve the equation $p^2 - 8p + 12 = 0$.

    c Hence find the two possible forms of the equation $x^2 - (p-2)x + (p-2) = 0$.

2 **a** Express $x^2 - 3x - 6$ in the form $(x-a)^2 + b$ stating the values of a and b.

    b Hence solve the equation $x^2 - 3x - 6 = 0$ giving your answer in surd form.

3 **a** Express $x^2 - 5x + 7$ in the form $(x-a)^2 + b$ stating the values of a and b.

    b Hence show that the equation $x^2 - 5x + 7 = 0$ has no real roots.

4 Solve the equations $x^2 - 2y = 2$
$$2x + y = 5$$

5 **a** Find the value of the discriminant of the expression $2x^2 - 5x + 2$.

    b Hence show that the equation $2x^2 - 5x + 2 = 0$ has real and distinct roots.

6 The quadratic equation $(2k-7)x^2 - (k-2)x + (k-3) = 0$ has real roots.

Show that $7k^2 - 48k + 80 \leq 0$

AQA MPC1 June 2013 (part question)

7 **i** Express $x^2 - 6x + 11$ in the form $(x-p)^2 + q$.

    ii **Use the result from part (i)** to show that the equation $x^2 - 6x + 11 = 0$ has no real solutions.

AQA MPC1 January 2013 (part question)

3 Algebraic Division

Introduction

Factorising quadratic polynomials can be done by sight with practice. Factorising polynomials where x^3 is the highest power cannot usually be done by sight. This chapter gives methods for dividing a polynomial by a linear expression and for deciding if a linear expression is a factor of that polynomial.

Objectives

By the end of this chapter, you should know how to...

▶ Divide a cubic polynomial by $x - a$ where a is a number.

▶ Use the remainder theorem.

▶ Use the factor theorem.

Recap

You need to remember how to...

▶ Factorise a quadratic expression.
▶ Expand expressions such as $(x - 5)(x^2 + 3x - 2)$.
▶ Do long division, such as $251 \div 36$, without using a calculator.
▶ Solve a quadratic equation and a linear equation simultaneously.
▶ Determine whether a quadratic equation has real and different roots, equal roots or no real roots.

3.1 Division of a polynomial by $x - a$

A fraction where both the numerator and the denominator are polynomials, is **proper** if the highest power of x in the numerator is less than the highest power of x in the denominator, for example $\dfrac{x+1}{x^2+x+2}$.

When the fraction is an **improper fraction** we can divide the numerator by the denominator using long division.

> **Note**
>
> $x^2 + 3x - 7$ is called the **dividend** and $x + 2$ is called the **divisor**.

The example shows how to divide $x + 2$ into $x^2 + 3x - 7$.

$$
\begin{array}{r}
x+1 \\
x+2 \overline{)\, x^2 + 3x - 7} \\
\underline{x^2 + 2x} \\
x - 7 \\
\underline{x + 2} \\
-9
\end{array}
$$

Start by dividing x into x^2; it goes x times.

Multiply $x + 2$ by x and then subtract this from $x^2 + 3x$.

Bring down the -7, divide x into x and repeat the process.

No more division by x can be done, so stop here.

$x + 1$ is called the **quotient** and the **remainder** is -9.

Compare with dividing 32 by 5: $32 \div 5 = 6$ with remainder 2.

Writing $\dfrac{32}{5}$ as a mixed number gives $6 + \dfrac{2}{5}$.

Similarly $\dfrac{x^2 + 3x - 7}{x + 2} = x + 1 - \dfrac{9}{x + 2}$

where $x+1$ is a linear polynomial and $\dfrac{9}{x+2}$ is a proper fraction.

You may have to do long division where the dividend has a 'missing' term, for example $2x^3 - x + 5$ has no term in x^2. In this case put in a term $0x^2$ (or at least leave a space where that term should be). Doing this makes sure that the terms always line up.

Example 1

Question

Divide $2x^3 - x + 5$ by $x + 3$, giving the quotient and the remainder.

Answer

$$\require{enclose}
\begin{array}{r}
2x^2 - 6x + 17 \\
x+3 \enclose{longdiv}{2x^3 + 0x^2 - x + 5} \\
\underline{2x^3 + 6x^2} \\
-6x^2 - x \\
\underline{-6x^2 - 18x} \\
17x + 5 \\
\underline{17x + 51} \\
-46
\end{array}$$

x into $2x^3$ goes $2x^2$ times.
Multiply $2x^2$ by $x+3$ and subtract.

x into $-6x^2$ goes $-6x$ times.

The quotient is $2x^2 - 6x + 17$ and the remainder is -46.

so $\dfrac{2x^3 - x + 5}{x+3} = 2x^2 - 6x + 17 - \dfrac{46}{x+3}$.

Exercise 1

Carry out each of the following divisions, giving the quotient and the remainder.

1. $(2x^2 + 5x - 3) \div (x + 2)$
2. $(x^2 - x + 4) \div (x + 1)$
3. $(4x^3 + x - 1) \div (2x - 1)$
4. $(2x^3 - x^2 + 2) \div (x - 2)$

3.2 The remainder theorem and the factor theorem

The remainder theorem

$x^3 - 7x^2 + 6x - 2$ divided by $x - 2$ gives a **quotient** and a **remainder**.

The relationship between these quantities can be written as
$x^3 - 7x^2 + 6x - 2 = (\text{quotient})(x - 2) + \text{remainder}$

Substituting 2 for x eliminates the term containing the quotient, giving just the remainder,

so $2^3 - 7(2^2) + 6(2) - 2 = \text{remainder} = -10$

This is an example of the general case, that is when a polynomial is divided by $(x - a)$ then

$\text{polynomial} = (\text{quotient})(x - a) + \text{remainder}$

$\Rightarrow \text{remainder} = $ the value of the polynomial when a is substituted for x

This result is called the **remainder theorem**.

Exercise 2

For the divisions in Questions 1 to 8, find just the remainder.

1 $x^3 - 2x + 4$, $x - 1$ **2** $x^3 + 3x^2 - 6x + 2$, $x + 2$

3 $2x^3 - x^2 + 2$, $x - 3$ **4** $x^4 - 3x^3 + 5x$, $2x - 1$

5 $9x^5 - 5x^2$, $3x + 1$ **6** $x^3 - 2x^2 + 6$, $x - a$

7 $x^2 + ax + b$, $x + c$ **8** $x^4 - 2x + 1$, $ax - 1$

9 $x^2 - 7x + a$ has a remainder 1 when divided by $x + 1$. Find a.

The factor theorem

It is not easy to see the factors of $x^3 - 3x^2 + 7x - 10$.

But if $(x - a)$ is a factor, then $x^3 - 3x^2 + 7x - 10$ can be written as $(x - a) \times (\text{a quadratic})$ as there is no remainder,

that is $x^3 - 3x^2 + 7x - 10 = (x - a) \times (\text{a quadratic})$ [1]

Substituting a for x in [1] we get

$$a^3 - 3a^2 + 7a - 10 = (0) \times (\text{a quadratic})$$

so $a^3 - 3a^2 + 7a - 10 = 0$

The same argument can be applied to any polynomial in x and it means that

> if $(x - a)$ is a factor of a polynomial in x, the result is zero when a is substituted for x in the polynomial.

Conversely, if the result is not zero, $x - a$ is not a factor.

This result is called the **factor theorem**.

The factor theorem can help to find factors of cubic polynomials.

Look at $x^3 - 3x^2 + 7x - 10$ again.

The coefficient of x^3 is 1 and the factors of 10 are 1, 2, 5 and 10. Therefore one or more of $(x \pm 1)$, $(x \pm 2)$ and $(x \pm 5)$ may be a factor. We can try each of these in turn using the factor theorem.

Try $(x - 1)$: substitute 1 for x in $x^3 - 3x^2 + 7x - 10$

$$\Rightarrow (1)^3 - 3(1)^2 + 7(1) - 10 = -5 \neq 0$$

so $(x - 1)$ is not a factor.

Try $(x + 1)$: substitute -1 for x in $x^3 - 3x^2 + 7x - 10$

$$\Rightarrow (-1)^3 - 3(-1)^2 + 7(-1) - 10 = -21 \neq 0$$

so $(x + 1)$ is not a factor.

Try $(x - 2)$: substitute 2 for x in $x^3 - 3x^2 + 7x - 10$

$$\Rightarrow (2)^3 - 3(2)^2 + 7(2) - 10 = 0$$

so $(x - 2)$ is a factor.

Therefore $x^3 - 3x^2 + 7x - 10 = (x - 2)(ax^2 + bx + c)$

The values of a, b and c can be found either by using long division to divide $x^3 - 3x^2 + 7x - 10$ by $(x - 2)$ or by inspection knowing how the terms in the expansion of the brackets are formed.

> **Note**
>
> In a cubic polynomial. the highest power of x is 3.

$$x^3 - 3x^2 + 7x - 10 = (x - 2)\,(ax^2 + bx + c)$$

Checking the pairs of products that give $7x$ (or $-3x^2$) finds the value of b:

$$x^3 - 3x^2 + 7x - 10 = (x - 2)\,(x^2 + bx + 5)$$

showing that $7x = 5x - 2bx$ so $b = -1$.

Therefore $x^3 - 3x^2 + 7x - 10 = (x - 2)(x^2 - x + 5)$.

There are no more factors because $x^2 - x + 5$ does not factorise.

Equating coefficients is another method that can be used to find the values of a, b and c:

as $\quad x^3 - 3x^2 + 7x - 10 = (x - 2)(ax^2 + bx + c)$, expanding the right-hand side gives

$$x^3 - 3x^2 + 7x - 10 = ax^3 + (b - 2a)x^2 + (c - 2b)x - 2c = 0$$

Because the left-hand side is another way of expressing the right-hand side, we can equate the coefficients of $x^3 \Rightarrow a = 1$

and equating the constants gives $-10 = -2c$ so $c = 5$

for x^2, $-3 = (b - 2a)$ so $b = -1$.

Therefore $x^3 - 3x^2 + 7x - 10 = (x - 2)(x^2 - x + 5)$.

3.3 The factors of $a^3 - b^3$ and $a^3 + b^3$

$a^3 - b^3 = 0$ when $a = b$, so $a - b$ is a factor of $a^3 - b^3$ therefore

$$a^3 - b^3 = (a - b)(a^2 + ab + b^2) \qquad\qquad\qquad\text{[1]}$$

$a^3 + b^3 = 0$ when $a = -b$, so $a + b$ is a factor of $a^3 + b^3$ therefore

$$a^3 + b^3 = (a + b)(a^2 - ab + b^2) \qquad\qquad\qquad\text{[2]}$$

These two facts can be used to factorise similar forms, for example $x^3 - 8 = x^3 - 2^3$, so replacing a by x and b by 2 in [1] gives $x^3 - 8 = (x - 2)(x^2 + 2x + 4)$.

Example 2

Question

Find the value of a for which $2x - 1$ is a factor of $4x^3 - 2x^2 + ax - 4$.

Answer

From the factor theorem, the value of the expression is zero when $x = \dfrac{1}{2}$ (the value of x for which $2x - 1 = 0$).

As $2x - 1$ is a factor of $4x^3 - 2x^2 + ax - 4$,

$$4\left(\frac{1}{2}\right)^3 - 2\left(\frac{1}{2}\right)^2 + a\left(\frac{1}{2}\right) - 4 = 0$$

so $\quad \dfrac{1}{2} - \dfrac{1}{2} + \dfrac{1}{2}a - 4 = 0$ therefore $\quad a = 8$.

Example 3

Question

The equation $4x^2 + px + q = 0$ has equal roots. When $4x^2 + px + q$ is divided by $x + 1$ the remainder is 1.

Find the values of p and q.

Answer

$4(-1)^2 + p(-1) + q = 1 \quad \Rightarrow \quad 4 - p + q = 1 \quad \Rightarrow \quad p = q + 3$ [1]

If $4x^2 + px + q = 0$ has an equal root then '$b^2 - 4ac$' $= 0$

so $\quad p^2 - 16q = 0$ [2]

Solving equations [1] and [2] simultaneously gives

$(q+3)^2 - 16q = 0 \quad \Rightarrow \quad q^2 - 10q + 9 = 0 \quad \Rightarrow \quad (q-9)(q-1) = 0$

therefore $q = 9$ and $p = 12$ or $q = 1$ and $p = 4$.

Exercise 3

1 State whether $x - 1$ is a factor of $x^3 - 7x + 6$.

2 Is $x + 1$ a factor of $x^3 - 2x^2 + 1$?

3 Find whether $x - 3$ or $2x + 1$ or both are factors of $4x^3 - 7x + 9$.

4 Show that $x - 3$ is a factor of $x^3 - 7x - 6$.

5 Factorise fully

    a $x^3 + 2x^2 - x - 2$ **b** $x^3 - x^2 - x - 2$ **c** $2x^3 - x^2 + 2x - 1$

6 $x - 4$ is a factor of $x^3 - ax + 16$. Find a.

7 Find one factor of $2x^3 + x^2 + 9x - 5$. Hence express $2x^3 + x^2 + 9x - 5$ in the form $(ax + b)(px^2 + qx + r)$ giving the values of a, b, p, q and r.

8 $(x + 1)$ and $(x + 2)$ are both factors of $2x^3 + bx^2 - 5x + c$. Find the values of b and c.

9 $x^3 - 4x^2 - 25$ has a factor $(x - a)$. Find the value of a.

10 When $5x^3 - px^2 + x - q$ is divided by $x - 2$, the remainder is 3.

    $(x - 1)$ is a factor of $5x^3 - px^2 + x - q$. Find p and q.

11 Show that the polynomial $x^3 - x^2 - x - 2$ has only one linear factor.

12 Show that $x^3 - 2x^2 - 9x + 4 = (x - 4)(ax^2 + bx + c)$ and find the values of a, b and c.

Summary

The remainder theorem

The remainder theorem states that when a polynomial in x is divided by $x - a$, the remainder equals the value of the polynomial when a is substituted for x.

The factor theorem

The factor theorem states that if $(x - a)$ is a factor of a polynomial in x, the result is zero when a is substituted for x in the polynomial.

Review

1 Divide $2x^2 + 3x + 4$ by $x - 2$, giving the quotient and the remainder.

2 Divide $3x^3 - 2x^2 - x - 3$ by $x + 1$ giving the quotient and the remainder.

3 Divide $3x^2 - 5x + 1$ by $x + 3$.

Hence express $\dfrac{3x^2 - 5x + 1}{x + 3}$ as a linear expression plus a proper fraction.

4 Divide $x^3 - 4x^2 + 5$ by $x - 1$, giving the quotient and the remainder.

Hence express $\dfrac{x^3 - 4x^2 + 5}{x - 1}$ as a quadratic expression plus a proper fraction.

5 If $x^2 - 7x + a$ has a remainder 1 when divided by $x + 1$, find a.

6 Is $x - 1$ a factor of $x^3 - 2x^2 + 1$? Explain your answer.

7 Show that $x - 3$ is a factor of $x^3 - 5x^2 + 7x - 3$. Hence factorise $x^3 - 5x^2 + 7x - 3$.

8 Factorise $2x^3 - x^2 - 2x + 1$.

9 Find the value of k given that $x - 2$ is a factor of $x^3 - x^2 + kx + 8$.

Assessment

1 Divide $x^3 - 4x^2 + 5 - 1$ by $x - 1$.

Hence show that $\dfrac{x^3 + 4x^2 + 5x - 1}{x - 1} = ax^2 + bx + c + \dfrac{d}{x - 1}$ and state the values of a, b, c and d.

2 **a** Show that $(x + 1)$ is a factor of $x^3 + x^2 - x - 1$.

   b Hence factorise $x^3 + x^2 - x - 1$ completely.

3 **a** Use the remainder theorem to find the remainder when $x^3 - 3x^2 + x - 3$ is divided by $x - 1$.

   b Use the factor theorem to show that $x - 3$ is a factor of $x^3 - 3x^2 + x - 3$.

   c Express $x^3 - 3x^2 + x - 3$ in the form $(x - 3)(ax^2 + bx + c)$.

4 **a** Both $(x - 1)$ and $(x + 2)$ are factors of $x^3 + ax^2 + bx - 6$. Find the values of a and b.

   b Express $x^3 + ax^2 + bx - 6$ as the product of three linear factors.

5 **a** The expression $x^2 - ax + 6$ has a remainder of 2 when divided by $(x - 1)$. Find the value of a.

   b Use the value of a found in part **a** to solve the equation $x^2 - ax + 6 = 0$.

6 The polynomial p(x) is given by

$$p(x) = x^3 + 7x^2 + 7x - 15$$

   a i Use the Factor Theorem to show that $x + 3$ is a factor of p(x).

      ii Express p(x) as the product of three linear factors.

   b Use the Remainder Theorem to find the remainder when p(x) is divided by $x - 2$.

   c Verify that p(-1) < p(0).

AQA MPC1 June 2010 (part question)

7 The polynomial p(x) is given by p(x) = $x^2(x - 3) + 20$.

   i Find the remainder when p(x) is divided by $x - 4$.

   ii Use the Factor Theorem to show that $x + 2$ is a factor of p(x).

   iii Express p(x) in the form $(x + 2)(x^2 + bx + c)$, where b and c are integers.

   iv Hence show that the equation p(x) = 0 has exactly one real root and state its value.

AQA MPC1 June 2015 (part question)

4 Functions and Graphs

Introduction

This chapter introduces the basic idea of a function and looks at how graphs of functions can be used to interpret the solutions of equations and inequalities.

Recap

You need to remember how to...

▶ Plot points on a graph.
▶ Solve quadratic equations.
▶ Factorise quadratic and cubic polynomials.
▶ Use the discriminant to determine the nature of the roots of a quadratic equation.

Objectives

By the end of this chapter, you should know how to...

▶ Define linear, quadratic and cubic functions and recognise the shapes of their graphs.
▶ Apply transformations to graphs and know the effect that transformations have on the equations of curves.
▶ Use the shapes of graphs to help solve equations and inequalities.

4.1 Functions

When any number is substituted for x in the expression $x^2 - 2x$, there is a single answer.

For example when $x = 3$, $x^2 - 2x = 3$.

$x^2 - 2x$ is an example of a function.

> A function is any expression involving one variable which, when a number is substituted for the variable, gives a single answer.

The notation 'f' is used to mean function. The notation f(x) means a 'function of x'.

f(3) means the value of the function when 3 is substituted for x.

Using the example above, f(x) = $x^2 - 2x$ and f(3) = $3^2 - 2(3) = 3$.

Linear functions

> The general form of the linear function is f(x) = $ax + b$
> where x is a variable and a and b are constants and $a \neq 0$.

The graph of a linear function is a straight line.

Two points are needed to draw a sketch of its graph.

For example, to sketch the graph of f(x) = $2x - 3$, two points are given by

$x = 0$, f(x) = -3 and $x = 1$, f(x) = -1.

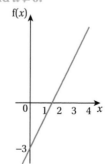

Quadratic functions

The general form of the quadratic function is $f(x) = ax^2 + bx + c$
where x is a variable, a, b and c are constants and $a \neq 0$.

When the graphs of quadratic functions are drawn for different values of a, b and c, the basic shape of the curve is always the same. This shape is called a **parabola**.

Every parabola has a line of symmetry which goes through the point where the curve has a least value or a greatest value. This point is called the **vertex**.

The vertex of a parabola can be below the x-axis, above the x-axis or touching the x-axis.

If the coefficient of x^2 is positive, that is $a > 0$,

then $f(x)$ has a least value and the parabola looks like this.

If the coefficient of x^2 is negative, that is $a < 0$,

then $f(x)$ has a greatest value and the curve is this way up.

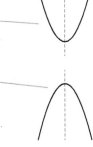

The value of x at the vertex is found by 'completing the square' on the left hand side of $f(x) = ax^2 + bx + c$.

This gives $f(x) = \left[\dfrac{4ac - b^2}{4a} \right] + a \left[x + \dfrac{b}{2a} \right]^2$

The first bracket is constant. The second bracket is squared, its value is zero when $x = -\dfrac{b}{2a}$ and greater than zero for all other values of x. Therefore

when a is positive, $f(x) = ax^2 + bx + c$ has a least value when $x = -\dfrac{b}{2a}$

when a is negative, $f(x) = ax^2 + bx + c$ has a greatest value when $x = -\dfrac{b}{2a}$

Example 1

Find the greatest or least value of the function given by $f(x) = 2x^2 - 7x - 4$ and sketch the graph of $f(x)$.

$f(x) = 2x^2 - 7x - 4 \implies a = 2$, $b = -7$ and $c = -4$.

As $a > 0$, $f(x)$ has a least value and this occurs when $x = -\dfrac{b}{2a} = \dfrac{7}{4}$

Therefore the least value of $f(x)$ is $f\left(\dfrac{7}{4}\right) = 2\left(\dfrac{7}{4}\right)^2 - 7\left(\dfrac{7}{4}\right) - 4 = -\dfrac{81}{8}$

This gives one point on the graph of $f(x)$ and the curve is symmetrical about this value of x.

To locate the curve more accurately we need another point so use $f(0)$ as it is easy to find.

$f(0) = -4$

'Sketch' a graph means that an accurate drawing is not needed. The axes do not need to be scaled, but the shape of the graph needs to be clear and points that show how the graph is placed on the axes need to be shown.

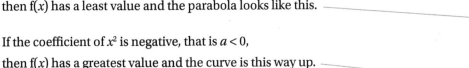

Example 2

Draw a sketch of the graph of $f(x) = (1 - 2x)(x + 3)$.

The coefficient of x^2 is negative, so $f(x)$ has a greatest value.

The curve cuts the x-axis when $f(x) = 0$.

When $f(x) = 0$, $(1 - 2x)(x + 3) = 0$

$\Rightarrow \quad x = \dfrac{1}{2}$ or $x = -3$

and this is where the curve cuts the x-axis.

The average of these values is $-\dfrac{5}{4}$, so the curve is symmetrical about $x = -\dfrac{5}{4}$.

This gives enough information to draw a quick sketch, but this method works only when the quadratic function factorises.

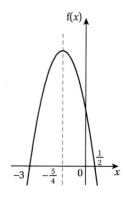

Cubic functions

The general form of a cubic function is
$f(x) = ax^3 + bx^2 + cx + d$
where a, b, c and d, are constants and $a \neq 0$.

Investigating the curve $y = ax^3 + bx^2 + cx + d$ for different values of a, b, c and d shows that the shape of the curve is

 when $a > 0$ and  when $a < 0$

Sometimes there are no turning points and the curve looks like this

 or

The graph of $f(x) = x^3$ looks like this.

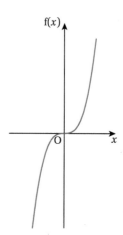

The graph of a cubic function always crosses the x-axis. The graph may cross the x-axis at three distinct points, it may cross the x-axis once and touch it once or cross the x-axis just once.

Example 3

Sketch the graph of $f(x) = (x-2)(x-1)(x+3)$.

Looking at the form of $f(x)$ shows that if the brackets are expanded, the coefficient of x^3 is 1.

The graph cuts the x-axis where $f(x) = 0$.

This is where $(x-2)(x-1)(x+3) = 0 \implies x = 2, x = 1$ and $x = -3$.

The graph cuts the vertical axis where $x = 0 \implies f(x) = 6$.

This gives enough information to sketch the graph.

Exercise 1

1. Find the greatest or least value of $f(x)$ where $f(x)$ is

   a $x^2 - 3x + 5$ **b** $2x^2 - 4x + 5$ **c** $3 - 2x - x^2$

2. Sketch the graph of each of the following quadratic functions, showing the greatest or least value and the coordinates of the vertex.

   a $x^2 - 2x + 5$ **b** $x^2 + 4x - 8$ **c** $2x^2 - 6x + 3$

   d $4 - 7x - x^2$ **e** $2 - 5x - 3x^2$

3. Write down the equation of the line of symmetry and draw a quick sketch of each of the following functions.

   a $(x-1)(x-3)$ **b** $(x+2)(x-4)$ **c** $(2x-1)(x-3)$

   d $(1+x)(2-x)$ **e** $x^2 - 9$ **f** $3x^2$

Find the values of x where the graph of $y = f(x)$ cuts the x-axis and sketch the curve when

4. $f(x) = x(x-1)(x+1)$ 5. $f(x) = (x-1)(x+1)(x-2)$

6. $f(x) = (x^2-1)(2-x)$ 7. $f(x) = x(x+1)(4-x)$

4.2 Graphical interpretation of equations

Graphical interpretation of quadratic equations

The graph of a function $f(x)$ is a curve or line. The equation $y = f(x)$ is called the equation of the curve or line.

When $f(x) = ax^2 + bx + c$, the equation $y = ax^2 + bx + c$ gives a parabola.

The curve cuts the x-axis when $y = 0$; these are the solutions of the quadratic equation $ax^2 + bx + c = 0$.

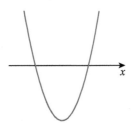

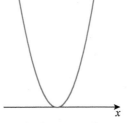

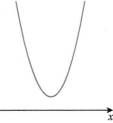

When $ax^2 + bx + c = 0$ has two distinct roots, the curve cuts the x-axis at two different values of x.

When $ax^2 + bx + c = 0$ has equal roots, the curve touches the x-axis at one value of x.

When $ax^2 + bx + c = 0$ has no real roots, the curve is all above (or below) the x-axis.

Graphical interpretation of cubic equations

When $y = ax^3 + bx^2 + cx + d$, the curve is a cubic curve.

The curve cuts the x-axis when $y = 0$; these are the solutions of the cubic equation $ax^3 + bx^2 + cx + d = 0$.

The shape of the curve shows that it always cuts the x-axis at least once.

Therefore $ax^3 + bx^2 + cx + d = 0$ always has at least one real root.

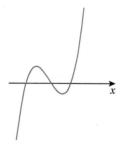 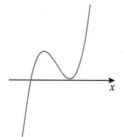 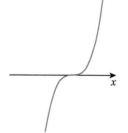 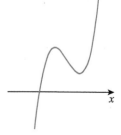

$ax^3 + bx^2 + cx + d = 0$ has three distinct roots. The curve cuts the x-axis at three different values of x.	$ax^3 + bx^2 + cx + d = 0$ has a pair of equal roots. The curve cuts the x-axis once and touches the x-axis at one value of x.	$ax^3 + bx^2 + cx + d = 0$ has three equal roots. The curve touches the x-axis and crosses it at the same value of x.	$ax^3 + bx^2 + cx + d = 0$ has only one real root. The curve cuts the x-axis once.

Intersection

Solving the simultaneous equations $y = x^2$ and $y = 2x + 3$ gives the values of x and y that satisfy both equations.

The points where the curve $y = x^2$ and the line $y = 2x + 3$ intersect have values of x and y that satisfy both equations.

Therefore the values of x and y at the points of intersection are the solutions of the simultaneous equations.

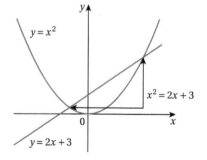

> In general, the points of intersection of the curves $y = f(x)$ and $y = g(x)$ are found by solving the equations $y = f(x)$ and $y = g(x)$ simultaneously.
>
> When there are no real solutions, the curves do not intersect.
>
> When there is a repeated root, the curves will touch (they may also intersect at another point).

Example 4

Question

The curve C has equation $y = x^2 - 3x + 9$.

a Express $x^2 - 3x + 9$ in the form $(x - a)^2 + b$.

b Show that C does not cross the x-axis.

c The line L has equation $y = 4x + 3$.

   Show that the values of x where C and L intersect are the solutions of the equation $x^2 - 7x + 6 = 0$.

d Sketch the graphs of C and L.

a $x^2 - 3x + 9 = \left(x - \dfrac{3}{2}\right)^2 + 9 - \dfrac{9}{4}$

$\qquad\qquad\quad = \left(x - \dfrac{3}{2}\right)^2 + \dfrac{27}{4}$

b The coefficient of x^2 is 1 so the curve has a least value.

$y = \left(x - \dfrac{3}{2}\right)^2 + \dfrac{27}{4}$ has a least value of $\dfrac{27}{4}$ which is positive.

Therefore C does not cross the x-axis.

(Alternatively, the discriminant of $x^2 - 3x + 9 = 0$ is $9 - 4(9) = -27$.

-27 is negative, therefore $x^2 - 3x + 9 = 0$ has no real roots so C does not cross the x-axis.)

c The curve C and the line L intersect where $x^2 - 3x + 9 = 4x + 3$

$\Rightarrow x^2 - 7x + 6 = 0.$

Therefore the values of x where C and L intersect are the solutions of the equation $x^2 - 7x + 6 = 0.$

d The curve $y = x^2 - 3x + 9$ has a least value of $\dfrac{27}{4}$ and crosses the y-axis where $x = 0$ when $y = 9$.

The line $y = 4x + 3$ crosses the x-axis where $y = 0$ when $x = -\dfrac{3}{4}$ and crosses the y-axis where $x = 0$ when $y = 3$.

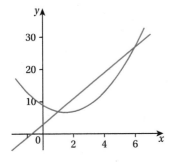

Exercise 2

1 **a** Show that the equation $x^2 - 7x + 13 = 0$ has no real roots.

    b The curve C has equation $y = x^2 - 7x + 13$ and the line L has equation $y = 7 - 2x$.

       Show that the value of x at the points where C and L intersect satisfy the equation $x^2 - 9x + 6 = 0$.

2 **a** Express $5 - 4x - x^2$ in the form $p - (q + x)^2$.

    b Hence give the values of x and y at the vertex of the parabola $y = 5 - 4x - x^2$.

    c Sketch the curve $y = 5 - 4x - x^2$.

3 **a** Factorise $x^3 - 1$.

    b Sketch the curve $y = x^3 - 1$.

4 **a** Factorise $x^3 - 3x - 2$.

    b Sketch the curve $y = x^3 - 3x - 2$.

5 Show that the curve C with equation $y = 2x^2 - 3x + 1$ and the line $y = x - 3$ do not intersect.

6 **a** The equation of a curve C is given by $y = (2 - x)(x^2 - x + 11)$.

       Sketch the curve.

    b The equation of a line L is given by $y = 2 - x$.

       State the number of points of intersection of C and L.

7 The equation of a curve C is given by $y = x^2 - 4x - 3$. The equation of a line L is given by $y = 2x - 12$.

    a Show that L touches C and find the the values of x and y at the point of contact.

    b Draw a sketch showing the curve C and the line L.

8 The equation of a curve C is given by $y = 3 - 2x - x^2$. The equation of a line L is given by $y = px + 2$.

    Show that the values of x at the points of intersection of C and L are given by the equation $x^2 + x(p + 2) - 3 = 0$.

9 The equation of a curve C is given by $y = x^2 + 12x + 20$. The equation of a line L is given by $y = px - 5$.

    a Find the equation that gives the values of x at the points of intersection of C and L.

    b Find the values of p for which L touches C.

4.3 Inequalities

An **inequality** compares two unequal quantities.

For example, comparing the two real numbers 3 and 8 shows that $8 > 3$.

The inequality remains true, that is the inequality sign is unchanged, when the same term is added or subtracted on both sides, for example

$$8 + 2 > 3 + 2 \quad \Rightarrow \quad 10 > 5$$

and $8 - 1 > 3 - 1 \quad \Rightarrow \quad 7 > 2$

The inequality sign is unchanged also when both sides are multiplied or divided by a positive quantity, for example

$$8 \times 4 > 3 \times 4 \quad \Rightarrow \quad 32 > 12$$

and $8 \div 2 > 3 \div 2 \quad \Rightarrow \quad 4 > 1\frac{1}{2}$

When both sides are multiplied or divided by a *negative* quantity the inequality is no longer true.

For example, multiplying $8 > 3$ by -1, the 8 becomes -8 and the 3 becomes -3. The correct inequality is now $-8 < -3$.

> When $a > b$
> $a + k > b + k$ for *all* values of k
> $ak > bk$ for *positive* values of k
> $ak < bk$ for *negative* values of k.

Solving linear inequalities

When an inequality contains an unknown quantity, the rules given above can be used to 'solve' it.

The solution of an equation is a value, or values, of the variable. The solution of an inequality is a range, or ranges, of values of the variable.

Example 5

Question

Find the set of values of x that satisfy the inequality $x - 5 < 2x + 1$.

Answer

$x - 5 < 2x + 1 \implies x < 2x + 6$ adding 5 to each side

$\implies -x < 6$ subtracting $2x$ from each side

$\implies x > -6$ multiplying both sides by -1

So $x - 5 < 2x + 1$ when $x > -6$.

Solving quadratic inequalities

A quadratic inequality involves a quadratic function, for example $x^2 - 3 > 2x$.

The solution is a range or ranges of values of the variable.

When the terms in the inequality can be collected and factorised, a graphical solution can be found.

Example 6

Question

Find the range(s) of values of x that satisfy the inequality $x^2 - 3 > 2x$.

Answer

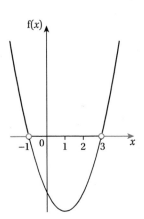

$x^2 - 3 > 2x \implies x^2 - 2x - 3 > 0$

$\implies (x - 3)(x + 1) > 0$

A sketch the graph of f(x) shows that f(x) > 0 where the graph is above the x-axis.

The values of x corresponding to these portions of the graph satisfy f(x) > 0.

Therefore the ranges of values of x which satisfy $x^2 - 3 > 2x$ are $x < -1$ and $x > 3$.

In this example there are two separate ranges.

But the solution of the inequality $(x - 3)(x + 1) < 0$ is the part of the graph below the x-axis *between* $x = -1$ and $x = 3$.

This is written as one range: $-1 < x < 3$.

Example 7

Question

The curve C has equation $y = ax^2 + 7x - 2$. Find the range of values of a for which C cuts the x-axis in two distinct places.

Answer

$y = ax^2 + 7x - 2$ cuts the x-axis when $ax^2 + 7x - 2 = 0$ has two distinct roots,

This is when '$b^2 - 4ac$' > 0 $\implies 49 + 8a > 0$

$$8a > -49$$

$$a > -\frac{49}{8} \text{ or } -6\frac{1}{8}$$

Exercise 3

Solve the following inequalities.

1 $x - 4 < 3 - x$ **2** $x + 3 < 3x - 5$ **3** $x < 4x + 9$

4 $7 - 3x < 13$ **5** $x > 5x + 2$ **6** $2x - 1 < x - 4$

7 $1 - 7x > x + 3$ **8** $2(3x - 5) > 6$ **9** $3(3 - 2x) < 2(3 + x)$

10 $(x - 2)(x - 1) > 0$ **11** $(x + 3)(x - 5) \geq 0$ **12** $(x - 2)(x + 4) < 0$

13 $(2x - 1)(x + 1) \geq 0$ **14** $x^2 - 4x > 3$ **15** $4x^2 < 1$

16 $(2 - x)(x + 4) \geq 0$ **17** $5x^2 > 3x + 2$ **18** $(3 - 2x)(x + 5) \leq 0$

19 $(x - 1)^2 > 9$ **20** $(x + 1)(x + 2) \leq 4$ **21** $(1 - x)(4 - x) > x + 11$

22 Find the range of values of k for which the equation $3x^2 - 7x + k = 0$ has two distinct real roots.

23 Find the range of values of b for which the equation $2x^2 + bx + 2 = 0$ has no real roots.

24 Find the range of values of k for which the equation $kx^2 - 2kx + 2 = 0$ has two distinct real roots.

25 The line $y = ax + 1$ and the curve $y = ax^2 - 49x + 5$ intersect at two distinct points. Find the range of values of a.

4.4 Transformations of graphs

A graph is **transformed** when it is moved (a **translation**) or reflected in the x-axis or the y-axis (a **reflection**) or it is stretched parallel to the x-axis or the y-axis (a **one-way stretch**).

Translations

The graph of the function $f(x) = 2^x$ is the curve $y = 2^x$.

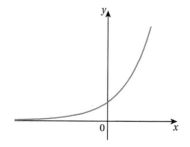

1 Look at the function g where $g(x) = f(x) + 2$.

Comparing $f(x) = 2^x$ with $g(x) = 2^x + 2$ shows that, for a particular value of x, the value of $g(x)$ is 2 units greater than the corresponding value of $f(x)$.

Therefore, for equal values of x, points on the curve $y = g(x)$ are two units above points on the curve $y = f(x)$.

Therefore the curve $y = 2^x + 2$ is a translation of the curve $y = 2^x$ by two units in the positive direction of the y-axis.

In general, for any function f, the curve $y = f(x) + a$ is the translation of the curve $y = f(x)$ by a units parallel to the y-axis. When $a > 0$ the translation is up the y-axis. When $a < 0$ the translation is down the y-axis.

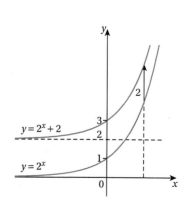

2 Look at the function g where $g(x) = 2^{x-2}$.

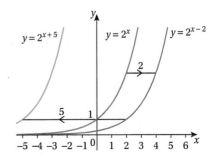

Comparing $f(x) = 2^x$ with $g(x) = 2^{x-2}$ shows that the values of $f(x)$ and $g(x)$ are the same when the value of x in $g(x)$ is 2 units greater than the value of x in $f(x)$, so $f(a) = g(a+2)$.

Therefore, for equal values of y, points on the curve $y = 2^{x-2}$ are 2 units to the *right* of points on the curve $y = 2^x$, so the curve $y = 2^{x-2}$ is a translation of the curve $y = 2^x$ by 2 units in the positive direction of the x-axis.

Also, comparing $f(x) = 2^x$ with $h(x) = 2^{x+5}$, shows that the values of $f(x)$ and $h(x)$ are the same when the value of x in $h(x)$ is five units less than the value of x in $f(x)$.

Therefore for equal values of y, points on the curve $y = 2^{x+5}$ are 5 units to the *left* of points on the curve $y = 2^x$.

In general, the curve $y = f(x + a)$ is a translation of the curve $y = f(x)$ by a units parallel to the x-axis.
If $a > 0$, the translation is in the negative direction of the x-axis and if $a < 0$, the translation is in the positive direction of the x-axis.

3 These two translations can be combined to give a single translation.

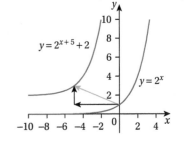

For example the curve $y = 2^{x+5} + 2$ is a translation of the curve $y = 2^x$ by 5 units in the negative direction of the x-axis and 2 units up the y-axis.

This translation can be described by the vector $\begin{bmatrix} -5 \\ 2 \end{bmatrix}$ where the top number gives the translation parallel to the x-axis and the bottom number gives the translation parallel to the y-axis.

In general the curve $y = f(x - a) + b$ is a translation
of the curve $y = f(x)$ by the vector $\begin{bmatrix} a \\ b \end{bmatrix}$.

Reflections

1 Look at the function $g(x) = -f(x) = -2^x$.

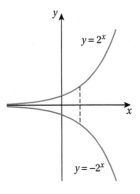

For a given value of x, $g(x)$ is equal to $-f(x)$.

Therefore for equal values of x, points on the curve $y = -2^x$ are the reflection in the x-axis of points on the curve $y = 2^x$, so the curve $y = -f(x)$ is the reflection in the x-axis of the curve $y = f(x)$.

In general, the curve $y = -f(x)$ is the reflection
of the curve $y = f(x)$ in the x-axis.

2 Look at the function $g(x) = f(-x) = 2^{-x}$.

Comparing $f(x) = 2^x$ with $g(x) = 2^{-x}$ shows that $f(x)$ and $g(x)$ are equal when the values given to x in $g(x)$ and $f(x)$ are equal but opposite in sign, therefore $g(a) = f(-a)$.

Therefore points with the same y-coordinates on the curves $y = 2^x$ and $y = 2^{-x}$, are symmetrical about $x = 0$, so the curve $y = 2^x$ is the reflection of the curve $y = 2^{-x}$ in the y-axis.

In general, the curve $y = f(-x)$ is the reflection of the curve $y = f(x)$ in the y-axis.

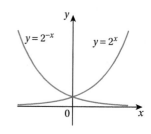

One-way stretches

Comparing points with the same x-coordinate on the curves $y = f(x)$ and $y = 3f(x)$, where $f(x) = (x-1)^2$ shows that the y-coordinate of the point on $y = 3(x-1)^2$ is 3 times the y-coordinate of the point on $y = (x-1)^2$.

Therefore the curve $y = 3(x-1)^2$ is $y = (x-1)^2$ stretched by a scale factor 3 parallel to the y-axis.

In general, the curve $y = af(x)$ is a one-way stretch of the curve $y = f(x)$ by a scale factor a parallel to the y-axis.

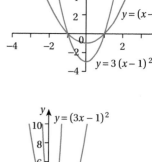

Comparing points with the same y-coordinate on the curves $y = f(x)$ and $y = f(3x)$ where $f(x) = (x-1)^2$ shows that the x-coordinate of the point on $y = (3x-1)^2$ is $\frac{1}{3}$ the x-coordinate of the point on $y = (x-1)^2$.

Therefore the curve $y = (3x-1)^2$ is the curve $y = (x-1)^2$ stretched by a scale factor $\frac{1}{3}$ parallel to the x-axis.

In general, the curve $y = f(ax)$ is a one-way stretch of the curve $y = f(x)$ by a factor $\frac{1}{a}$ parallel to the x-axis.

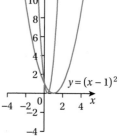

Example 8

The function f is given by $f(x) = x^3$. The diagram shows a sketch of the curve $y = x^3$.

On the same set of axes sketch the curves $y = g(x)$ and $y = h(x)$ where
$$g(x) = x^3 + 1, \quad h(x) = (x+1)^3.$$

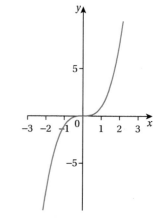

$f(x) = x^3 \Rightarrow g(x) = f(x) + 1$ so the curve $y = g(x)$ is a translation of the curve $y = x^3$ one unit in the direction of the positive y-axis.

$f(x) = x^3 \Rightarrow h(x) = f(x+1)$ so the curve $y = h(x)$ is a translation of the curve $y = x^3$ one unit in the direction of the negative x-axis.

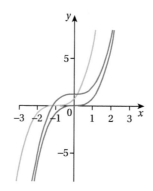

Example 9

The curve $y = x^3$ is translated by the vector $\begin{bmatrix} 3 \\ -1 \end{bmatrix}$ to the curve $y = g(x)$.
Write down an expression for g(x).

The curve $y = f(x - a) + b$ is a translation of the curve $y = f(x)$ by the vector $\begin{bmatrix} a \\ b \end{bmatrix}$.
Therefore $g(x) = (x - 3)^3 - 1$.

Exercise 4

1. On the same set of axes sketch the graphs of $f(x) = x^2$, $f(x) = (x + 1)^2$, $f(x) = -x^2$ and $f(x) = 2 + x^2$.

2. The equation of a line L is $y = 2x - 1$.
   a. The line L is transformed to the line $y = x - \frac{1}{2}$. Describe the transformation.
   b. The line L is transformed to the line $y = 6x - 3$. Describe the transformation.

3. The curve with equation $y = 3^{x+2} - 5$ is reflected in the y-axis to give the curve $y = g(x)$. Write down an expression for g(x).

4. The curve with equation $y = 3^{x+2} - 5$ is stretched by a scale factor 2 in the direction of the y-axis to give the curve $y = g(x)$. Write down an expression for g(x).

5. The curve with equation $y = \sqrt{2x-1}$ is translated by the vector $\begin{bmatrix} 3 \\ 5 \end{bmatrix}$ to give the curve $y = g(x)$. Write down an expression for g(x).

6. The curve with equation $y = \frac{1}{x}$ is transformed to give the curve with equation $y = g(x)$. Describe the transformation when g(x) is given by
   a. $\frac{1}{2x}$ b. $\frac{2}{x}$ c. $\frac{1}{x+2}$ d. $\frac{1}{x} + 2$

7. The curve with equation $y = \sqrt{4 - x^2}$ is stretched by a scale factor 2 in the direction of the x-axis to give the curve with equation $y = g(x)$.
   Write down an expression for g(x).

8. The curve with equation $y = \frac{2}{1-x}$ is reflected in the y-axis to give the curve with equation $y = g(x)$.
   Write down an expression for g(x).

9. The curve with equation $y = \frac{1}{\sqrt{8x^3 - 1}}$ is transformed to the curve with equation $y = \frac{1}{\sqrt{x^3 - 1}}$. Describe the transformation.

10. The curve $y = \frac{\sqrt{x^2 + 5}}{x}$ is translated by the vector $\begin{bmatrix} -1 \\ 5 \end{bmatrix}$ to give the curve with equation $y = g(x)$.
    Write down an expression for g(x).

Summary

Functions

The general form of a linear function is $f(x) = ax + b$, $a \neq 0$.

The general form of the quadratic function is $f(x) = ax^2 + bx + c$, $a \neq 0$.

When a is positive, $f(x) = ax^2 + bx + c$ has a least value when $x = -\dfrac{b}{2a}$.

When a is negative, $f(x) = ax^2 + bx + c$ has a greatest value when $x = -\dfrac{b}{2a}$.

The general form of a cubic function is $f(x) = ax^3 + bx^2 + cx + d$, $a \neq 0$.

Intersection

The points of intersection of the curves $y = f(x)$ and $y = g(x)$ are found by solving the equations $y = f(x)$ and $y = g(x)$ simultaneously.

If there are no real solutions then the curves do not intersect.

If there is a repeated root, the curves will touch (they may also intersect at another point).

Inequalities

When $a > b$ $a + k > b + k$ for *all* values of k

$ak > bk$ for *positive* values of k

$ak < bk$ for *negative* values of k.

Transformations

The curve $y = f(x - a) + b$ is a translation of the curve $y = f(x)$ by the vector $\begin{bmatrix} a \\ b \end{bmatrix}$.

The curve $y = -f(x)$ is the reflection of the curve $y = f(x)$ in the x-axis.

The curve $y = f(-x)$ is the reflection of the curve $y = f(x)$ in the y-axis.

The curve $y = af(x)$ is a one-way stretch of the curve $y = f(x)$ by a scale factor a parallel to the y-axis.

The curve $y = f(ax)$ is a one-way stretch of the curve $y = f(x)$ by a factor $\dfrac{1}{a}$ parallel to the x-axis.

Review

1 Find the greatest or least value of each of the following functions and state the value of x at which they occur.

    a $f(x) = x^2 - 3x + 5$ **b** $f(x) = 2x^2 - 7x + 1$ **c** $f(x) = (x - 1)(x + 5)$

2 **a**   Factorise $x^3 + 1$.   **b**   Sketch the curve $y = x^3 + 1$.

3 **a**   Show that the values of x where the line $y = x + 7$ intersects the curve
        $y = 2x^2 + 8x + 3$ satisfy the equation $2x^2 + 7x - 4 = 0$.

    b   Hence find the values of x and y at the points of intersection of the curve
        $y = 2x^2 + 8x + 3$ and the line $y = x + 7$.

P1 4

4 The curve C_1 has equation $y = x^2 + 1$ and the curve C_2 has equation $y = x^3 + x$.

    a Write down the equation satisfied by x at the point where C_1 and C_2 intersect.

    b Solve this equation and hence show that the curves intersect at only one point.

5 The curve C has equation $y = x^2 + x + 1$. The line L has equation $y = 3x$.

    Explain why L touches C but does not intersect C.

6 Solve the inequalities

    a $2x + 1 < 4 - x$ **b** $x - 5 > 1 - 3x$

    c $(x - 3)(x + 2) > 0$ **d** $(2x - 3)(3x + 2) < 0$

    e $x^2 - 3 < 10$ **f** $(x - 3)^2 > 2$

7 The graph of the function $f(x) = \dfrac{1}{x^2}$ is transformed to the graph of the function $g(x)$.

    Describe the transformation when

    a $g(x) = \dfrac{2}{x^2}$ **b** $g(x) = \dfrac{1}{2x^2}$ **c** $g(x) = \dfrac{1}{x^2} - 3$ **d** $g(x) = -\dfrac{1}{x^2}$

8 The curve with equation $y = \dfrac{x}{1-x}$ is transformed to the curve with equation $y = g(x)$.

    Give an expression for $g(x)$ when the transformation is

    a a reflection in the y-axis

    b a one way stretch parallel to the x-axis by a scale factor of 3

    c a translation by the vector $\begin{bmatrix} -2 \\ 3 \end{bmatrix}$.

Assessment

1 The function $f(x)$ is given by $f(x) = x^2 + 7x + 2$.

    a Express $f(x)$ in the form $a + (x + b)^2$.

    b Write down the equation of the line of symmetry of the curve C with equation $y = x^2 + 7x + 2$.

    c Write down the coordinates of the vertex of the curve C.

    d The curve C is transformed by the vector $\begin{bmatrix} 3.5 \\ 0 \end{bmatrix}$ to the curve C_1. Sketch the curve C_1.

2 The curve C_1 has equation $y = x^2 - 1$ and the curve C_2 has equation $y = x^3 + 2x - 1$.

    a Show that equation satisfied by x at the point where C_1 and C_2 intersect is $x^3 - x^2 + 2x = 0$.

    b Solve this equation and hence find the coordinates of the point where C_1 and C_2 intersect.

    c Describe the transformation that maps the curve C_1 to the curve with equation $y = x^2$.

Functions and Graphs 53

3 The curve C has equation $y = x^2 + 4x - 2$. The line L has equation $y = k(2x - 1)$.

  a Show that the x-coordinates of any points of intersection of C and L satisfy the equation $x^2 + (4 - 2k)x + (k - 2) = 0$.

  b For each value of k, the line L touches the curve C at one point. Show that $k^2 - 5k + 6 = 0$.

  c For each value of k, find the coordinates of the point where the line L touches the curve C.

4 The curve C_1 has equation $y = x^3 - x$.

  a Factorise $x^3 - x$ completely.

  b Explain why C_1 intersects the x-axis at three distinct points.

  c The curve C_2 has equation $y = x^3 + (k - 2)x^2 + k$.

    Show that the x-coordinates of the points of intersection of C_1 and C_2 satisfy the equation $(k - 2)x^2 + x + k = 0$.

  d The curves C_1 and C_2 intersect in two distinct points. Show that $4k^2 - 8k - 1 < 0$.

  e Solve the inequality $4k^2 + 4k + 1 > 0$.

5 The curve C has equation $y = \sqrt{2x^2 - 1}$

  a The curve C is stretched parallel to the y-axis with scale factor 2 to give the curve with equation $y = g(x)$.

    Write down an expression for $g(x)$.

  b The curve C is reflected in the x-axis to give the curve with equation $y = h(x)$.

    Write down an expression for $h(x)$.

6 **a** Express $x^2 + 3x + 2$ in the form $(x + p)^2 + q$, where p and q are rational numbers.

  b A curve has equation $y = x^2 + 3x + 2$.

    i Use the result from part **a** to write down the coordinates of the vertex of the curve.

    ii State the equation of the line of symmetry of the curve.

  c The curve with equation $y = x^2 + 3x + 2$ is translated by the vector $\begin{bmatrix} 2 \\ 4 \end{bmatrix}$

    Find the equation of the resulting curve in the form $y = x^2 + bx + c$.

AQA MPC1 June 2015

7 **a** **i** Express $2x^2 + 6x + 5$ in the form $2(x + p)^2 + q$, where p and q are rational numbers.

    ii Hence write down the minimum value of $2x^2 + 6x + 5$.

  b The point A has coordinates $(-3, 5)$ and the point B has coordinates $(x, 3x + 9)$.

    i Show that $AB^2 = 5(2x^2 + 6x + 5)$

    ii Use your result from part **a ii** to find the minimum value of the length AB as x varies, giving your answer in the form $\dfrac{1}{2}\sqrt{n}$, where n is an integer.

AQA MPC1 June 2013

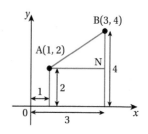

5 Coordinate Geometry

Introduction

Coordinate geometry is the name given to the graphical analysis of geometric properties.

Graphical methods are useful for investigating the geometric properties of many curves and surfaces. Straight lines and plane figures bounded by straight lines are studied in this chapter.

Recap

You need to remember how to...

▶ Use Pythagoras' Theorem.
▶ Simplify surds.
▶ Use properties of triangles and circles.
▶ Use properties of similar triangles.

Objectives

By the end of this chapter, you should know how to...

▶ Find and use formulae to give the distance between two points and the midpoint and the gradient of the line between two points.
▶ Use formulae to find the equation of a straight line.

- -

5.1 Lines joining two points

Cartesian coordinates

Cartesian coordinates use a fixed point O, called the **origin**, and a pair of perpendicular lines through O to locate a point on a plane. One line is drawn horizontally and is called the x-axis. The other line is drawn vertically and is called the y-axis. These lines are called Cartesian axes.

Three types of points are used in coordinate geometry:

1 Fixed points whose coordinates are known, e.g. the point (1, 2).

2 Fixed points whose coordinates are not known numerically. These points are referred to as (x_1, y_1), (x_2, y_2), ... etc. or (a, b), etc.

3 Points which are not fixed. These are general points and we refer to them as (x, y), (X, Y), etc.

The letters A, B, ... are used for fixed points and the letters P, Q, ... for general points.

The axes usually use identical scales to avoid distorting the shape of figures.

The distance between two points

The distance between points A and B can be found by using Pythagoras' theorem:

$$AB^2 = AN^2 + BN^2$$

For A(1, 2) and B(3, 4), the length of the line is given by

$$AB^2 = (3-1)^2 + (4-2)^2$$
$$= 8$$

Therefore $AB = \sqrt{8} = 2\sqrt{2}$.

In the same way the distance between any two points $A(x_1, y_1)$ and $B(x_2, y_2)$ can be found.

From the diagram, $AB^2 = AN^2 + BN^2$

$$= (x_2 - x_1)^2 + (y_2 - y_1)^2$$
$$\Rightarrow AB = \sqrt{(x_2 - x_1)^2 + (y_2 - y_1)^2}$$

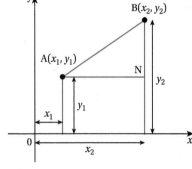

so

the distance between $A(x_1, y_1)$ and $B(x_2, y_2)$ is given by

$$\mathbf{AB} = \sqrt{(x_2 - x_1)^2 + (y_2 - y_1)^2}$$

This formula is true when some, or all, of the coordinates are negative.

Example 1

Question

Find the distance between A(−2, 2) and B(3, −1).

Answer

$$AB = \sqrt{(x_2 - x_1)^2 + (y_2 - y_1)^2}$$
$$= \sqrt{(3 - \{-2\})^2 + (-1-2)^2}$$
$$= \sqrt{5^2 + (-3)^2}$$
$$= \sqrt{34}$$

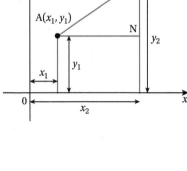

From the diagram, BN = 3 + 2 = 5 and AN = 2 + 1 = 3

$$\Rightarrow AB^2 = 5^2 + 3^2 = 34 \Rightarrow AB = \sqrt{34}$$

This confirms that the formula used above also works when some of the coordinates are negative.

The midpoint of the line joining two given points

Look at the line joining the points A(1, 1) and B(5, 3).

As M is the midpoint of AB, then S is the midpoint of CD.

The x-coordinate of M is half way between 1 and 5, so $\dfrac{(1+5)}{2} = 3$.

The y-coordinate of M is half way between 1 and 3, so $\dfrac{(1+3)}{2} = 2$.

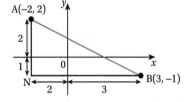

Therefore M is the point (3, 2).

In general, if $A(x_1, y_1)$ and $B(x_2, y_2)$ are two points, then the coordinates of M, the midpoint of AB, can be found in the same way.

The x-coordinate of M is half way between x_1 and x_2

so it is $\dfrac{1}{2}(x_1 + x_2)$.

The y-coordinate of M is half way between y_1 and y_2

so it is $\dfrac{1}{2}(y_1 + y_2)$.

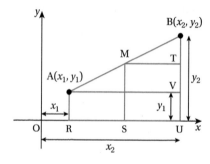

> The coordinates of the midpoint of the line joining $A(x_1, y_1)$ and $B(x_2, y_2)$
> are $\left(\dfrac{1}{2}(x_1 + x_2), \dfrac{1}{2}(y_1 + y_2)\right)$.

Example 2

Question

Find the coordinates of the midpoint of the line joining $A(-3, -2)$ and $B(1, 3)$.

Answer

The coordinates of M are $\left[\dfrac{1}{2}(x_1 + x_2), \dfrac{1}{2}(y_1 + y_2)\right]$

$= \left[\dfrac{1}{2}(-3+1), \dfrac{1}{2}(-2+3)\right] = \left(-1, \dfrac{1}{2}\right)$

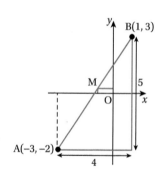

Alternatively, from the diagram, M is half-way from A to B horizontally and vertically,

therefore at M $\quad x = -3 + \dfrac{1}{2}(4) = -1$ and $y = -2 + \dfrac{1}{2}(5) = \dfrac{1}{2}$

This confirms that the formula works when some of the coordinates are negative.

Exercise 1

1 Find the distance between the points

    a $A(1, 2)$ and $B(4, 6)$ **b** $C(3, 1)$ and $D(2, 0)$ **c** $E(4, 2)$ and $F(2, 5)$

2 Find the coordinates of the midpoints of the lines joining the points in Question 1.

3 Find

    i the length

    ii the coordinates of the midpoint of the line joining

        a $A(-1, -4)$, $B(2, 6)$ **b** $C(0, 0)$, $D(-1, -2)$ **c** $E(-1, -4)$, $F(-3, -2)$

4 Find the distance from the origin to the point $(7, 4)$.

5 Find the length of the line joining the point $(-3, 2)$ to the origin.

6 Find the coordinates of the midpoint of the line from the point $(4, -8)$ to the origin.

7 Show, by using Pythagoras' Theorem, that the lines joining $A(1, 6)$, $B(-1, 4)$ and $C(2, 1)$ form a right-angled triangle.

8 The points A, B and C have coordinates (7, 3), (−4, 1) and (−3, −2) respectively.

   a Show that triangle ABC is isosceles.

   b Find the midpoint of BC.

9 The vertices of a triangle are A(0, 2), B(1, 5) and C(−1, 4). Find

   a the perimeter of the triangle

   b the coordinates of D where D is the midpoint of BC

   c the length of AD.

10 Show that the lines OA and OB are perpendicular where A and B are the points (4, 3) and (3, −4) respectively.

11 The point M is the midpoint of the line joining A to B. The coordinates of A and M are (5, 7) and (0, 2) respectively. Find the coordinates of B.

5.2 Gradient

The gradient of a straight line is defined as the increase in *y* divided by the increase in *x* between one point and another point on the line.

The diagram shows the line passing through A(1, 2) and B(4, 3).

From A to B, the increase in *y* is 1,

               the increase in *x* is 3.

Therefore the gradient of AB is $\frac{1}{3}$.

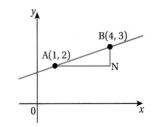

The gradient of a line can be found from *any* two points on the line.

Moving from A to B $\dfrac{\text{increase in } y}{\text{increase in } x} = \dfrac{-2}{4} = -\dfrac{1}{2}$

Alternatively, moving from B to A $\dfrac{\text{increase in } y}{\text{increase in } x} = \dfrac{2}{-4} = -\dfrac{1}{2}$

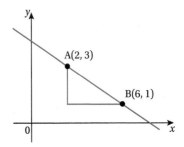

This shows that it does not matter in which order the two points are used, provided that they are used in the *same* order when calculating the increases in *x* and in *y*.

These two examples show that the gradient of a line can be positive or negative.

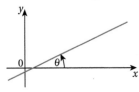

A positive gradient shows that the line makes an acute angle with the positive sense of the *x*-axis.

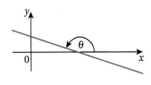

A negative gradient shows that the line makes an obtuse angle with the positive sense of the *x*-axis.

In general, the gradient of the line passing through $A(x_1, y_1)$ and $B(x_2, y_2)$ is

$$\frac{\text{the increase in } y}{\text{the increase in } x} = \frac{y_2 - y_1}{x_2 - x_1}$$

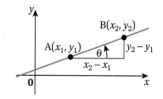

The gradient of a straight line is the increase in y divided by the increase in x from one point on the line to another, so

gradient measures the increase in y per unit increase in x, that is the rate of increase of y with respect to x.

Parallel lines

If l_1 and l_2 are **parallel** lines, they are inclined at the same angle to the positive direction of the x-axis,

so

parallel lines have equal gradients.

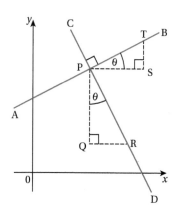

Perpendicular lines

The lines AB and CD are **perpendicular** with gradients m_1 and m_2 respectively.

The line AB makes an angle θ with the x-axis so the line CD also makes an angle θ with the y-axis.

Therefore triangles PQR and PST are similar.

The gradient of AB is $\dfrac{ST}{PS} = m_1$

and the gradient of CD is $\dfrac{-PQ}{QR} = m_2 \Rightarrow \dfrac{PQ}{QR} = -m_2$.

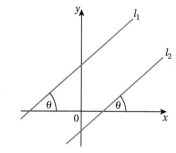

But $\dfrac{ST}{PS} = \dfrac{QR}{PQ}$ (triangles PQR and PST are similar)

therefore $m_1 = -\dfrac{1}{m_2}$ or $m_1 m_2 = -1$.

The product of the gradients of perpendicular lines is -1, therefore when one line has gradient m, any perpendicular line has gradient $-\dfrac{1}{m}$.

Example 3

Find, by comparing gradients, find whether or not the following three points are **collinear** (meaning they lie on the same straight line).

$$A\left(\frac{2}{3}, 1\right), B\left(1, \frac{1}{2}\right), C(2, -1)$$

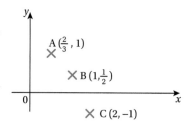

Answer

The gradient of AB is $\dfrac{1-\dfrac{1}{2}}{\dfrac{2}{3}-1}=-\dfrac{3}{2}$.

The gradient of BC is $\dfrac{-1-\dfrac{1}{2}}{2-1}=-\dfrac{3}{2}$.

As the gradients of AB and BC are the same, A B and C are collinear.

> **Note**
>
> The diagram is not necessary but it gives a check that the answer is reasonable.

Exercise 2

1 Find the gradient of the line through each pair of points.

a $(0, 0), (1, 3)$ **b** $(1, 4), (3, 7)$ **c** $(5, 4), (2, 3)$

d $(-1, 4), (3, 7)$ **e** $(-1, -3), (-2, 1)$ **f** $(-1, -6), (0, 0)$

g $(-2, 5), (1, -2)$ **h** $(3, -2), (-1, 4)$ **i** $(h, k), (0, 0)$

2 Find whether the given points are collinear.

a $(0, -1), (1, 1), (2, 3)$ **b** $(0, 2), (2, 5), (3, 7)$

c $(-1, 4), (2, 1), (-2, 5)$ **d** $(0, -3), (1, -4), \left(-\dfrac{1}{2}, -\dfrac{5}{2}\right)$

3 Find whether AB and CD are parallel, perpendicular or neither.

a A$(0, -1)$, B$(1, 1)$, C$(1, 5)$, D$(-1, 1)$

b A$(1, 1)$, B$(3, 2)$, C$(-1, 1)$, D$(0, -1)$

c A$(3, 3)$, B$(-3, 1)$, C$(-1, -1)$, D$(1, -7)$

d A$(2, -5)$, B$(0, 1)$, C$(-2, 2)$, D$(3, -7)$

e A$(2, 6)$, B$(-1, -9)$, C$(2, 11)$, D$(0, 1)$

5.3 The equation of a straight line

The Cartesian axes give a way of defining the position of any point in a plane. This plane is called the *xy*-plane.

In general x and y are independent variables. This means that they can each take any value independently of the value of the other unless some restriction is placed on them.

When the value of x is restricted to 2 but the value of y is not restricted, the condition gives a set of points which form a straight line parallel to the *y*-axis and passing through P, Q and R as shown.

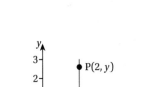

In the *xy*-plane, the equation $x = 2$ defines the line shown in the diagram; $x = 2$ is called *the equation of this line* or simply *the line* $x = 2$.

A straight line can be defined in many ways; for example, a line passes through the origin and has a gradient of $\dfrac{1}{2}$.

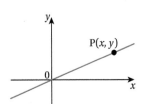

The point P(x, y) is on this line if and only if the gradient of OP is $\dfrac{1}{2}$.

In terms of x and y, the gradient of OP is $\dfrac{y}{x}$, so the statement above can be written in the form

P(x, y) is on the line $\Leftrightarrow$ $\dfrac{y}{x}=\dfrac{1}{2}$ $\Rightarrow$ $2y = x$.

The symbol $\Leftrightarrow$ means 'if and only if' or 'implies and is implied by'.

Therefore the coordinates of points on the line satisfy the relationship $2y = x$, and the coordinates of points that are not on the line do not satisfy this relationship.

$2y = x$ is called the equation of the line.

The equation of a line (straight or curved) is a relationship between the x and y-coordinates of all points on the line.

This relationship is not satisfied by any other point on the plane.

Example 4

Question

Find the equation of the line through the points A(1, –2) and B(–2, 4).

Answer

$P(x, y)$ is on the line if and only if the gradient of PA is equal to the gradient of AB (or PB).

The gradient of PA is $\dfrac{y-(-2)}{x-1} = \dfrac{y+2}{x-1}$.

The gradient of AB is $\dfrac{-2-4}{1-(-2)} = -2$.

Therefore the coordinates of P satisfy the equation $\dfrac{y+2}{x-1} = -2$

giving $y + 2x = 0$.

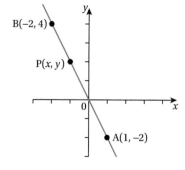

The general form of the equation of a line

Look at the more general case of a line whose gradient is m and which cuts the y-axis at a directed distance c from the origin.

c is called the **intercept** on the y-axis.

$P(x, y)$ is on this line if and only if the gradient of AP is m.

Therefore the coordinates of P satisfy the equation $\dfrac{y-c}{x-0} = m$.

Therefore $y = mx + c$.

This is the general form for the equation of a straight line.

> An equation of the form $y = mx + c$ represents a straight line
> with gradient m and intercept c on the y-axis.

Because the value of m and/or c may be a fraction, this equation can be rearranged and expressed as $ax + by + c = 0$, therefore

> $ax + by + c = 0$ where a, b and c are constants,
> is the equation of a straight line.

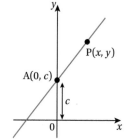

Note

In this form c is not the intercept.

Example 5

Question

a Write down the gradient of the line $3x - 4y + 2 = 0$.

b Find the equation of the line through the origin which is perpendicular to the given line.

a Rearranging $3x - 4y + 2 = 0$ to the general form gives $y = \dfrac{3}{4}x + \dfrac{1}{2}$.

Comparing with $y = mx + c$ we can 'read' the gradient (m) and the intercept on the y-axis.

The gradient of the line is $\dfrac{3}{4}$.

b The gradient of a perpendicular line is $-\dfrac{1}{m}$, that is $-\dfrac{4}{3}$. It passes through the origin so the intercept on the y-axis is 0.

Therefore its equation is $y = -\dfrac{4}{3}x + 0 \quad \Rightarrow \quad 3y + 4x = 0$.

Example 6

Sketch the line $x - 2y + 3 = 0$.

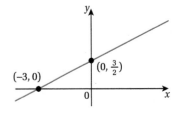

This line can be located accurately in the xy-plane when we know two points on the line. We will use the intercepts on the axes as these can be found easily

$$x = 0 \quad \Rightarrow \quad y = \dfrac{3}{2} \text{ and } y = 0 \quad \Rightarrow \quad x = -3.$$

[The diagram is a sketch, not an accurate plot, but it shows reasonably accurately the position of the line in the plane.]

Exercise 3

1 Write down the equation of the line that passes through the origin and has gradient

    a 2   **b** −1   **c** $\dfrac{1}{3}$ **d** $-\dfrac{1}{4}$ **e** 0

2 Write down the equation of the line passing through the given point with the given gradient.

    a $(0, 1)$, $\dfrac{1}{2}$ **b** $(0, 0)$, $-\dfrac{2}{3}$ **c** $(-1, -4)$, 4

    Draw a sketch showing all these lines on the same set of axes.

3 Write down the equation of the line passing through the points

    a $(0, 1), (2, 1)$ **b** $(1, 4), (3, 0)$ **c** $(-1, 3), (-4, -3)$

4 Write down the equation of the line that passes through the origin and is perpendicular to

    a $y = 2x + 3$ **b** $3x + 2y - 4 = 0$ **c** $x - 2y + 3 = 0$

5 Write down the equation of the line that passes through $(2, 1)$ and is perpendicular to

    a $3x + y - 2 = 0$ **b** $2x - 4y - 1 = 0$

6 Write down the equation of the line that passes through $(3, -2)$ and is parallel to

    a $5x - y + 3 = 0$ **b** $x + 7y - 5 = 0$

7 $A(1, 5)$ and $B(4, 9)$ are two adjacent vertices of a square. Find the equation of the line on which the side BC of the square lies. How long are the sides of this square?

Finding the equation of a straight line

Straight lines are important in graphical analysis. This section gives two ways in which the equation of a straight line can be found. Each gives a formula that can be used to write down the equation of a particular line.

The equation of a line with gradient m and passing through the point (x_1, y_1)

The point P (x, y) is a point on the line if and only if the gradient of AP is m

therefore $\dfrac{y - y_1}{x - x_1} = m$

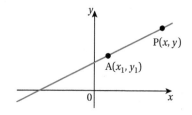

$\Rightarrow \quad y - y_1 = m(x - x_1)$ [1]

In some cases it is often easier to work from a diagram than to apply a formula.

The equation of the line passing through (x_1, y_1) and (x_2, y_2)

In the equation $y = mx + c$, m is the gradient of AB,

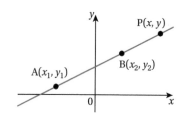

$\Rightarrow \quad m = \dfrac{y_2 - y_1}{x_2 - x_1}$

So the equation of the line through A and B is

$$y - y_1 = \left[\dfrac{y_2 - y_1}{x_2 - x_1} \right] (x - x_1)$$ [2]

Example 7

Question

Find the equation of the line with gradient $-\dfrac{1}{3}$ that passes through $(2, -1)$.

Answer

Using [1] with $m = -\dfrac{1}{3}$, $x_1 = 2$ and $y_1 = -1$ gives

$$y - (-1) = -\dfrac{1}{3}(x - 2)$$

$\Rightarrow \qquad x + 3y + 1 = 0$

Alternatively, the equation of this line can be found from the general form of the equation of a straight line, $y = mx + c$.

Using $y = mx + c$ and $m = -\dfrac{1}{3}$ gives $y = -\dfrac{1}{3}x + c$.

The point $(2, -1)$ lies on this line so its coordinates satisfy the equation,

$\Rightarrow \qquad -1 = -\dfrac{1}{3}(2) + c \quad \Rightarrow \quad c = -\dfrac{1}{3}$

Therefore $\quad y = -\dfrac{1}{3}x - \dfrac{1}{3}$

$\Rightarrow \qquad x + 3y + 1 = 0$

Example 8

Question

Find the equation of the line through the points $(1, -2)$ and $(3, 5)$.

Answer

Using formula [2] with $x_1 = 1$, $y_1 = -2$, $x_2 = 3$ and $y_2 = 5$ gives

$$y - (-2) = \frac{5 - (-2)}{3 - 1}(x - 1)$$

$\Rightarrow \quad 7x - 2y - 11 = 0$

Example 9

Question

Find the equation of the line through $(1, -2)$ which is perpendicular to the line $3x - 7y + 2 = 0$ and passes through $(1, 2)$.

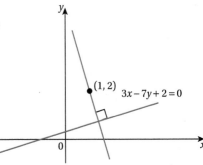

Answer

Expressing $3x - 7y + 2 = 0$ in general form gives $y = \frac{3}{7}x + \frac{2}{7}$.

Hence $3x - 7y + 2 = 0$ has gradient $\frac{3}{7}$.

So the perpendicular line has a gradient of $-\frac{7}{3}$ and it passes through $(1, 2)$.

Using $y - y_1 = m(x - x_1)$

$\Rightarrow \quad y - 2 = \frac{-7}{3}(x - 1) \quad \Rightarrow \quad 7x + 3y - 13 = 0$

Example 9 shows that the line perpendicular to $3x - 7y + 2 = 0$ passing through $(1, 2)$ has equation $7x + 3y - 13 = 0$

The coefficients of x and y have been transposed and the sign between the x and y terms has changed. This is a particular example of the general fact that

> given a line with equation $ax + by + c = 0$ then the equation of any
> perpendicular line is $bx - ay + k = 0$.

This property of perpendicular lines can be used to shorten the working of problems, for example to find the equation of the line passing through $(2, -6)$ which is perpendicular to the line $5x - y + 3 = 0$, then the required line has an equation of the form $x + 5y + k = 0$. Use the fact that the coordinates $(2, -6)$ satisfy this equation to find the value of k.

Exercise 4

1 Find the equation of the line with the given gradient that passes through the given point.

    a $3, (4, 9)$ **b** $-5, (2, -4)$ **c** $\frac{1}{4}, (4, 0)$

    d $0, (-1, 5)$ **e** $-\frac{2}{5}, \left(\frac{1}{2}, 4\right)$ **f** $-\frac{3}{8}, \left(\frac{22}{5}, -\frac{5}{2}\right)$

2 Find the equation of the line passing through the points

    a $(0, 1), (2, 4)$ **b** $(-1, 2), (1, 5)$ **c** $(3, -1), (3, 2)$

3 State which of the following pairs of lines are perpendicular.

   a $x - 2y + 4 = 0$ and $2x + y = 3$ **b** $x + 3y = 6$ and $3x + y + 2 = 0$

   c $x + 3y - 2 = 0$ and $y = 3x + 2$ **d** $y + 2x + 1 = 0$ and $x = 2y - 4$

4 Find the equation of the line through the point $(5, 2)$ that is perpendicular to the line $x - y + 2 = 0$.

5 Find the equation of the perpendicular bisector of the line joining

   a $(0, 0), (2, 4)$ **b** $(3, -1), (-5, 2)$ **c** $(5, -1), (0, 7)$

6 Find the equation of the line through the origin which is parallel to the line $4x + 2y - 5 = 0$.

7 The line $4x - 5y + 20 = 0$ cuts the x-axis at A and the y-axis at B.

   Find the equation of the line through O and the midpoint of the line AB.

8 Find the equation of the line through O that is perpendicular to the line AB defined in Question 7.

9 Find the equation of the perpendicular line from $(5, 3)$ to the line $2x - y + 4 = 0$.

10 The points A(1, 4) and B(5, 7) are two adjacent vertices of a parallelogram ABCD. The point C(7, 10) is another vertex of the parallelogram. Find the equation of the side CD.

5.4 Intersection

The point where two lines (or curves) cut is called a point of intersection.

If A is the point of intersection of the lines

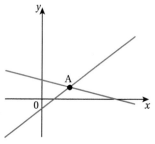

$$y - 3x + 1 = 0 \qquad\qquad [1]$$
$$\text{and} \qquad y + x - 2 = 0 \qquad\qquad [2]$$

then the coordinates of A satisfy both of these equations. Therefore A can be found by solving [1] and [2] simultaneously:

$$[2] - [1] \implies 4x - 3 = 0 \implies x = \frac{3}{4} \text{ and } y = \frac{5}{4}$$

Therefore $\left(\dfrac{3}{4}, \dfrac{5}{4}\right)$ is the point of intersection.

Exercise 5

1 Show that the triangle whose vertices are $(1, 1)$, $(3, 2)$ and $(2, -1)$ is isosceles.

2 Find the area of the triangular region enclosed by the x and y axes and the line $2x - y - 1 = 0$.

3 Find the coordinates of the vertices of the triangular region enclosed by the lines $y = 0$, $y = x + 5$ and $x + 2y - 6 = 0$.

4 Write down the equation of the perpendicular bisector of the line joining the points $(2, -3)$ and $\left(-\dfrac{1}{2}, 3\dfrac{1}{2}\right)$.

5 Find the equation of the line through A(5, 2) which is perpendicular to the line $y = 3x - 5$. Hence find the coordinates of the point where the two lines meet.

6 Find expressions in a and b for the coordinates of the point where the perpendicular from the point (a, b) to the line $x + 2y - 4 = 0$ meets the line $x + 2y - 4 = 0$.

7 The coordinates of a point P are $(t + 1, 2t - 1)$. Sketch the positions of P when $t = -1, 0, 1$ and 2. Show that these points are collinear and write down the equation of the line on which they lie.

8 Find the equation of the perpendicular bisector of the line joining the points (a, b) and (2a, − 3b).

9 The equations of two sides of a square are $y = 3x - 1$ and $x + 3y - 6 = 0$.

The point $(0, -1)$ is one vertex of the square. Find the coordinates of the other vertices.

10 The lines $y = 2x$, $2x + y - 12 = 0$ and $y = 2$ enclose a triangular region of the xy-plane. Find the coordinates of the vertices of this region.

Summary

Lines joining two points

The distance between points $A(x_1, y_1)$ and $B(x_2, y_2)$ is given by
$$AB = \sqrt{(x_2 - x_1)^2 + (y_2 - y_1)^2}$$

The coordinates of the midpoint of the line joining $A(x_1, y_1)$ and $B(x_2, y_2)$ are
$$\left(\frac{1}{2}(x_1 + x_2), \frac{1}{2}(y_1 + y_2) \right)$$

Gradient

The gradient of the line through $A(x_1, y_1)$ and $B(x_2, y_2)$ is
$$\frac{\text{the increase in } y}{\text{the increase in } x} = \frac{y_2 - y_1}{x_2 - x_1}$$

Parallel lines have equal gradients.

The product of the gradients of perpendicular lines is −1.

The equation of a straight line

The equation $y = mx + c$ gives a straight line with gradient m and y-intercept c. The equation of a straight line can also be expressed as $ax + by + c = 0$ where a, b and c are constants, but in this form c is not the y-intercept.

The formula $y - y_1 = m(x - x_1)$ or the formula $y - y_1 = \left[\dfrac{y_2 - y_1}{x_2 - x_1} \right](x - x_1)$ can be used to find the equation of a straight line.

Given a line with equation $ax + by + c = 0$, the equation of any perpendicular line is $bx - ay + k = 0$

Review

In Questions 1 to 4 state the letter that gives the correct answer.

1 The distance between the points $(3, -4)$ and $(-7, 2)$ is

    a $2\sqrt{3}$ **b** 16 **c** $2\sqrt{34}$ **d** $2\sqrt{5}$ **e** 6

2 The midpoint of the line joining $(-1, -3)$ to $(3, -5)$ is

    a $(1, 1)$ **b** $(0, 0)$ **c** $(2, -8)$ **d** $(1, -4)$ **e** $(1, -1)$

3 The gradient of the line joining $(1, 4)$ and $(-2, 5)$ is

    a $\dfrac{1}{3}$ **b** $-\dfrac{1}{3}$ **c** 3 **d** -3 **e** 1.3

4 The gradient of the line perpendicular to the line joining $(-1, 5)$ and $(2, -3)$ is

    a $\dfrac{3}{8}$ **b** $-2\dfrac{2}{3}$ **c** $\dfrac{1}{2}$ **d** 2 **e** $2\dfrac{2}{3}$

5 **a** Find the gradient of the straight line $2x + 3y = 5$.

    b Find the equation of the line through $(-2, 1)$ that is perpendicular to $2x + 3y = -9$.

    c Find the coordinates of the point where these two lines intersect.

6 The vertices of triangle ABC are A $(-3, 1)$, B $(10, -8)$ and C $(1, 4)$.

    a Find the equation of the line passing through A and B.

    b Show that CA and CB are perpendicular.

7 The straight line through the point P $(2, 1)$ and Q $(4, p)$ has gradient $-\dfrac{5}{12}$. Find the value of p.

Assessment

1 The line L has equation $2x + 3y - 4 = 0$.

    a Find the gradient of L.

    The point C has coordinates $(3, -1)$.

    b Find the equation of a line through C parallel to L.

    c Find the equation of the line through C perpendicular to L.

2 The vertices of triangle ABC are A $(-3, 1)$, B $(10, -8)$ and C $(1, 4)$ respectively.

    a Find the equation of the line passing through A and B.

    b Find the gradients of the lines CA and CB. Hence show that CA and CB are perpendicular.

    c The point D is the midpoint of AB. Write down the coordinates of D.

3 The line $3x + y + 2 = 0$ crosses the x-axis at the point A and crosses the y-axis at the point B.

    a Write down the coordinates of the point A and B.

    b Find the area of the triangle OAB where O is the origin.

    c Find the equation of the line through O and the midpoint of AB.

4 The coordinates of the point A are (1, 12) and the coordinates of the point B are $(-2, p)$.

The equation of the line through A and B is $y = 3x + 9$.

a Find the value of p.

b Find the gradient of AB.

c Find the coordinates of the midpoint of AB.

d Find the equation of the line through the point (1, 1) that is perpendicular to AB.

5 The line L_1 has equation $y = 5x - 2$. The point A with coordinates $(3, p)$ lies on the line L_1.

a Find the value of p.

The line L_2 has equation $2x - 3y + 1 = 0$.

b The lines L_1 and L_2 intersect at the point C. Find the coordinates of the point C.

c Find the length of the line AC.

6 The line AB has equation $7x + 3y = 13$.

a Find the gradient of AB.

b The point C has coordinates $(-1, 3)$.

   i Find an equation of the line which passes through the point C and which is parallel to AB.

   ii The point $\left(1\frac{1}{2}, -1\right)$ is the mid-point of AC. Find the coordinates of the point A.

c The line AB intersects the line with equation $3x + 2y = 12$ at the point B. Find the coordinates of B.

AQA MPC1 June 2011

7 The point A has coordinates $(-1, 2)$ and the point B has coordinates $(3, -5)$.

a **i** Find the gradient of AB.

   ii Hence find an equation of the line AB, giving your answer in the form $px + qy = r$, where p, q and r are integers.

b The midpoint of AB is M.

   i Find the coordinates of M.

   ii Find an equation of the line which passes through M and which is perpendicular to AB.

c The point C has coordinates $(k, 2k + 3)$. Given that the distance from A to C is $\sqrt{13}$, find the two possible values of the constant k.

AQA MPC1 June 2014

6 Differentiation

Introduction

As a car accelerates, the speed of the car changes as time changes. Differentiation is a way of measuring how one quantity changes as another quantity changes. This chapter also shows other situations that may involve finding conditions that will minimize or maximize a quantity.

Objectives

By the end of this chapter, you should...

▶ Understand and be able to use the rule for differentiating simple functions.

▶ Understand the idea of stationary points.

▶ Be able to use methods for distinguishing between different types of stationary points.

Recap

You need to remember how to...

▶ Work with surds and indices.
▶ Solve quadratic equations.
▶ Find the equation of a straight line.
▶ Use the formulae for the volume of a cuboid and the volume of a cylinder.
▶ Use Pythagoras' Theorem.

Applications

When a rocket is launched its speed increases rapidly. The rate at which the speed increases can be found when the distance of the rocket from the launch pad is expressed as a function of the time since launch.

A rectangular box is made from a sheet of cardboard with a fixed length and width. The volume of such a box can be optimized when the volume is expressed as a function of one of its dimensions.

6.1 Chords, tangents, normals and gradients

The points A and B are any two points on a curve.

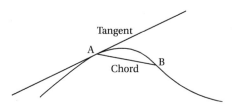

The line joining A and B is called a **chord**.

The line that touches the curve at A is called the **tangent** at A.

The word *touch* has an exact meaning. A line that meets a curve at a point and carries on without crossing to the other side of the curve at that point, *touches* the curve at the *point of contact*.

The line perpendicular to the tangent at A is called the **normal** at A.

Gradient defines the direction of a line (lines can be straight or curved).

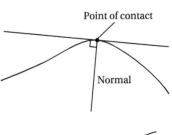

Point of contact

Normal

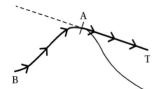

However when we reach A the direction we are pointing in is along the tangent AT, so

the gradient of the curve at A is the same as the gradient of the tangent to the curve at A.

The point B is a point on the curve, fairly close to A, so the gradient of the chord AB gives an *approximate* value for the gradient of the tangent at A.

As B gets nearer to A, the chord AB gets closer to the tangent at A so the approximation becomes more accurate.

Therefore,

as $B \rightarrow A$

the gradient of chord AB $\rightarrow$ the gradient of the tangent at A

or

$\displaystyle \lim_{\text{as } B \rightarrow A} (\text{gradient of chord AB}) = \text{gradient of tangent at A}$

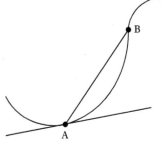

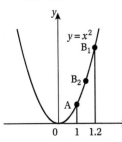

This example shows how to find the gradient at the point A(1, 1) on the curve with equation $y = x^2$.

Using the points B_1, B_2 ... that are getting closer to the point A(1, 1) and taking the x-coordinates of these points as 1.2, 1.1, 1.05, 1.01, 1.001, then calculating the gradient of the chord joining A to each position of B gives this table.

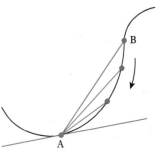

> **Note**
>
> The symbol $\rightarrow$ means 'tends to' or 'approaches'.

x	1.2	1.1	1.05	1.01	1.001
Gradient of chord AB	2.2	2.1	2.05	2.01	2.001

The numbers in the last row of the table show that, as B gets nearer to A, the gradient of the chord gets nearer to 2, so

$\displaystyle \lim_{\text{as } B \rightarrow A} (\text{gradient of chord AB}) = 2$

This method is too long to use every time the gradient at a point on the curve is needed, so a general method is required. To find the gradient at any point on the curve $y = x^2$ a general point A(x, y) is used and a variable small change in the value of x between A and B.

The symbol, δ, is used to mean a small change.

When δ is a prefix to any letter representing a variable quantity, it means a small increase in that quantity.

For example δx means a small increase in x

δy means a small increase in y

δt means a small increase in t.

(δ is only a prefix. It has no independent value and cannot be treated as a factor.)

Using *any* point A(x, y) on the curve $y = x^2$ and a point B on $y = x^2$ where the x-coordinate of B is $x + \delta x$,

gives the y-coordinate of B as $(x + \delta x)^2 = x^2 + 2x\delta x + (\delta x)^2$.

The gradient of chord AB is given by $\dfrac{\text{increase in } y \text{ from A to B}}{\text{increase in } x \text{ from A to B}}$

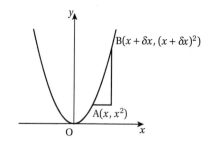

$$\dfrac{(x+\delta x)^2 - x^2}{(x+\delta x) - x}$$

$$= \dfrac{2x\delta x + (\delta x)^2}{\delta x}$$

$$= 2x + \delta x$$

As B $\rightarrow$ A, $\delta x \rightarrow 0$

therefore the gradient of the curve at A $= \underset{\text{B}\rightarrow\text{A}}{\text{limit}}(\text{gradient of chord AB})$

$$= \underset{\text{as } \delta x \rightarrow 0}{\text{limit}}(2x + \delta x)$$

$$= 2x$$

This result can now be used to find the gradient at any particular point on the curve with equation $y = x^2$.

For example at the point where $x = 3$, the gradient is $2(3) = 6$ and at the point $(4, 16)$, the gradient is $2(4) = 8$.

6.2 Differentiation

The process of finding a general expression for the gradient of a curve at any point is called **differentiation**.

The general expression for the gradient of a curve $y = \text{f}(x)$ is also a function so it is called the **gradient function**.

For the curve $y = x^2$, the gradient function is $2x$.

The gradient function is derived from the given function, it is also called the **derived function** or the **derivative**.

The example above shows that the derivative is found from (the increase in y) ÷ (the increase in x) so it also gives the rate at which the value of y changes with respect to x.

The method used above, in which the limit of the gradient of a chord is used to find the derived function, is called differentiating from first principles. It is the method by which the gradient of each new type of function is found.

Notation

We now need a short way to write 'the derivative of x^2 is $2x$'.

One notation uses the equation of the curve.

For example when $y = x^2$, the derived function is written as $\dfrac{dy}{dx} = 2x$ (spoken as 'dy by dx').

The complete symbol $\dfrac{d}{dx}$ means 'the derivative with respect to x of'.

So, $\dfrac{dy}{dx}$ means 'the derivative with respect to x of y'

and $\dfrac{d}{dx}(x^2 - x)$ means 'the derivative with respect to x of $(x^2 - x)$'.

Another notation uses the function of x.

For example when $f(x) = x^2$ the derived function is written as $f'(x) = 2x$

where f' means 'the gradient function' or 'the derived function'.

Note

d has no independent meaning and must never be thought of or used as a factor.

Differentiating x^n

The table shows the results of differentiating some equations from first principles.

y	x^2	x^3	x^4	x^{-1}
$\dfrac{dy}{dx}$	$2x$	$3x^2$	$4x^3$	$-x^{-2}$ $\left(\text{or } -\dfrac{1}{x^2}\right)$

From this table it appears that to differentiate a power of x, multiply by that power and then reduce the power by 1,

so $$\textbf{when } y = x^n \textbf{, then } \dfrac{dy}{dx} = nx^{n-1}$$

The rule works for all numerical powers of x, including those that are fractional or negative.

For example $\dfrac{d}{dx}(x^7) = 7x^6 \qquad \dfrac{d}{dx}(x^3) = 3x^2$

and $\dfrac{d}{dx}(x^{-2}) = -2x^{-3} \qquad \dfrac{d}{dx}(x^{\frac{3}{2}}) = \dfrac{3}{2}x^{\frac{1}{2}}$

Example 1

Question

Differentiate with respect to x

a $x^{-\frac{1}{3}}$

b $\sqrt[4]{(x^3)}$

Answer

a Use $\dfrac{d}{dx}(x^n) = nx^{n-1}$, where $n = -\dfrac{1}{3}$

$\dfrac{dy}{dx} = -\dfrac{1}{3}x^{-\frac{1}{3}-1} = -\dfrac{1}{3}x^{-\frac{4}{3}} = -\dfrac{1}{3x^{\frac{4}{3}}}$

b $\sqrt[4]{(x^3)}$ can be written $x^{\frac{3}{4}}$, so $n = \dfrac{3}{4}$

$\dfrac{d}{dx} = (x^{\frac{3}{4}}) = \dfrac{3}{4}x^{\frac{3}{4}-1} = \dfrac{3}{4}x^{-\frac{1}{4}}$ or $\dfrac{3}{4\sqrt[4]{x}}$

Exercise 1

Differentiate with respect to x.

1 x^5 **2** x^{-3} **3** $x^{\frac{4}{3}}$ **4** $\dfrac{1}{x}$ **5** x^{10} **6** $\dfrac{1}{x^2}$ **7** $\sqrt{x^3}$ **8** $x^{-\frac{1}{2}}$

9 $\dfrac{1}{x^4}$ **10** $x^{\frac{1}{3}}$ **11** $x^{-\frac{1}{4}}$ **12** x **13** $\sqrt{x^7}$ **14** $\dfrac{1}{x^7}$ **15** $x^{\frac{1}{7}}$ **16** $\sqrt{(x^2)^3}$

Differentiating constants and multiples of x

Any line with equation $y = c$ is a horizontal straight line whose gradient is zero.

Therefore $y = c \Rightarrow \dfrac{dy}{dx} = 0$.

Any line with equation $y = kx$ has gradient k,

$$y = kx \Rightarrow \frac{dy}{dx} = k$$

When y is a constant multiple of a function of x, so $y = af(x)$ then $\dfrac{dy}{dx} = af'(x)$.

For example when $\qquad y = 3x^5, \quad \dfrac{dy}{dx} = 3 \times 5x^4 = 15x^4$

and when $\qquad y = 4x^{-2}, \quad \dfrac{dy}{dx} = 4 \times -2x^{-3} = -8x^{-3}$

In general, if a is a constant

$$\frac{d}{dx} ax^n = anx^{n-1}$$

A function of x which contains the sum or difference of a number of separate terms can be differentiated term by term, applying the basic rule to each term.

For example, if $y = x^4 + \dfrac{1}{x} - 6x$

then $\qquad \dfrac{dy}{dx} = \dfrac{d}{dx}(x^4) + \dfrac{d}{dx}(x^{-1}) - \dfrac{d}{dx}(6x) = 4x^3 - \dfrac{1}{x^2} - 6$

Exercise 2

Differentiate each of the following functions with respect to x.

1 $x^3 - x^2 + 5x - 6$ **2** $3x^2 + 7 - \dfrac{4}{x}$ **3** $\sqrt{x} + \dfrac{1}{\sqrt{x}}$

4 $2x^4 - 4x^2$ **5** $x^3 - 2x^2 - 8x$ **6** $x^2 + 5\sqrt{x}$

7 $x^{-\frac{3}{4}} - x^{\frac{3}{4}} + x$ **8** $3x^3 - 4x^2 + 9x - 10$ **9** $x^{\frac{3}{2}} - x^{\frac{1}{2}} + x^{-\frac{1}{2}}$

10 $\sqrt{x} + \sqrt{x^3}$ **11** $\dfrac{1}{x^2} - \dfrac{1}{x^3}$ **12** $\dfrac{1}{\sqrt{x}} - \dfrac{2}{x}$

13 $x^{-\frac{1}{2}} + 3x^{\frac{3}{2}}$ **14** $x^{\frac{1}{4}} - x^{\frac{1}{5}}$ **15** $\dfrac{4}{x^3} + \dfrac{x^3}{4}$

16 $\dfrac{4}{x} + \dfrac{5}{x^2} - \dfrac{6}{x^3}$ **17** $3\sqrt{x} - 3x$ **18** $x - 2x^{-1} - 3x^{-3}$

19 $x\sqrt{x} - x^2\sqrt{x}$ **20** $\dfrac{\sqrt{x}}{x^2} + \dfrac{x^2}{\sqrt{x}}$

Differentiating products and fractions

The rules given above can be used to the differentiate expressions containing products or quotients only when they can be multiplied out or divided into separate terms.

Example 2

Question

Find $\dfrac{dy}{dx}$ when $y = (x-3)(x^2 + 7x - 1)$.

Answer

$$y = (x-3)(x^2 + 7x - 1) = x^3 + 4x^2 - 22x + 3$$

$$\Rightarrow \dfrac{dy}{dx} = 3x^2 + 8x - 22$$

Example 3

Question

Find $\dfrac{dt}{dz}$ when $t = \dfrac{6z^2 + z - 4}{2z}$.

Answer

$$t = \dfrac{6z^2 + z - 4}{2z} = \dfrac{6z^2}{2z} + \dfrac{z}{2z} - \dfrac{4}{2z}$$

$$= 3z + \dfrac{1}{2} - \dfrac{2}{z}$$

$$\Rightarrow \dfrac{dt}{dz} = 3 + 0 - 2(-z^{-2})$$

$$= 3 + \dfrac{2}{z^2}$$

Exercise 3

Differentiate each of the following equations with respect to the variable concerned.

1. $y = (x+1)^2$
2. $z = x^{-2}(2-x)$
3. $y = (3x-4)(x+5)$
4. $y = (4-z)^2$
5. $s = \dfrac{t^{-1} + 3t^2}{2t^2}$
6. $s = \dfrac{t^2 + t}{2t}$
7. $y = \left(\dfrac{1}{x}\right)(x^2 + 1)$
8. $y = \dfrac{z^3 - z}{\sqrt{z}}$
9. $y = 2x(3x^2 - 4)$
10. $s = (t+2)(t-2)$
11. $s = \dfrac{t^3 - 2t^2 + 7t}{t^2}$
12. $y = \dfrac{\sqrt{x} + 7}{x^2}$

6.3 Gradients of tangents and normals

When the equation of a curve is known, and the gradient function can be found, the gradient, m, at a given point A on that curve can be calculated. This is also the gradient of the tangent to the curve at A.

The normal at A is perpendicular to the tangent at A, therefore its gradient is $-\dfrac{1}{m}$.

Example 4

The equation of a curve is $s = 6 - 3t - 4t^2 - t^3$.

Find the gradient of the tangent and of the normal to the curve at the point $(-2, 4)$.

$s = 6 - 3t - 4t^2 - t^3 \implies \dfrac{ds}{dt} = 0 - 3 - 8t - 3t^2$

At the point $(-2, 4)$, $\dfrac{ds}{dt} = -3 - 8(-2) - 3(-2)^2 = 1$.

Therefore the gradient of the tangent at $(-2, 4)$ is 1 and the gradient of the normal is $-\dfrac{1}{1} = -1$.

Example 5

Find the coordinates of the points on the curve $y = 2x^3 - 3x^2 - 8x + 7$ where the gradient is 4.

$y = 2x^3 - 3x^2 - 8x + 7 \implies \dfrac{dy}{dx} = 6x^2 - 6x - 8$

When the gradient is 4, $\dfrac{dy}{dx} = 4$

therefore $6x^2 - 6x - 8 = 4 \implies 6x^2 - 6x - 12 = 0 \implies x^2 - x - 2 = 0$.

Therefore $(x - 2)(x + 1) = 0 \implies x = 2$ or $x = -1$.

When $x = 2$, $y = 16 - 12 - 16 + 7 = -5$.

when $x = -1$, $y = -2 - 3 + 8 + 7 = 10$.

Therefore the gradient is 4 at the points $(2, -5)$ and $(-1, 10)$.

Exercise 4

In Questions 1 to 12, find the gradient of the tangent and the gradient of the normal at the point given on the curve.

1 $y = x^2 + 4$ where $x = 1$

2 $y = \dfrac{3}{x}$ where $x = -3$

3 $y = \sqrt{z}$ where $z = 4$

4 $s = 2t^3$ where $t = -1$

5 $v = 2 - \dfrac{1}{u}$ where $u = 1$

6 $y = (x + 3)(x - 4)$ where $x = 3$

7 $y = z^3 - z$ where $z = 2$

8 $s = t + 3t^2$ where $t = -2$

9 $z = x^2 - \dfrac{2}{x}$ where $x = 1$

10 $y = \sqrt{x} + \dfrac{1}{\sqrt{x}}$ where $x = 9$

11 $s = \sqrt{t}(1 + \sqrt{t})$ where $t = 4$

12 $y = \dfrac{x^2 - 4}{x}$ where $x = -2$

In Questions 13 to 20, find the coordinates of the point(s) on the curve where the gradient has the value given.

13 $y = 3 - \dfrac{2}{x}; \dfrac{1}{2}$

14 $z = x^2 - x^3; -1$

15 $s = t^3 - 12t + 9; 15$

16 $v = u + \dfrac{1}{u}; 0$

17 $s = (t + 3)(t - 5); 0$

18 $y = \dfrac{1}{x^2}; \dfrac{1}{4}$

19 $y = (2x - 5)(x + 1); -3$

20 $y = z^3 - 3z; 0$

Equations of tangents and normals

The tangent is a line touching a curve at a given point A on the curve so the gradient of the tangent can be found at that point. Therefore the equation of the tangent can be found using $y - y_1 = m(x - x_1)$.

The equation of a normal can be found in the same way.

Example 6

Question

Find the equation of the normal to the curve $y = \dfrac{4}{x}$ at the point where $x = 1$.

Answer

$y = \dfrac{4}{x} \quad \Rightarrow \quad \dfrac{dy}{dx} = -\dfrac{4}{x^2}$

When $x = 1$, $y = 4$ and $\dfrac{dy}{dx} = -4$.

The gradient of the tangent at $(1, 4)$ is -4, therefore the gradient of the normal at $(1, 4)$ is $-\dfrac{1}{-4} = \dfrac{1}{4}$.

The equation of the normal is given by $y - y_1 = m(x - x_1)$

giving $\quad y - 4 = \dfrac{1}{4}(x - 1) \quad \Rightarrow \quad 4y = x + 15$.

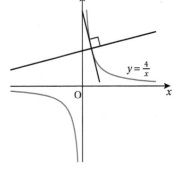

Example 7

Question

Find the equation of the tangent to the curve $y = x^2 - 6x + 5$ at each of the points where the curve crosses the x-axis. Find also the coordinates of the point where these tangents meet.

Answer

The curve crosses the x-axis where $y = 0$,

that is where $x^2 - 6x + 5 = 0 \quad \Rightarrow \quad (x - 5)(x - 1) = 0$

$\Rightarrow \quad x = 5 \quad$ and $\quad x = 1$

Therefore the curve crosses the x-axis at $(5, 0)$ and $(1, 0)$

$$y = x^2 - 6x + 5 \quad \Rightarrow \quad \dfrac{dy}{dx} = 2x - 6$$

At $(5, 0)$, the gradient of the tangent is given by $\dfrac{dy}{dx} = 10 - 6 = 4$

therefore the equation of this tangent is $y - 0 = 4(x - 5) \Rightarrow y = 4x - 20$.

At $(1, 0)$ the gradient of the tangent is given by $\dfrac{dy}{dx} = 2 - 6 = -4$.

Therefore the equation of the tangent is $y - 0 = -4(x - 1) \Rightarrow y + 4x = 4$.

The two tangents meet at the point P so at P,

$$y + 4x = 4 \text{ and } y - 4x = -20$$

Solving these equations simultaneously gives $2y = -16 \Rightarrow y = -8$.

Using $y = -8$ in $y + 4x = 4$ gives $-8 + 4x = 4 \Rightarrow x = 3$.

Therefore the tangents meet at $(3, -8)$.

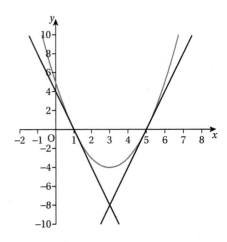

Exercise 5

In Questions 1 to 6 find, at the given point on the curve

a the equation of the tangent   **b** the equation of the normal.

1 $y = x^2 - 4$ where $x = 1$ **2** $y = x^2 + 4x - 2$ where $x = 0$

3 $y = \dfrac{1}{x}$ where $x = -1$ **4** $y = x^2 + 5$ where $x = 0$

5 $y = x^2 - 5x + 7$ where $x = 2$ **6** $y = (x - 2)(x^2 - 1)$ where $x = -2$

7 Find the equation of the normal to the curve $y = x^2 + 4x - 3$ at the point where the curve cuts the y-axis.

8 Find the equation of the tangent to the curve $y = x^2 - 3x - 4$ at the point where this curve cuts the line $x = 5$.

9 Find the equation of the tangent to the curve $y = (2x - 3)(x - 1)$ at each of the points where this curve cuts the x-axis. Find the point of intersection of these tangents.

10 Find the equation of the normal to the curve $y = x^2 - 6x + 5$ at each of the points where the curve cuts the x-axis.

11 Find the equation of the tangent to the curve $y = 3x^2 + 5x - 1$ at each of the points of intersection of the curve and the line $y = x - 1$.

12 Find the equations of the tangent to the curve $y = x^2 + 5x - 3$ at the points where the line $y = x + 2$ crosses the curve.

13 Find the coordinates of the point on the curve $y = 2x^2$ where the gradient is 8. Hence find the equation of the tangent to $y = 2x^2$ whose gradient is 8.

14 Find the coordinates of the point on the curve $y = 3x^2 - 1$ where the gradient is 3.

15 Find the equation of the tangent to the curve $y = 4x^2 + 3x$ whose gradient is -1.

16 Find the equation of the normal to the curve $y = 2x^2 - 2x + 1$ whose gradient is $\dfrac{1}{2}$.

17 Find the value of k for which $y = 2x + k$ is a tangent to the curve $y = 2x^2 - 3$.

18 Find the equation of the tangent to the curve $y = (x - 5)(2x + 1)$ that is parallel to the x-axis.

19 Find the coordinates of the point(s) on the curve $y = x^2 - 5x + 3$ where the gradient of the normal is $\dfrac{1}{3}$.

20 A curve has the equation $y = x^3 - px + q$. The tangent to this curve at the point $(2, -8)$ is parallel to the x-axis. Find the values of p and q.

Find also the coordinates of the other point where the tangent is parallel to the x-axis.

6.4 Increasing and decreasing functions

The derived function, $f'(x)$, expresses the rate at which the function $f(x)$ increases with respect to x.

At a particular point,

> when $f'(x)$ is positive, $f(x)$ is increasing as x increases, but if $f'(x)$ is negative then $f(x)$ is decreasing as x increases.

Example 8

Question

Show that the function given by $f(x) = 2x - x^2$ is

a an increasing function for values of x less than 1

b a decreasing function for values of x greater than 1.

Answer

$f(x) = 2x - x^2 \Rightarrow f'(x) = 2 - 2x$.

a $f(x)$ is increasing when $f'(x) > 0$,
that is when $2 - 2x > 0 \Rightarrow 1 > x$ so $x < 1$.

b $f(x)$ is decreasing when $f'(x) < 0$,
that is when $2 - 2x < 0 \Rightarrow 1 < x$ so $x > 1$.
Note: A sketch of $f(x)$ confirms this.

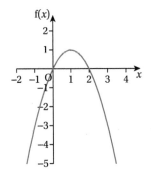

6.5 Stationary points

There may be points where $f'(x)$ is zero, that is where $f(x)$ is momentarily neither
increasing nor decreasing with respect to x.

The value of $f(x)$ at such a point is called a **stationary value**

therefore $f'(x) = 0 \Rightarrow y$ has a stationary value.

The diagram shows a general curve with equation $y = f(x)$.

At the points A and B, $f(x)$, and therefore y, is neither increasing
nor decreasing with respect to x.

So the values of y at A and B are stationary values,

therefore $\dfrac{dy}{dx} = 0 \Rightarrow y$ has a stationary value.

The point on a curve where y has a stationary value is called a
stationary point. At any stationary point, the gradient of the tangent to the
curve is zero, so the tangent is parallel to the x-axis.

To sum up

at a stationary point $\begin{cases} y, \text{ or } f(x) \text{ has a stationary value} \\ \dfrac{dy}{dx}, \text{ or } f'(x), \text{ is zero} \\ \text{the tangent is parallel to the } x\text{-axis.} \end{cases}$

Example 9

Question

Find the stationary values of the function $x^3 - 4x^2 + 7$.

Answer

$f(x) = x^3 - 4x^2 + 7 \Rightarrow f'(x) = 3x^2 - 8x$

At stationary points, $f'(x) = 0$ so $3x^2 - 8x = 0$

$\Rightarrow x(3x - 8) = 0 \Rightarrow x = 0$ and $x = \dfrac{8}{3}$.

Therefore there are stationary points where $x = 0$ and $x = \dfrac{8}{3}$.

When $x = 0$, $f(x) = 0 - 0 + 7 = 7$.

When $x = \dfrac{8}{3}$, $f(x) = \left(\dfrac{8}{3}\right)^3 - 4\left(\dfrac{8}{3}\right)^2 + 7 = -2\dfrac{13}{27}$.

Therefore the stationary points of $x^3 - 4x^2 - 5$ are $(0, 7)$ and $\left(\dfrac{8}{3}, -2\dfrac{13}{27}\right)$.

Exercise 6

1 Show that $f(x) = x^2$ is an increasing function for $x > 0$.

2 Show that $f(x) = x^3$ is an increasing function of x for all values of x.

In Questions 3 to 8 find the value(s) of x at which the following functions have stationary values.

3 $x^2 + 7$ **4** $2x^2 - 3x - 2$ **5** $x^3 - 4x^2 + 6$

6 $4x^3 - 3x - 9$ **7** $x^3 - 2x^2 + 11$ **8** $x^3 - 3x - 5$

In Questions 9 to 14 find the value(s) of x for which y has a stationary value.

9 $y = x^2 - 8x + 1$ **10** $y = x + \dfrac{9}{x}$ **11** $y = 2x^3 + x^2 - 8x + 1$

12 $y = 9x^3 - 25x$ **13** $y = 2x^3 + 9x^2 - 24x + 7$ **14** $y = 3x^3 - 12x + 19$

In Questions 15 to 20 find the coordinates of the stationary points on the curves.

15 $y = \dfrac{x^2 + 9}{2x}$ **16** $y = x^3 - 2x^2 + x - 7$ **17** $y = (x - 3)(x + 2)$

18 $y = x^{\frac{3}{2}} - x^{\frac{1}{2}}$ **19** $y = \sqrt{x} + \dfrac{1}{\sqrt{x}}$ **20** $y = x + \dfrac{9}{x}$

6.6 Maximum and minimum points

Close to a stationary point, a curve can have any one of the shapes shown in the diagram.

Moving through A from left to right shows that the curve is rising, then turns at A and begins to fall, so the gradient changes from positive, to zero at A, and then becomes negative.

At A there is a **turning point**.

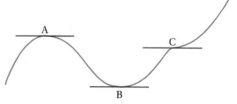

The value of y at A is called a **maximum value** and A is called a **maximum point**.

Moving through B from left to right the curve is falling, then turns at B and begins to rise, so the gradient changes from negative, to zero at B, and then becomes positive.

At B there is a **turning point**.

The value of y at B is called a **minimum value** and B is called a **minimum point**.

> **The tangent is always horizontal at a turning point.**

At C the curve does not turn. The gradient goes from positive, to zero at C and then becomes positive again, therefore the gradient does not change sign at C.

C is not a turning point but, because there is a change in the sense in which the curve is turning (from clockwise to anti-clockwise), C is called a point of inflection.

Note

A maximum value of y is *not always the greatest value of y on the curve*. The terms maximum and minimum apply only to the behaviour of the curve close to a stationary point; the point can be called a local maximum or a local minimum.

Investigating the nature of stationary points

This section shows two ways of distinguishing between the different types of stationary point.

Method 1

This method uses the sign of the gradient at points close to, and on either side of, the stationary point.

For a maximum point, A

$\dfrac{dy}{dx}$ at A_1 is +ve, $\dfrac{dy}{dx}$ at A_2 is −ve

For a minimum point, B

$\dfrac{dy}{dx}$ at B_1 is −ve, $\dfrac{dy}{dx}$ at B_2 is +ve

For a point of inflection, C

$\dfrac{dy}{dx}$ at C_1 is +ve, $\dfrac{dy}{dx}$ at C_2 is +ve

These conclusions are summarised in the table:

Sign of $\dfrac{dy}{dx}$	Passing through a maximum point $+\ 0\ -$	Passing through a minimum point $-\ 0\ +$	Passing through a point of inflection $\frac{dy}{dx}$ does not change sign
Gradient	/ ‾ \	\ _ /	/ – /

Method 2

This method looks at how $\dfrac{dy}{dx}$ changes with respect to x going through a stationary point.

The rate at which $\dfrac{dy}{dx}$ increases with respect to x can be written $\dfrac{d}{dx}\left(\dfrac{dy}{dx}\right)$ but this notation is shortened to $\dfrac{d^2y}{dx^2}$ (spoken as 'd 2 y by dx squared').

$\dfrac{d^2y}{dx^2}$ is the *second derivative* of y with respect to x. If $\dfrac{dy}{dx}$ is increasing as x increases then $\dfrac{d^2y}{dx^2}$ is +ve.

Looking at the behaviour of $\dfrac{dy}{dx}$ at each stationary point shows:

for the maximum point A; at A_1 $\dfrac{dy}{dx}$ is +ve and at A_2 $\dfrac{dy}{dx}$ is −ve so,

passing through A, $\dfrac{dy}{dx}$ goes from + to −, so $\dfrac{dy}{dx}$ decreases

$\Rightarrow$ at A, $\dfrac{d^2y}{dx^2}$ is negative

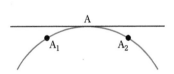

for the minimum point B; at B_1 $\dfrac{dy}{dx}$ is −ve and at B_2 $\dfrac{dy}{dx}$ is +ve so,

passing though B, $\dfrac{dy}{dx}$ goes from − to +, so $\dfrac{dy}{dx}$ increases

$\Rightarrow$ at B, $\dfrac{d^2y}{dx^2}$ is positive.

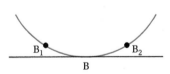

This table sums up method 2:

	Maximum	Minimum
Sign of $\dfrac{d^2y}{dx^2}$	negative (or zero)	positive (or zero)

This method is easy to use but $\dfrac{d^2y}{dx^2}$ can be zero at a turning point and it can also be zero at other points.

Therefore this method fails when $\dfrac{d^2y}{dx^2}=0$ at stationary point so the first method must be used.

Example 10

Find the stationary points on the curve $y = 4x^3 + 3x^2 - 6x - 1$ and determine the nature of each one.

$y = 4x^3 + 3x^2 - 6x - 1$

$\Rightarrow \dfrac{dy}{dx} = 12x^2 + 6x - 6$

At stationary points, $\dfrac{dy}{dx} = 0$ so $12x^2 + 6x - 6 = 0$

$$\Rightarrow \quad 6(2x - 1)(x + 1) = 0$$

Therefore there are stationary points where $x = \dfrac{1}{2}$ and $x = -1$.

When $x = \dfrac{1}{2}$, $y = -2\dfrac{3}{4}$ and when $x = -1$, $y = 4$

so the stationary points are $\left(\dfrac{1}{2}, -2\dfrac{3}{4} \right)$ and $(-1, 4)$.

Differentiating $\dfrac{dy}{dx}$ with respect to x gives $\dfrac{d^2 y}{dx^2} = 24x + 6$.

When $x = \dfrac{1}{2}$, $\qquad \dfrac{d^2 y}{dx^2} = 12 + 6$ which is positive

$\Rightarrow \left(\dfrac{1}{2}, -2\dfrac{3}{4} \right)$ is a minimum point.

When $x = -1$, $\qquad \dfrac{d^2 y}{dx^2} = -24 + 6$ which is negative

$\Rightarrow (-1, 4)$ is a maximum point.

Exercise 7

In Questions 1 to 15 find the stationary points on the following curves and distinguish between them.

1 $y = 2x - x^2$

2 $y = 3x - x^3$

3 $y = \dfrac{9}{x} + x$

4 $y = x^2(x - 5)$

5 $y = x^2$

6 $y = x + \dfrac{1}{2x^2}$

7 $y = 2x^2 - x^4$

8 $y = x^4$

9 $y = (2x + 1)(x - 3)$

10 $y = x^5 - 5x$

11 $y = x^2(x^2 - 8)$

12 $y = x^2 + \dfrac{16}{x^2}$

13 $x + \dfrac{1}{x}$

14 $3 - x + x^2$

15 $x^2(3x^2 - 2x - 3)$

16 Show that the curve with equation $y = x^5 + x^3 + 4x - 3$ has no stationary points.

(Hint: replace x^2 by w in the derivative.)

Applications

The example below shows how a variable (in this case the volume of a box) can be optimised.

Example 11

An open box is made from a square sheet of cardboard, with sides half a metre long. Squares are cut from each corner. The sides are then folded up to make the box. Find the maximum volume of the box.

The volume of the box depends on the unknown length of the side of the square cut from each corner so use x metres for this length. The side of the cardboard sheet is $\frac{1}{2}$ m, so this means that $0 < x < \frac{1}{4}$.

Using metres throughout,

the base of the box is a square of side $\left(\frac{1}{2} - 2x\right)$

and the height of the box is x,

therefore the volume, C, of the box is given by

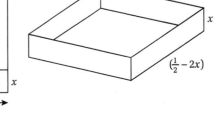

$$C = x\left(\frac{1}{2} - 2x\right)^2 = \frac{1}{4}x - 2x^2 + 4x^3 \text{ for } 0 < x < \frac{1}{4}$$

$$\Rightarrow \quad \frac{dC}{dx} = \frac{1}{4} - 4x + 12x^2$$

At a stationary value of C, $\frac{dC}{dx} = 0$

therefore $12x^2 - 4x + \frac{1}{4} = 0 \quad \Rightarrow \quad 48x^2 - 16x + 1 = 0$

$$\Rightarrow (4x - 1)(12x - 1) = 0 \quad \Rightarrow \quad x = \frac{1}{4} \quad \text{or} \quad x = \frac{1}{12}$$

Therefore there are stationary values of C when $x = \frac{1}{4}$ and when $x = \frac{1}{12}$.

It is not possible to make a box if $x = \frac{1}{4}$ so only check that $x = \frac{1}{12}$ gives a maximum volume.

$$\frac{d^2C}{dx^2} = -4 + 24x \text{ which is negative when } x = \frac{1}{12}.$$

Therefore C has a maximum value of $\frac{1}{12}\left(\frac{1}{2} - \frac{1}{6}\right)^2 = \frac{1}{108}$ so the maximum capacity of the box is $\frac{1}{108}$ m³.

Example 12

The function $ax^2 + bx + c$ has a gradient function $4x + 2$ and a stationary value of 1. Find the value of a, b and c.

$$f(x) = ax^2 + bx + c \quad \Rightarrow \quad f'(x) = 2ax + b$$

Given $f'(x) = 4x + 2 \quad \Rightarrow \quad 2ax + b$ is identical to $4x + 2$.

Therefore $a = 2$ and $b = 2$.

The stationary value of $f(x)$ occurs when $f'(x) = 0$ that is when

$$4x + 2 = 0 \quad \Rightarrow \quad x = -\frac{1}{2}$$

So the stationary value of $f(x)$ is $2\left(-\frac{1}{2}\right)^2 + 2\left(-\frac{1}{2}\right) + c = -\frac{1}{2} + c$.

The stationary value of $f(x)$ is given as 1 therefore $-\frac{1}{2} + c = 1 \quad \Rightarrow \quad c = \frac{3}{2}$.

Example 13

A cylinder has a radius r metres and a height h metres. The sum of the radius and height is 2 m.

Find an expression for the volume, V cubic metres, of the cylinder in terms of r only. Hence find the maximum volume.

$V = \pi r^2 h$ and $r + h = 2$

Therefore $\quad V = \pi r^2 (2 - r) = \pi (2r^2 - r^3)$

For maximum volume, $\dfrac{dV}{dr} = 0$,

so $\quad \pi(4r - 3r^2) = 0 \Rightarrow \pi r(4 - 3r) = 0$.

Therefore there are stationary values of V when $r = 0$ and $r = \dfrac{4}{3}$.

When $r = 0$, $V = 0$ so no cylinder exists. Therefore check the sign of $\dfrac{d^2V}{dr^2}$ only for $r = \dfrac{4}{3}$.

$\dfrac{d^2V}{dr^2} = \pi(4 - 6r)$ which is negative when $r = \dfrac{4}{3}$.

Therefore the maximum value of V occurs when $r = \dfrac{4}{3}$ and is $\pi\left(\dfrac{4}{3}\right)^2\left(2 - \dfrac{4}{3}\right)$.

Therefore the maximum volume is $\dfrac{32\pi}{27}$ m³.

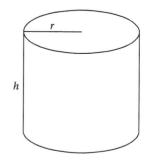

Exercise 8

1 A farmer wants to use an 80 m length of fencing to make three sides of a rectangle against an existing straight fence, which forms the longer side of the rectangle. The shorter side of the rectangle is x m long.

  a Find the length of the longer side of the rectangle in terms of x.

  b Show that the area of the enclosure is given by $A = 2x(40 - x)$

  c Hence find the maximum area that he can enclose and give its dimensions.

2 An open rectangular cardboard box is to be made with a square base and a volume of 4 m³.

  a The length of the base is x m. Find an expression for the height of the box in terms of x.

  b Show that the area of cardboard used to make the box is given by $A = x^2 + \dfrac{16}{x}$.

  c Hence find the dimensions of the box which contains the minimum area of cardboard.

3 The diagram shows a cylinder cut from a solid sphere of radius 3 m.

  a Given that the cylinder has a height of $2h$ m, find its radius in terms of h.

  b Hence show that the volume, V cubic metres, of the cylinder is given by

$$V = 2\pi h(9 - h^2)$$

  c Find the maximum volume of the cylinder as h varies.

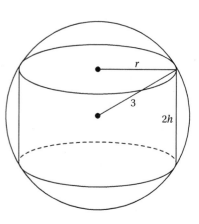

4 A rectangle has a perimeter of 20 cm.

    a The length of the rectangle is cm. Find an expression for the width of the rectangle in terms of x.

    b Find an expression for the area of the rectangle and hence find the lengths of the sides when the area is as large a possible.

5 A rectangle whose length and width vary has a constant area of 35 cm^2.

  Find the lengths of the sides when the perimeter is a minimum.

6 The curve $y = ax^2 + bx + c$ crosses the y-axis at the point $(0, 3)$ and has a stationary point at $(1, 2)$. Find the values of a, b and c.

Summary

Chords, tangents, normals and gradients

The tangent to a curve at a point A touches the curve at A.

The normal to a curve at a point A is perpendicular to the tangent at A.

The gradient of a curve at a point A on the curve is the gradient of the tangent to the curve at A.

Differentiation

When $y = x^n$, $\dfrac{\mathrm{d}y}{\mathrm{d}x} = nx^{n-1}$ for all values of n.

When $y = k$, $\dfrac{\mathrm{d}y}{\mathrm{d}x} = 0$.

When $y = ax^n$, $\dfrac{\mathrm{d}y}{\mathrm{d}x} = anx^{n-1}$.

When $y = \mathrm{f}(x) + \mathrm{g}(x)$, $\dfrac{\mathrm{d}y}{\mathrm{d}x} = \mathrm{f}'(x) + \mathrm{g}'(x)$.

Stationary points

$$\text{At a stationary point} \begin{cases} y, \text{ or } \mathrm{f}(x) \text{ has a stationary value} \\ \dfrac{\mathrm{d}y}{\mathrm{d}x}, \text{ or } \mathrm{f}'(x), \text{ is zero} \\ \text{the tangent is parallel to the } x\text{-axis.} \end{cases}$$

The turning point on a curve may be a maximum or a minimum point.

To distinguish between turning points use either

    at a maximum point the gradient changes from positive to zero to negative

    at a minimum point the gradient changes from negative to zero to positive

or

    at a maximum point $\dfrac{\mathrm{d}^2 y}{\mathrm{d}x^2}$ is negative (or zero)

    at a minimum point $\dfrac{\mathrm{d}^2 y}{\mathrm{d}x^2}$ is positive (or zero)

When $\dfrac{\mathrm{d}^2 y}{\mathrm{d}x^2}$ is zero use the first method.

Review

1 Find the derivative of

   a $x^{-3} - x^3 + 7$ **b** $x^{\frac{1}{2}} - x^{-\frac{1}{2}}$ **c** $\dfrac{1}{x^2} + \dfrac{2}{x^3}$

2 Differentiate with respect to x

   a $y = x^{\frac{3}{2}} - x^{\frac{2}{3}} + x^{-\frac{1}{3}}$ **b** $y = \sqrt{x} - \dfrac{1}{x} + \dfrac{1}{x^3}$ **c** $\dfrac{1}{x^{\frac{3}{4}}} - \dfrac{1}{x^{\frac{1}{4}}}$

3 Find the gradient of the curve $y = 2x^3 - 3x^2 + 5x - 1$ at the point

   a $(0, -1)$ **b** $(1, 3)$ **c** $(-1, -11)$

4 Find the gradient of the given curve at the given point.

   a $y = x^2 + x - 9; x = 2$ **b** $y = x^2 - 4x; x = 5$

5 The equation of a curve is $y = (x - 3)(x + 4)$. Find the gradient of the curve

   a at the point where the curve crosses the y-axis

   b at each of the points where the curve crosses the x-axis.

6 Find the coordinates of the point(s) on the curve $y = 3x^3 - x + 8$ at which the gradient is

   a 8   **b** 0

7 Find the equation of the normal to the curve $y = 1 - x^2$ at the point where the curve crosses the positive x-axis.

Find also the coordinates of the point where the normal meets the curve again.

8 Find the equations of the tangents to the curve $y = x^3 - 6x^2 + 12x + 2$ that have a gradient of 3.

9 Find the equation of the normal to the curve $y = x^2 - 6$ which has a gradient of $-\dfrac{1}{2}$.

10 Find the coordinates of the turning points on the curve $y = x^3 - 12x$ and determine their nature.

11 Find the stationary values of the function $x + \dfrac{1}{x}$.

12 The perimeter of a rectangle is 10 m.

   a The length of the rectangle is x m. Find an expression in terms of x for the width of the rectangle.

   b Hence show that the area A m^2 of the rectangle is given by $A = x(5 - x)$.

   c Show that the area of the rectangle is greatest when it is square.

13 The shape of a door is a rectangle with a semicircle on the top edge of the rectangle.

The diameter of the semicircle is equal to the width of the rectangle.

The perimeter of the door is 7 m, and the radius of the semicircle is r m.

   a Express the height of the rectangle in terms of r.

   b Show that the area of the door has a maximum value when the radius is $\dfrac{7}{4 + \pi}$.

Assessment

1 The point (x, y) is on a curve for which $x > 0$ and the gradient of the curve is given by $\dfrac{dy}{dx} = \dfrac{4}{x^2} - 3x + 5$.

  a Show that $\dfrac{dy}{dx} = 0$ when $x = 2$.

  b Find the value of $\dfrac{d^2 y}{dx^2}$ when $x = 2$.

  The point P(2, 3) lies on the curve.

  c State whether P is a maximum or a minimum point. Give a reason for your answer.

2 The equation of a curve is $y = x^4 - 2x^3 + 3x - 7$.

  a Find $\dfrac{dy}{dx}$.   **b** Find $\dfrac{d^2 y}{dx^2}$.

  c The point P is on the curve and the x coordinate of P is 1. Find the equation of the tangent to the curve at the point P.

3 The equation of a curve is $y = \dfrac{1}{x^2} + x^{\frac{1}{2}}$ for $x > 0$. The point P(1, 2) is on the curve.

  a Find the value of $\dfrac{dy}{dx}$ at the point P.

  b Show that the equation of the normal to the curve at the point P is $2x - 3y + 4 = 0$.

4 An open tank is made with a square base and vertical sides and holds 32 cubic metres of water.

  The side of the square base is x m long and the length of the vertical sides is y m.

  a Show that $x^2 y = 32$. Hence find the length of the vertical sides in terms of x.

  b Show that the area A m² of the sheet metal used to make the tank is given by $x^2 - \dfrac{128}{x}$.

  c Find the dimensions of the tank when the area of sheet metal used to make it has a minimum value.

5 Triangle ABC has a right angle at C. The shape of the triangle can vary but the sides BC and CA have a fixed total length of 10 cm.

  a Given that AC $= x$ cm, find an expression the length of BC.

  b Show that the area, A cm² of the triangle is given by $A = \dfrac{1}{2} x(10 - x)$.

  c Hence find the maximum area of the triangle.

6 The curve with equation $y = x^5 - 3x^2 + x + 5$ is sketched below. The point O is at the origin and the curve passes through the points A(−1, 0) and B(1, 4).

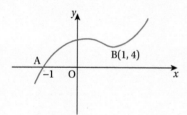

a Given that $y = x^5 - 3x^2 + x + 5$, find:

   i $\dfrac{dy}{dx}$;

   ii $\dfrac{d^2y}{dx^2}$

b Find an equation of the tangent to the curve at the point A(−1, 0).

c Verify that the point B, where $x = 1$, is a minimum point of the curve.

AQA MPC1 January 2012 (part question)

7 A bird flies from a tree. At time t seconds, the bird's height, y metres, above the horizontal ground is given by

$$y = \frac{1}{8}t^4 - t^2 + 5, \quad 0 \le t \le 4$$

a Find $\dfrac{dy}{dt}$.

b **i** Find the rate of change of height of the bird in metres per second when $t = 1$.

   ii Determine, with a reason, whether the bird's height above the horizontal ground is increasing or decreasing when $t = 1$.

c **i** Find the value of $\dfrac{d^2y}{dt^2}$ when $t = 2$.

   ii Given that y has a stationary value when $t = 2$, state whether this is a maximum value or a minimum value.

AQA MPC1 January 2013

7 Integration

Introduction

When x^2 is differentiated with respect to x the derivative is $2x$.

Reversing this, when the derivative of an unknown function is $2x$ then the unknown function could be x^2.

This process of finding a function from its derivative, which reverses the operation of differentiating, is called *integration*.

Recap

You need to remember how to...

▶ Work with surds and indices.
▶ Expand brackets such as $(x-4)(x^2+3x-1)$.
▶ Differentiate powers of x.
▶ Understand the meaning of rotational symmetry.

Objectives

By the end of this chapter, you should know how to...

▶ Explain the meaning of an indefinite and a definite integral and how to find them.
▶ Find areas bounded by straight lines and a curve by using integration.
▶ Find an approximate value of a definite integral using the trapezium rule.

- -

7.1 Indefinite integration

The derivative of x^2 is $2x$, but it is also the derivative of x^2+3, x^2-9, and the derivative of $x^2 + $ any constant.

Therefore the result of integrating $2x$ is not a unique function but is of the form

$$x^2 + c \text{ where } c \text{ is any constant}$$

c is called the **constant of integration**.

This is written $\quad \int 2x \, dx = x^2 + c$

where $\int \ldots dx$ **means 'the integral of ... with respect to x'.**

Integrating *any* function reverses the process of differentiating so, for any function $f(x)$, we have

$$\int \frac{d}{dx} f(x) \, dx = f(x) + c$$

For example, because differentiating x^3 with respect to x gives $3x^2$

then $\int 3x^2 \, dx = x^3 + c$

and it follows that $\int x^2 \, dx = \dfrac{1}{3}x^3 + c$

It is not necessary to write $\dfrac{1}{3}c$ in the second form, as c represents *any* constant in either expression.

In general, the derivative of x^{n+1} is $(n+1)x^n$ so $\int x^n \, dx = \dfrac{1}{n+1}x^{n+1} + c$.

Therefore

> to integrate a power of x, *increase* that power by 1 and *divide* by the new power.

This rule can be used to integrate any power of x *except* -1 because $-1+1=0$ and dividing by 0 is meaningless.

Because integrating a function of x involves adding an unknown constant, it is called indefinite integration.

Integrating a sum or difference of functions

A function can be differentiated term by term. Therefore, because integration reverses differentiation, integration also can be done term by term.

To integrate products or quotients of functions, first express them as sums or differences of functions.

For example $\quad \displaystyle\int \frac{2x-1}{\sqrt{x}}\,dx = \int \frac{2x}{\sqrt{x}} - \frac{1}{\sqrt{x}}\,dx = \int 2x^{\frac{1}{2}}\,dx - \int x^{-\frac{1}{2}}\,dx$

Example 1

Question

Find the integral of $1 + x^7 + \dfrac{1}{x^2} - \sqrt{x}$ with respect to x.

Answer

$$\int\left(1 + x^7 + \frac{1}{x^2} - \sqrt{x}\right)dx = \int(1 + x^7 + x^{-2} - x^{\frac{1}{2}})\,dx$$

$$= \int 1\,dx + \int x^7\,dx + \int x^{-2}\,dx - \int x^{\frac{1}{2}}\,dx$$

$$= x + \frac{1}{8}x^8 + \frac{1}{-1}x^{-1} - \frac{1}{\frac{3}{2}}x^{\frac{3}{2}} + c$$

$$= x + \frac{1}{8}x^8 - \frac{1}{x} - \frac{2}{3}x^{\frac{3}{2}} + c$$

Finding the constant of integration

The value of the constant of integration can be found when more information about a curve is known, for example a point on the curve.

Example 2

Question

The gradient of a curve is given by $\dfrac{dy}{dx} = 2x+4$ and (1, 3) is a point on the curve. Find the equation of the curve.

Answer

$$\frac{dy}{dx} = 2x+4 \quad \Rightarrow \quad y = x^2 + 4x + c$$

(1, 3) is a point on the curve, therefore $3 = 1 + 4 + c \Rightarrow c = -2$.

So $y = x^2 + 4x - 2$.

Exercise 1

Integrate with respect to x

1 x^5

2 $\dfrac{1}{x^5}$

3 $\sqrt[4]{x}$

4 x^{-3}

5 $\dfrac{1}{x^{\frac{5}{2}}}$

6 $x^{-\frac{1}{2}}$

7 x^1

8 $\dfrac{1}{\sqrt[3]{x}}$

9 $1 + x^2$

10 $2x - \sqrt{x}$

11 $1 + \dfrac{1}{x^2}$

12 $x(1 + x)$

13 $(2 - 3x)(1 + 5x)$

14 $\dfrac{1 + x}{\sqrt{x}}$

15 $\dfrac{1 - 2x}{x^3}$

16 $\dfrac{1 + x + x^3}{\sqrt{x}}$

17 $(1 - x)^2$

18 $x(1 + x)(1 - x)$

19 $\dfrac{1 - \sqrt{x}}{x^2}$

20 The gradient of a curve is given by $\dfrac{dy}{dx} = 3x^2 - 2x$. The point $(0, 1)$ is on the curve. Find the equation of the curve.

21 The point $(4, 1)$ is on a curve. The gradient of the curve is given by $\dfrac{dy}{dx} = \sqrt{x}$. Find the equation of the curve.

22 The gradient of a curve is given by $\dfrac{dy}{dx} = \dfrac{x - 3}{\sqrt{x}}$. The point $(9, 2)$ is on the curve. Find the equation of the curve.

23 At the point $(1, 5)$ the gradient of a curve is 4. Given that $\dfrac{d^2y}{dx^2} = 6x - 2$, find the equation of the curve.

7.2 Using integration to find an area

The area shown in the diagram is bounded by the curve $y = f(x)$, the x-axis and the lines $x = a$ and $x = b$.

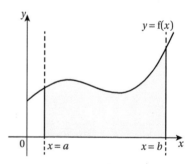

There are several ways in which this area can be estimated, for example by counting squares on graph paper. A better method is to divide the area into thin vertical strips and treat each strip, or *element*, as being approximately rectangular.

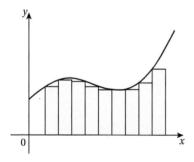
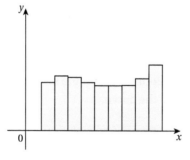

The sum of the areas of the rectangular strips gives an approximate value for the required area. The approximation gets better as the strips get thinner.

Every strip has one end on the x-axis, one end on the curve and two vertical sides, so all the strips have the same type of boundaries.

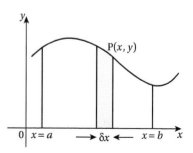

Look at a typical element bounded on the left by the **ordinate** through a general point P(x, y). (An ordinate is the length of a line from a point $x = a$ on the x-axis to a curve and parallel to the y-axis, that is the ordinate is the y-coordinate of the point on a curve where $x = a$.)

The width of the element represents a small increase in the value of x and so can be called δx.

Also, if A represents the part of the area up to the ordinate through P, then the area of the element represents a small increase in the value of A and so can be called δA.

A typical strip is approximately a rectangle of height y and width δx.

Therefore, for any element $\delta A \approx y\, \delta x$ [1]

The approximate area can now be found by adding the areas of all the strips from $x = a$ to $x = b$.

The notation for this is $\displaystyle\sum_{x=a}^{x=b} \delta A$ where $\displaystyle\sum$ means 'the sum of'.

so,   total area $\approx \displaystyle\sum_{x=a}^{x=b} \delta A$

$\Rightarrow$ total area $\approx \displaystyle\sum_{x=a}^{x=b} y\, \delta x$

As δx gets smaller the accuracy of the results increases until, in the limiting case,

total area $= \displaystyle\lim_{\delta x \to 0} \sum_{x=a}^{x=b} y\, \delta x$

The equation $\delta A \approx y\delta x$ can also be written in the form $\dfrac{\delta A}{\delta x} \approx y$.

This form too becomes more accurate as δx gets smaller, giving $\displaystyle\lim_{\delta x \to 0}\dfrac{\delta A}{\delta x} = y$.

But $\displaystyle\lim_{\delta x \to 0}\dfrac{\delta A}{\delta x}$ is $\dfrac{dA}{dx}$ so $\dfrac{dA}{dx} = y$

Hence $A = \displaystyle\int y\, dx.$

The boundary values of x defining the total area are $x = a$ and $x = b$ and we indicate this by writing

total area $= \displaystyle\int_a^b y\, dx$

The total area can therefore be found in two ways, either as the limit of a sum or by integration,

therefore $\displaystyle\lim_{\delta x \to 0}\sum_{x=a}^{x=b} y\, \delta x = \int_a^b y\, dx$

and this shows that integration is a process of summation.

Definite integration

This section shows how to calculate the value of expressions like $\displaystyle\int_a^b y\, dx.$

For example, using the method above find the area bounded by the x-axis, the lines $x = a$ and $x = b$ and the curve $y = 3x^2$ gives $A = \displaystyle\int 3x^2\, dx$, so $A = x^3 + c.$

From this area function we can find the value of A corresponding to a particular value of x.

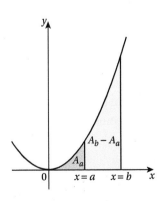

Hence using $x = a$ gives $A_a = a^3 + c$

and using $x = b$ gives $A_b = b^3 + c$.

Then the area between $x = a$ and $x = b$ is given by $A_b - A_a$ where

$$A_b - A_a = (b^3 + c) - (a^3 + c) = b^3 - a^3$$

$A_b - A_a$ is called the definite integral from a to b of $3x^2$ and is denoted by

$$\int_a^b 3x\,dx, \text{ so } \int_a^b 3x^2\,dx = (x^3)_{x=b} - (x^3)_{x=a}$$

The right hand side of this equation is usually written in the form $\left[x^3\right]_a^b$ where a and b are called the boundary values or limits of integration; b is the upper limit and a is the lower limit.

$\int_a^b y\,dx$ is called the definite integral from a to b of y with respect to x.

Whenever a definite integral is calculated, the constant of integration disappears.

(A definite integral can be found in this way only if the function to be integrated is defined for every value of x from a to b, for example $\int_{-1}^1 \dfrac{1}{x^2}dx$ cannot be found directly as $\dfrac{1}{x^2}$ is undefined when $x = 0$.)

Example 3

Question

Evaluate $\displaystyle\int_1^4 \dfrac{1}{x^2}dx$.

Answer

$$\int_1^4 \frac{1}{x^2}dx = \int_1^4 x^{-2}\,dx$$

$$= \left[-x^{-1}\right]_1^4 = \{-4^{-1}\} - \{-1^{-1}\} = -\frac{1}{4} + 1 = \frac{3}{4}$$

Exercise 2

Evaluate each of the following definite integrals.

1. $\displaystyle\int_0^2 x^3\,dx$

2. $\displaystyle\int_1^2 \sqrt{x^5}\,dx$

3. $\displaystyle\int_2^4 (x^2 + 4)\,dx$

4. $\displaystyle\int_4^9 \sqrt{x}\,dx$

5. $\displaystyle\int_0^3 (x^2 + 2x - 1)\,dx$

6. $\displaystyle\int_0^2 (x^3 - 3x)\,dx$

7. $\displaystyle\int_{-1}^0 (1 - x)^2\,dx$

8. $\displaystyle\int_1^2 \dfrac{3x + 1}{\sqrt{x}}dx$

9. $\displaystyle\int_{-1}^0 (2 + 3x)^2\,dx$

Finding area by definite integration

The area bounded by a curve $y = f(x)$, the lines $x = a$, $x = b$, and the x-axis, can be found from the definite integral $\int_a^b f(x)\,dx$. It helps to draw a diagram showing a typical element.

Example 4

Question

Find the area in the first quadrant bounded by the x and y axes and the curve $y = 1 - x^2$.

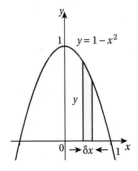

Answer

The area starts at the y-axis where $x = 0$ and ends where the curve crosses the x-axis.

$$\text{Area} = \lim_{\delta x \to 0} \sum_{x=0}^{x=1} y\, \delta x = \int_0^1 (1-x^2)\,dx = \left[x - \frac{x^3}{3} \right]_0^1 = \left(1 - \frac{1}{3} \right) - (0-0) = \frac{2}{3}$$

The area is $\dfrac{2}{3}$ of a square unit.

The meaning of a negative result

Look at the area bounded by $y = 4x^3$, the x-axis and the lines

a $x = 1$ and $x = 2$ **b** $x = -2$ and $x = -1$.

This curve has rotational symmetry about the origin so the two shaded areas are equal.

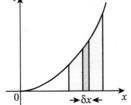

a $A = \lim_{\delta x \to 0} \sum_{x=1}^{x=2} y\, \delta x = \int_1^2 y\, dx$

$= \int_1^2 4x^3\, dx$

$= \left[x^4 \right]_1^2 = 16 - 1 = 15$

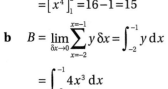

b $B = \lim_{\delta x \to 0} \sum_{x=-2}^{x=-1} y\, \delta x = \int_{-2}^{-1} y\, dx$

$= \int_{-2}^{-1} 4x^3\, dx$

$= \left[x^4 \right]_{-2}^{-1} = 1 - 16 = -15$

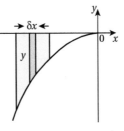

This integral has a negative value because, from -2 to -1, the value of y which gives the length of the strip, is negative.

Area cannot be negative; the minus sign means that area A is below the x-axis. The actual area is 15 square units.

Be careful with problems involving a curve that crosses the x-axis between the boundary values.

Example 5

Question

The diagram shows the curve $y = x(x-1)(x-2)$ and the x-axis.

a   Find the area marked A on the diagram.

b   Find the area marked B on the diagram.

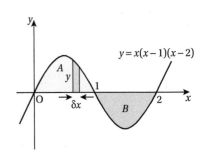

a $\text{Area } A = \int_0^1 y\,dx = \int_0^1 (x^3 - 3x^2 + 2x)\,dx$

$$= \left[\frac{x^4}{4} - x^3 + x^2 \right]_0^1$$

$$= \frac{1}{4}$$

b $\text{Area } B = \int_1^2 (x^3 - 3x^2 + 2x)\,dx = \left[\frac{x^4}{4} - x^3 + x^2 \right]_1^2$

$$= (4 - 8 + 4) - \left(\frac{1}{4} - 1 + 1 \right)$$

$$= -\frac{1}{4}$$

> **Note**
>
> The minus sign refers only to the *position* of area B relative to the x-axis. The actual area is $\frac{1}{4}$ of a square unit.

Finding compound areas

Example 6

Find the area between the curve $y = x^2$ and the line $y = 3x$.

The area can be found from

(area of the triangle bounded by the line, the x-axis and $x = 3$) − (area between the curve. the x-axis and $x = 3$).

$$\Rightarrow \quad \frac{1}{2} \times 3 \times 9 - \int_0^3 x^2\,dx = 13\frac{1}{2} - \left[\frac{1}{3} x^3 \right]_0^3$$

$$= 13\frac{1}{2} - 9 = 4\frac{1}{2}$$

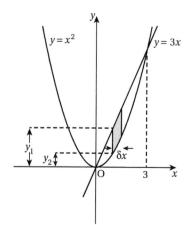

Exercise 3

In Questions 1 to 9, find the area with the given boundaries.

1. The x-axis, the curve $y = x^2 + 3$ and the lines $x = 1$, $x = 2$.
2. The curve $y = \sqrt{x}$, the x-axis and the lines $x = 4$, $x = 9$.
3. The x-axis, the lines $x = -1$, $x = 1$, and the curve $y = x^2 + 1$.
4. The curve $y = x^2 + x$, the x-axis and the line $x = 3$.
5. The positive x and y axes and the curve $y = 4 - x^2$.
6. The lines $x = 2$, $x = 4$, the x-axis and the curve $y = x^3$.
7. The curve $y = 4 - x^2$, the positive y-axis and the negative x-axis.
8. The x-axis, the lines $x = 1$ and $x = 2$, and the curve $y = \frac{1}{2} x^3 + 2x$.
9. The x-axis and the lines $x = 1$, $x = 5$, and $y = 2x$.

   Check the result by sketching the area and finding it by calculation.

10 Find the area below the x-axis and above the curve $y = x^2 - 1$.

11 Find the area bounded by the curve $y = 1 - x^3$, the x-axis and the lines $x = 2$, $x = 3$.

12 Find the area between the x and y axes and the curve $y = (x - 1)^2$.

13 Sketch the curve $y = x(x^2 - 1)$, showing where it crosses the x-axis. Find

    a the area enclosed above the x-axis and below the curve

    b the area enclosed below the x-axis and above the curve.

14 Repeat Question 4 for the curve $y = x(4 - x^2)$.

15 Evaluate

    a $\displaystyle\int_0^2 (x-2)\,\mathrm{d}x$ **b** $\displaystyle\int_2^4 (x-2)\,\mathrm{d}x$ **c** $\displaystyle\int_0^4 (x-2)\,\mathrm{d}x$

    Interpret your results with a sketch.

16   **a** Find, by integration, the area bounded by the x-axis, the line $x = 2$ and the curve $y = x^2$.

    b Hence find the area bounded by the y-axis, the line $y = 4$ and the curve $y = x^2$.

17 Find the area bounded by the curve $y = 1 - x^2$ and the line $y = 1 - x$.

18 Evaluate the area between the line $y = x - 1$ and the curve $y = x(1 - x)$.

19 Evaluate the area between the line $y = x - 1$ and the curve $y = (2x + 1)(x - 1)$.

20 Calculate the area of the region bounded by the curve $y = (x + 1)(x - 2)$ and the line $y = x$.

7.3 The trapezium rule

The definite integral $\displaystyle\int_a^b \mathrm{f}(x)\,\mathrm{d}x$ can be used to find the area between the curve $y = \mathrm{f}(x)$, the x-axis and the lines $x = a$ and $x = b$ but it is not always possible to find a function whose derivative is $\mathrm{f}(x)$. In this case the definite integral, and hence the exact value of an area, cannot be found.

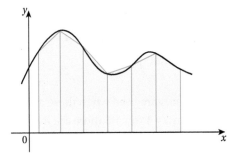

However the area can be divided into a *finite* number of strips. The sum of the areas of these strips gives an approximate value for the area and hence an approximate value of the definite integral.

When the area in the diagram is divided into vertical strips as shown, each strip is approximately a trapezium.

When the width of the strip and its two vertical sides are known, the area of the strip can be found using the formula

$$\text{area} = \frac{1}{2}\ (\text{sum of parallel sides}) \times \text{width}$$

The sum of the areas of all the strips then gives an approximate value for the area under the curve.

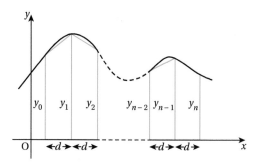 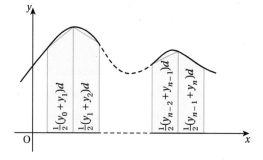

With *n* strips *all with the same width, d,* and with the vertical edges of the strips labelled $y_0, y_1, y_2, \ldots, y_{n-1}, y_n$,

then the sum of the areas of all the strips is

$$\frac{1}{2}(y_0+y_1)(d)+\frac{1}{2}(y_1+y_2)(d)+\frac{1}{2}(y_2+y_3)(d)+\cdots+\frac{1}{2}(y_{n-2}+y_{n-1})(d)+\frac{1}{2}(y_{n-1}+y_n)(d)$$

Therefore the area, *A*, under the curve is given approximately by

$$A \approx \frac{1}{2}(d)[y_0+2y_1+2y_2+\cdots+2y_{n-1}+y_n]$$

This formula is known as the trapezium rule.

An easy way to remember the formula in terms of ordinates is

half width of strip × (first + last + twice all the others)

Be careful not to confuse the number of strips and the number of ordinates – they are not the same.

Example 7

Question

Use the trapezium rule and four strips to find an approximate value for the definite integral $\int_1^5 x^3$.

Answer

Five ordinates are used when there are four strips whose widths must all be the same. From $x=1$ to $x=5$ there are four units so the width of each strip must be 1 unit. Hence the five ordinates are where $x=1$, $x=2$, $x=3$, $x=4$ and $x=5$.

Using the trapezium rule,

$$y_0=1^3=1, y_1=2^3=8, y_2=3^3=27, y_3=64, y_4=125$$

The area, *A*, is given by

$$A \approx \frac{1}{2}(1)[1+125+2\{8+27+64\}]=162$$

The area is approximately 162 square units.

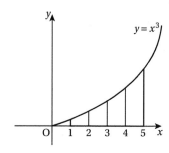

Exercise 4

In Questions 1 to 6 to estimate the value of each definite integral, using the trapezium rule with five ordinates.

Give answers correct to 3 significant figures where necessary.

1 $\displaystyle\int_0^4 x^2\,dx$

2 $\displaystyle\int_0^4 \frac{1}{x^2}\,dx$

3 $\displaystyle\int_0^4 (x-1)(x-2)\,dx$

4 $\displaystyle\int_5^9 \sqrt{x}\,dx$

5 $\displaystyle\int_5^9 \sqrt{x^2-9}\,dx$

6 $\displaystyle\int_{0.6}^1 \frac{1}{x}\,dx$

Summary

The indefinite integral of x^n with respect to x is given by $\displaystyle\int x^n\,dx = \frac{1}{n+1}x^{n+1}+c$ so to integrate a power of x, *increase* that power by 1 and *divide* by the new power.

Integration can be done term by term, so $\displaystyle\int (f(x)+g(x))\,dx = \int f(x)\,dx + \int g(x)\,dx$.
The area between a curve $y = f(x)$, the x-axis and the ordinates where $x = a$ and $x = b$ is given by $\displaystyle x = \int_a^b y\,dx$.

The trapezium rule gives an approximate value for the definite integral $\displaystyle\int_a^b f(x)\,dx$ where

$$\int_a^b f(x)\,dx \approx \frac{1}{2}d[y_0 + 2y_1 + \ldots + 2y_{n-1} + y_n]$$

where d is the width of each strip and the values of y are the lengths of the parallel sides of the trapeziums, that is, the y-coordinates of the points on the curve at the edge of each strip.

Review

Integrate the functions in Questions 1 to 6 with respect to x.

1 $x^2 - \dfrac{1}{x^2}$

2 $\sqrt[3]{x}$

3 $\sqrt{x} + \dfrac{1}{\sqrt{x}}$

4 $\dfrac{x\sqrt{x}-1}{\sqrt{x}}$

5 $\dfrac{x^3-1}{x^2}$

6 $\dfrac{x^2-1}{\sqrt{x}}$

Evaluate the definite integrals in Questions 7 to 9.

7 $\displaystyle\int_3^6 (6-x)^2\,dx$

8 $\displaystyle\int_{-1}^8 \frac{3y}{\sqrt[3]{8y}}\,dy$

9 $\displaystyle\int_1^{32}\left(\sqrt[5]{x} - \frac{1}{\sqrt[5]{x}}\right)dx$

10 Find the area bounded by the x and y axes and the curve $y = 1 - x^3$.

11 Find the area bounded by the curve $y = x^2 - 4$ and the x-axis.

12 Find the area between the curve $y = 4 - x^2$ and the line $y = 4 - x$.

13 a Find an approximate value for the area between the x-axis and the curve $y = (x-1)(x-4)$, using the trapezium rule with four ordinates.

   b Evaluate $\displaystyle\int_1^4 (x-1)(x-4)\,dx$.

Assessment

1 **a** Show that $\dfrac{\left(x^2-3\right)^2}{x^4}$ can be expressed as $1-\dfrac{6}{x^2}+\dfrac{9}{x^4}$.

 b Hence find $\displaystyle\int\left(\dfrac{\left(x^2-3\right)^2}{x^4}\right)dx.$ **c** Hence find $\displaystyle\int_1^2 \dfrac{\left(x^2-3\right)^2}{x^4}\,dx.$

2 The sketch shows the curve with equation $y=6-x-x^2$.

 a Find $\displaystyle\int_{-1}^1 (6-x-x^2)\,dx.$

 b Hence find the area of the shaded region bounded by the curve $y=6-x-x^2$ and the line AB.

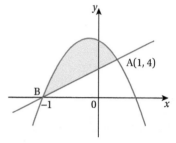

A(1, 4)

B

−1   0   x

3 The point P(4, 1) lies on the curve with equation $y=f(x)$ where
$f'(x)=\sqrt{x}-\dfrac{1}{\sqrt{x}}+1.$

 a Find the equation of the normal to the curve at the point P.

 b Find the equation of the curve.

4 **a** Use the trapezium rule with three ordinates to estimate the value of
$\displaystyle\int_0^5 (3+x)\,dx.$

 b Find the value of $\displaystyle\int_0^5 (3+x)\,dx.$

 c Explain the connection between the results of **a** and **b**.

5 The equation of a curve C is $y=2^x$.

 a Describe the transformation that transforms the curve C to the curve with equation $y=2^{\frac{x}{2}}$.

 b Use the trapezium rule with four ordinates (three strips) to estimate the area between the curve $y=2^{\frac{x}{2}}$, the x-axis and the lines $x=1$ and $x=4$.

6 The curve with equation $y = x^5 - 3x^2 + x + 5$ is sketched below. The point O is at the origin and the curve passes through the points A(−1, 0) and B(1, 4).

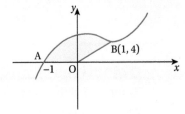

i Find $\int_{-1}^{1} (x^5 - 3x^2 + x + 5)\,dx$

ii Hence find the area of the shaded region bounded by the curve between A and B and the line segments AO and OB.

AQA MPC1 January 2012 (part question)

7 a Use the trapezium rule with five ordinates (four strips) to find an approximate value for

$$\int_0^4 \frac{2^x}{x+1}\,dx$$

giving your answer to three significant figures.

b State how you could obtain a better approximation to the value of the integral using the trapezium rule.

AQA MPC2 January 2012

8 Sequences and Series

Introduction

Look at this set of numbers: 2, 4, 6, 8, 10, ...

This set of numbers, *in the order given*, has a pattern. There is also a clear rule for getting the next number and as many following numbers as needed. Sets like these are called *sequences* and each member of the set is a term of the sequence.

Now look at this sum: $2 + 4 + 6 + 8 + 10 + ...$

The terms of the sequence, in order again, are added. Sums like these are called *series*.

Objectives

By the end of this chapter, you should know how to...

► Decide if a sequence is convergent or divergent.
► Distinguish between a finite series and an infinite series.
► Understand and use the Σ notation.
► Define an arithmetic series and a geometric series.
► Find the sum of an arithmetic series and the sum of a geometric series including the sum to infinity.
► Expand $(a + b)^n$ as a series when n is a positive integer.

Recap

You need to remember how to...

► Expand brackets such as $(1 + 2x)(x^2 - 3x - 7)$.
► Solve a pair of simultaneous equations.
► Solve a quadratic equation.
► Solve a linear inequality.

Applications

To calculate 2.01^6 without a calculator involves several long multiplications. By expressing 2.01^6 as $(2 + 0.01)^6$, it is possible to expand $(2 + 0.01)^6$ as the sum of terms involving powers of 0.01. This makes the arithmetic simpler as powers of 2 and 0.01 are easy to calculate.

8.1 Defining a sequence

The terms in a **sequence** are denoted by $u_1, u_2, ..., u_n, ...$ where u_n is the nth term.

Each term in the sequence 1, 2, 4, 8, 16, ... is a power of 2 so the sequence can be written as $\qquad$ $2^0, 2^1, 2^2, 2^3, 2^4, ...$

All the terms are of the form 2^r, so 2^r is a general term.

However, 2^r is *not* the nth term, as $u_1 = 2^0$, $u_2 = 2^1$, $u_3 = 2^2$, ...

This shows that nth term, u_n is given by $u^n = 2^{n-1}$.

$u^n = 2^{n-1}$ can now be used to find any term of the sequence, for example, the ninth term, u_9, is given by $u_9 = 2^{9-1} = 256$.

Therefore the rule $u_n = 2^{n-1}$ for $n = 1, 2, 3, \ldots$ enables the whole sequence to be generated and so it defines the sequence completely.

When a sequence is defined by $u_n = f(n)$, any term of the sequence can be found.

There is another way to define a sequence.

Look at the sequence 2, 4, 6, 8, 10, …

The obvious way to describe this sequence is 'starting with 2, add 2 to each term to get the next term', or, using notation, $u_1 = 2$, $u_{n+1} = u_n + 2$.

This is also a definition of the sequence because it can be generated as follows.

$$u_1 = 2, \ u_2 = u_1 + 2 = 4, \ u_3 = u_2 + 2 = 6, \text{ and so on.}$$

Any sequence can be generated when one term is known together with a relationship of the form $u_{n+1} = f(u_n)$.

The relationship between u_n and u_{n+1} is called a **recurrence relation**.

Example 1

Write down the first four terms of the sequence defined by

a $u_n = \dfrac{n}{n+1}$

b $u_1 = 2, \quad u_{n+1} = \dfrac{u_n}{u_n + 1}$

a $u_n = \dfrac{n}{n+1} \Rightarrow u_1 = \dfrac{1}{1+1} = \dfrac{1}{2}$

$$u_2 = \dfrac{2}{2+1} = \dfrac{2}{3}$$

$$u_3 = \dfrac{3}{3+1} = \dfrac{3}{4}$$

$$u_4 = \dfrac{4}{4+1} = \dfrac{4}{5}$$

b $u_1 = 2$, and $u_{n+1} = \dfrac{u_n}{u_n + 1}$

$$\Rightarrow \quad u_2 = \dfrac{u_1}{u_1 + 2} = \dfrac{2}{2+1} = \dfrac{2}{3}$$

$$u_3 = \dfrac{\frac{2}{3}}{\frac{2}{3}+1} = \dfrac{2}{5}$$

$$u_4 = \dfrac{\frac{2}{5}}{\frac{2}{5}+1} = \dfrac{2}{7}$$

The behaviour of u_n as $n \to \infty$

Look at the sequence $\dfrac{1}{2}, \dfrac{2}{3}, \dfrac{3}{4}, \dfrac{4}{5}, \ldots$

All the terms are less than 1, and the values of the terms are increasing as n increases. As the sequence progresses, the value of the terms is getting closer to 1. Expressing this in symbols gives

$$u_n \to 1 \text{ as } n \to \infty \quad \text{or} \quad \lim_{n \to \infty} u_n = 1$$

and the sequence is called **convergent**.

We can illustrate this on a graph by plotting values of u_n against values of n.

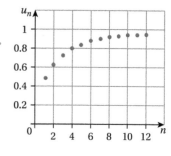

> **Any sequence whose terms approach one finite value, L, converges to L. A sequence that is not convergent is called divergent.**

When a sequence is defined by a recurrence relation, $u_{n+1} = f(u_n)$, then *if* the sequence converges to L,

both u_n and u_{n+1} approach L as $n \to \infty$. Therefore replacing u_{n+1} and u_n by L in $u_{n+1} = f(u_n)$ gives the equation $L = f(L)$.

The solutions of the equation $L = f(L)$, if they exist, do not always give a reasonable value for $\lim_{n \to \infty} u_n$. They need to be looked at together with the first few terms of the sequence.

For example, when $u_1 = 3$ and $u_{n+1} = 2u_n$, the first few terms of the sequence are 3, 6, 12, 24, ... so the sequence clearly does not converge. However the equation $L = 2L$ gives $L = 0$ which is not a reasonable answer.

Exercise 1

1 Write down the first six terms of each sequence and state whether it is convergent or divergent.

  a $u_n = \dfrac{1}{n^n}$

  b $u_n = (-1)^n 2^n$

  c $u_n = \dfrac{(-1)^n}{2^n}$

2 Write down the first six terms of the sequence given by the recurrence relation. Give answers correct to 2 decimal places where necessary.

  a $u_{n+1} = \dfrac{1 + u_n}{1 + 2u_n}; \ u_1 = 3$

  b $u_{n+1} = 2 - (u_n)^2 - u_n; \ u_1 = -4$

  c $u_n = \sqrt{2 - u_{n-1}}; \ u_1 = 0.5$

  d $u_n = \dfrac{1}{5}(5 - \{u_{n-1}\}^2); \ u_1 = 1$

3 A sequence is generated by the recurrence relation $u_{n+1} = \dfrac{2}{2 - u_n}$.

  Describe whether the terms converge or diverge as n increases when

  a $u_1 = 1$ **b** $u_1 = 0.5$ **c** $u_1 = -2$

4 Repeat Question 3 for the sequence defined by $u_{n+1} = \dfrac{1}{4}(2 - u_n)$ when

  a $u_1 = 1$ **b** $u_1 = -1$ **c** $u_1 = 2$

  Find the value to which the terms of the sequence converge for those sequences that do converge.

8.2 Series

A **series** is formed when the terms of a sequence are added.

For example $1 + 2 + 4 + 8 + 16 + \ldots$ is a series.

When the series stops after a finite number of terms it is called a **finite series**.

For example $1 + 2 + 4 + 8 + 16 + 32 + 64$ is a finite series of seven terms.

If the series continues indefinitely it is called an **infinite series**.

For example $1 + \dfrac{1}{2} + \dfrac{1}{4} + \dfrac{1}{8} + \dfrac{1}{16} + \dfrac{1}{32} + \cdots + \dfrac{1}{1024} + \cdots$ is an infinite series.

Look again at the series $1 + 2 + 4 + 8 + 16 + 32 + 64$.

Each term is a power of 2 so this series can be written as

$$2^0 + 2^1 + 2^2 + 2^3 + 2^4 + 2^5 + 2^6$$

All the terms of this series are of the form 2^r with $r = 0$ giving the first term, so 2^{n-1} is the nth term.

Therefore the series can be defined as the sum of terms of the form 2^{n-1}, where n takes all integral values in order from 1 to 7 inclusive.

Using Σ as a symbol for 'the sum of terms such as' the series can be defined simply as $\Sigma 2^{n-1}$, n taking all integer values from 1 to 7 inclusive, or, even more simply as $\displaystyle\sum_{n=1}^{7} 2^{n-1}$.

Placing the lowest and highest value that n takes below and above Σ respectively, shows that n takes all integer values between and including these extreme values.

So $\displaystyle\sum_{n=1}^{10} n^3$ means 'the sum of all terms of the form n^3, where n takes all integer values from 1 to 10 inclusive', that is $\displaystyle\sum_{n=1}^{10} n^3 = 1^3 + 2^3 + 3^3 + 4^3 + 5^3 + 6^3 + 7^3 + 8^3 + 9^3 + 10^3$.

The infinite series $1 + \dfrac{1}{2} + \dfrac{1}{4} + \dfrac{1}{8} + \dfrac{1}{16} + \cdots$ can also be written in the sigma notation. The continuing dots after the last written term show that the series is infinite, that is there is *no* last term.

Each term of this series is a power of $\dfrac{1}{2}$. The first term is 1 or $\left(\dfrac{1}{2}\right)^0$ so the nth term can be written $\left(\dfrac{1}{2}\right)^{n-1}$.

There is no last term of this series, so there is no upper limit for the value of n.

Therefore $1 + \dfrac{1}{2} + \dfrac{1}{4} + \dfrac{1}{8} + \dfrac{1}{16} + \cdots$ can be written as $\displaystyle\sum_{n=1}^{\infty} \left(\dfrac{1}{2}\right)^{n-1}$.

Writing a series in sigma notation, apart from the obvious advantage of being brief, means that a particular term of a series can be written without having to write down all the earlier terms.

For, example, in the series $\displaystyle\sum_{n=1}^{10} (2n+5)$,

the first term is the value of $2n + 5$ when $n = 1$, this is $2 \times 1 + 5 = 7$,

the last term is the value of $2n + 5$ when $n = 10$, this is 25,

the fourth term is the value of $2n + 5$ when $n = 4$, this is 13.

> **Note**
>
> When a finite series is written out, it should always end with the last term even if intermediate terms are omitted, for example $3 + 6 + 9 \ldots + 99$.

> **Note**
>
> When a given series is rewritten in sigma notation, check that the first few values of n gives the correct first few terms of the series.

Example 2

Write the following series in the sigma notation,

a $-2+4-8+16+\ldots-128$ **b** $1-x+x^2-x^3+\ldots$

a $-2+4-8+16+\ldots-128=-2+(2)^2-(2)^3+(2)^4-\ldots-(2)^7$

So the nth term is $\pm 2^n$, and is positive when n is even and negative when n is odd.

Because $(-1)^n$ is positive when n is even and negative when n is odd, the nth term can be written $(-1)^n 2^n$.

Hence $-2+4-8+16+\ldots-128=\sum_{n=1}^{7}(-1)^n 2^n$.

b $1-x+x^2-x^3+\ldots=x^0-x+x^2-x^3+\ldots$

The nth term of this series is $\pm x^{n-1}$. The nth term is positive when n is odd and negative when n is even.

$(-1)^{n-1}$ is positive when n is odd and negative when n is even.

Therefore nth term is $(-1)^{n-1}x^{n-1}$.

Hence $1-x+x^2-x^3+\ldots=\sum_{n=1}^{\infty}(-1)^{n-1}x^{n-1}$.

Exercise 2

1 Write the following series in the sigma notation:

a $1+8+27+64+125$ **b** $2+4+6+8+\ldots+20$

c $\dfrac{1}{2}+\dfrac{1}{3}+\dfrac{1}{4}+\dfrac{1}{5}+\ldots+\dfrac{1}{50}$ **d** $1+\dfrac{1}{3}+\dfrac{1}{9}+\dfrac{1}{27}+\ldots$

e $-4-1+2+5\ldots+17$ **f** $8+4+2+1+\dfrac{1}{2}+\ldots$

2 Write down the first three terms and, where there is one, the last term of each of the following series:

a $\sum_{n=1}^{\infty}\dfrac{1}{n}$ **b** $\sum_{n=1}^{5}n(n-1)$

c $\sum_{n=1}^{20}\dfrac{n+2}{(n+1)(2n+1)}$ **d** $\sum_{n=1}^{\infty}\dfrac{1}{(n^2+1)}$

e $\sum_{n=1}^{8}n(n+1)(n+2)$ **f** $\sum_{n=1}^{\infty}a^n(-1)^{n+1}$

3 For the following series, write down the term indicated, and the number of terms in the series.

a $\sum_{n=1}^{9}2^n$, 3rd term **b** $\sum_{n=1}^{8}(2n+3)$, 5th term

c $\sum_{n=1}^{\infty}\dfrac{1}{(n+1)(n+2)}$, 20th term **d** $\sum_{n=1}^{\infty}\left(\dfrac{1}{2}\right)^n$, nth term

e $8+4+0-4-8-12\ldots-80$, 15th term **f** $\dfrac{1}{16}+\dfrac{1}{8}+\dfrac{1}{4}+\dfrac{1}{2}+\ldots+32$, 7th term

8.3 Arithmetic series

Look at the series $5 + 8 + 11 + 14 + 17 + \ldots + 29$.

Each term of this series is 3 greater than the previous term, so the series can be written as

$$5 + (5 + 3) + (5 + 2 \times 3) + (5 + 3 \times 3) + (5 + 4 \times 3) + \ldots + (5 + 8 \times 3)$$

This series is an example of an **arithmetic series** which is a series where any term differs from the term before it by the same constant. This constant is called the **common difference**.

The common difference can be positive or negative. For example, the first six terms of an arithmetic series whose first term is 8 and whose common difference is -3, are 8, 5, 2, -1, -4, -7.

In general, if an arithmetic series has a first term a, and a common difference d, the first four terms are

$$a, (a + d), (a + 2d), (a + 3d), \text{ and the } n\text{th term, } u_n, \text{ is } a + (n - 1)d.$$

So

an arithmetic series with n terms can be written as

$$a + (a + d) + (a + 2d) + \cdots + [a + (n-1)d]$$

Example 3

Question

The eighth term of an arithmetic series is 11 and the 15th term is 21.

Find

a   the common difference

b   the first term of the series

c   the nth term.

Answer

a   If the first term of the series is a and the common difference is d, then the eighth term is $a + 7d$,

Therefore $\qquad\qquad\qquad\qquad a + 7d = 11 \qquad\qquad\qquad\qquad$ [1]

The 15th term is $a + 14d$, $\quad \Rightarrow \quad a + 14d = 21 \qquad\qquad\qquad$ [2]

[2] – [1] gives $7d = 10 \qquad \Rightarrow \quad d = \dfrac{10}{7}$

The common differences is $\dfrac{10}{7}$.

b   From [2], $a = 1$, so the first term is 1.

c   Hence the nth term is $a + (n-1)d = 1 + (n-1)\dfrac{10}{7} = \dfrac{1}{7}(10n - 3)$.

Example 4

Question

The nth term of an arithmetic series is $12 - 4n$. Find the first term and the common difference.

Answer

The nth term is $12 - 4n$ so the first term ($n = 1$) is 8 and the second term ($n = 2$) is 4.

Therefore the common difference is -4.

The sum of an arithmetic series

The sum of the first ten even numbers is an arithmetic series.

Writing it first in normal, then in reverse order gives

$$S = 2 \ +4 \ +6 \ +8 \ +...+18+20$$

Adding gives
$$\frac{S = 20 + 18 + 16 + 14 + ... + 4 \ + 2}{2S = 22 + 22 + 22 + 22 + ... + 22 + 22}$$

There are ten terms in this series, therefore

$$2S = 10 \times 22 \quad \Rightarrow \quad S = 110$$

Using this method with a general arithmetic series gives formulae for the sum, which may be quoted and used.

If S_n is the sum of the first n terms of an arithmetic series with last term l,

then $\qquad S_n = a + (a+d) + (a+2d) + ... + (l-d) + l$

reversing $\qquad S_n = l + (l-d) + (l-2d) + ... + (a+d) + a$

adding $\qquad \overline{2S_n = (a+l) + (a+l) + (a+l) + ... + (a+l) + (a+l)}$

There are n terms so $2S_n = n(a+l)$

$$\Rightarrow \quad S_n = \frac{1}{2}n(a+l)$$

Because the nth term, l, is equal to $a + (n-1)d$, then $S_n = \frac{1}{2}n[a + a + (n-1)d]$.

Therefore $\quad S_n = \frac{1}{2}n[2a + (n-1)d]$.

Either of these formulae can be used to find the sum of the first n terms of an arithmetic series.

The sum of the first *n* natural numbers

$1 + 2 + 3 + ... + n$ is an arithmetic series with $a = 1$ and $d = 1$ so

$$\sum_{r=1}^{n} r = \frac{1}{2}n(n+1)$$

Note

This result can be quoted.

Example 5

Question

Find the sum of the following series

a an arithmetic series of eleven terms whose first term is 1 and whose last term is 6

b $\displaystyle\sum_{r=1}^{8}\left(2 - \frac{2r}{3}\right)$.

Answer

a The first and last terms and the number of terms are given so use

$$S_n = \frac{1}{2}n(a+l) \Rightarrow S_{11} = \frac{11}{2}(1+6) = \frac{77}{2}$$

b $\displaystyle\sum_{r=1}^{8}\left(2 - \frac{2r}{3}\right) = \frac{4}{3} + \frac{2}{3} + 0 - \frac{2}{3} - ... - \frac{10}{3}$

This is an arithmetic series with 8 terms where $a = \frac{4}{3}, d = -\frac{2}{3}$.

Using $S_n = \frac{1}{2}n(2a + (n-1)d)$ gives $S_8 = 4\left[\frac{8}{3} + 7\left(-\frac{2}{3}\right)\right] = -8$.

Example 6

Question

The sum of the first ten terms of an arithmetic series is 50. The 5th term is three times the 2nd term.

Find the first term and the sum of the first 20 terms.

Answer

Using $S = \frac{1}{2}n[2a+(n-1)d]$ gives $\qquad S_{10} = 50 = 5(2a+9d)$ $\qquad$ [1]

Using $u_n = a+(n-1)d$ gives $\qquad u_5 = a+4d$ and $u_2 = a+d$

Therefore $\qquad\qquad\qquad\qquad a+4d = 3(a+d)$ $\qquad\qquad$ [2]

Solving [1] and [2] simultaneously gives $\quad d=1$ and $a = \frac{1}{2}$.

So the first term is $\frac{1}{2}$ and the sum of the first 20 terms is S_{20} where

$S_{20} = 10(1+19 \times 1) = 200$.

Example 7

Question

The sum of the first n terms of a series is given by $S_n = n(n+3)$.

Find the fourth term of the series and show that the terms are in arithmetic sequence.

Answer

The terms of the series are $a_1, a_2, a_3, ..., a_n$ so $S_n = a_1 + a_2 + ... + a_n = n(n+3)$.

Therefore $\quad S_4 = a_1 + a_2 + a_3 + a_4 = 4 \times 7 = 28$ $\qquad\qquad$ [1]

and $\qquad\quad S_3 = a_1 + a_2 + a_3 = 3 \times 6 = 18$ $\qquad\qquad\qquad$ [2]

[1] – [2] gives $a_4 = 10$.

$\qquad\qquad S_n = a_1 + a_2 + ... + a_{n-1} + a_n = n(n+3)$ $\qquad\qquad$ [1]

and $\qquad\quad S_{n-1} = a_1 + a_2 + ... + a_{n-1} = (n-1)(n+2)$ $\qquad$ [2]

[1] – [2] gives $a_n = n(n+3) - (n-1)(n+2) = 2n+2$

Replacing n by $n-1$ gives the $(n-1)$th term

so $\qquad\qquad a_{n-1} = 2(n-1)+2 = 2n$

Then $\qquad\quad a_n - a_{n-1} = (2n+2) - 2n = 2$

Therefore there is a common difference of 2 between successive terms, showing that the series is an arithmetic series.

Exercise 3

1. Write down the fifth term and the nth term of the following arithmetic series.

    a $\displaystyle\sum_{n=1}^{n}(2n-1)$ $\qquad\qquad$ **b** $\displaystyle\sum_{n=1}^{n}4(n-1)$ $\qquad\qquad$ **c** $\displaystyle\sum_{n=0}^{n}(3n+3)$

    d first term 5, common difference 3

    e first term 6, common difference −2

    f first term p, common difference q

    g first term 10, last term 30, 11 terms

    h $1, 5, ...$ $\qquad\qquad$ **i** $2, 1\frac{1}{2}, ...$ $\qquad\qquad$ **j** $-4, -1, ...$

2 Find the sum of the first ten terms of each arithmetic series given in Question 1.

3 The 9th term of an arithmetic series is 8 and the 4th term is 20. Find the first term and the common difference.

4 The 6th term of an arithmetic series is twice the 3rd term and the first term is 3. Find the common difference and the 10th term.

5 The nth term of an arithmetic series is $\frac{1}{2}(3-n)$. Write down the first three terms and the 20th term.

6 Find the sum, to the number of terms indicated, of each of the following arithmetic series.

a $1+2\frac{1}{2}+\ldots$, 6 terms

b $3+5+\ldots$, 8 terms

c the first twenty odd integers

d $a_1+a_2+a_3+\ldots+a_8$ where $a_n=2n+1$

e $4+6+8+\ldots+20$

f $\displaystyle\sum_{n=1}^{3n}(3-4n)$

g $S_n=n^2-3n$, 8 terms

h $S_n=2n(n+3)$, m terms

7 The sum of the first n terms of an arithmetic series is S_n where $S_n=n^2-3n$. Write down the fourth term and the nth term.

8 The sum of the first n terms of a series is given by S_n where $S_n=n(3n-4)$. Show that the series is an arithmetic series.

9 In an arithmetic series, the 8th term is twice the 4th term and the 20th term is 40.

Find the common difference and the sum of the terms from the 8th to the 20th inclusive.

10 How many terms of the arithmetic series, $1+3+5+\ldots$ are needed to make a sum of 1521?

11 Find the least number of terms of the arithmetic series, $1+3+5+\ldots$, that are needed to make a sum greater than 4000.

12 The sum of the first n terms of a series is S_n where $S_n=2n^2-n$.

a Prove that the series is an arithmetic series, stating the first term and the common difference.

b Find the sum of the terms from the 3rd term to the 12th term inclusive.

13 In an arithmetic series the 6th term is half the 4th term and the 3rd term is 15.

a Find the first term and the common difference.

b How many terms are needed to give a sum that is less than 65?

8.4 Geometric series

Consider the sequence 12, 6, 3, 1.5, 0.75, 0.375, ...

Each term of this sequence is half the term before it so the sequence can be written

$$12,\ 12\left(\frac{1}{2}\right),\ 12\left(\frac{1}{2}\right)^2,\ 12\left(\frac{1}{2}\right)^3,\ 12\left(\frac{1}{2}\right)^4,\ 12\left(\frac{1}{2}\right)^5,\ \ldots$$

A sequence like this is called a **geometric series** which is a series where each term is the same constant multiple of the term before it. This constant multiplying factor is called the **common ratio**, and it can have any value.

Therefore, if a geometric series has a first term of 3 and a common ratio of –2, the first four terms are

$$3, 3(-2), 3(-2)^2, 3(-2)^3 \implies 3, -6, 12, -24$$

In general if a geometric series has a first term a, and a common ratio r, the first four terms are a, ar, ar^2, ar^3 and the nth term, u_n, is ar^{n-1}.

Therefore, a geometric series with n terms can be written $a + ar + ar^2 + \ldots + ar^{n-1} + \ldots$

The sum of a geometric series

Look at the sum of the first eight terms, S_8, of the geometric series with first term 1 and common ratio 3,

so $\quad S_8 = 1 + 1(3) + 1(3)^2 + 1(3)^3 + \ldots + 1(3)^7$

$\implies 3S_8 = \quad\quad 3 + 3^2 + 3^3 + \ldots + 3^7 + 3^8$

Hence $\quad S_8 - 3S_8 = 1 + 0 + 0 + 0 + \ldots + 0 - 3^8$.

Therefore $\quad S_8(1-3) = 1 - 3^8 \implies S_8 = \dfrac{1-3^8}{1-3} = \dfrac{3^8-1}{2}$.

Now look at the sum, S_n, of the first n terms of a geometric series with first term a and common ration r,

so $\quad\quad\quad\quad\quad\quad S_n = a + ar + \ldots + ar^{n-2} + ar^{n-1}$ [1]

Multiplying by r gives $\quad rS^n = \quad ar + ar^2 + \ldots + ar^{n-1} + ar^n$ [2]

[1] – [2] $\implies S_n - rS_n = a - ar^n$

$\quad\quad\quad \implies S_n(1-r) = a(1-r^n)$

$\quad\quad\quad \implies S_n = \dfrac{a(1-r^n)}{1-r}$

When $r > 1$ the formula can be written $\dfrac{a(r^n-1)}{r-1}$.

Example 8

The 5th term of a geometric series is 8, the third term is 4, and the sum of the first 10 terms is positive. Find the first term, the common ratio, and the sum of the first 10 terms.

For a first term a and common ratio r, the nth term is ar^{n-1}.

Therefore when $n = 5$, $\quad ar^4 = 8$ [1]

and when $n = 3$, $\quad\quad ar^2 = 4$ [2]

[1] ÷ [2] $\implies r^2 = 2$

Therefore $\quad r = \pm\sqrt{2}$ and $a = 2$.

Using $S_n = \dfrac{a(r^n-1)}{r-1}$ gives,

when $r = \sqrt{2}, S_{10} = \dfrac{2\left(\left(\sqrt{2}\right)^{10} - 1\right)}{\sqrt{2}-1} = \dfrac{62}{\sqrt{2}-1}$

(continued)

(continued)

when $r = -\sqrt{2}$, $S_{10} = \dfrac{2\left((-\sqrt{2})^{10} - 1\right)}{-\sqrt{2} - 1} = \dfrac{-62}{\sqrt{2} + 1}$

But $S_{10} > 0$, so $r = \sqrt{2}$ and $S_{10} = \dfrac{62}{\sqrt{2} - 1} = 62(\sqrt{2} + 1)$

Example 9

The first term of a geometric series is 3 and the common ratio is $\dfrac{1}{2}$.

a   Write down the sixth term of the series.

b   Find the sum of the first 5 terms.

a $a = 3$ and $r = \dfrac{1}{2}$ so using nth term $= ar^{n-1}$,

the sixth term is $3\left(\dfrac{1}{2}\right)^5 = \dfrac{3}{32}$

b   Using $S_n = \dfrac{a(1 - r^n)}{1 - r}$, $S_5 = \dfrac{3\left(1 - \left(\frac{1}{2}\right)^5\right)}{1 - \frac{1}{2}} = \dfrac{3\left(\frac{31}{32}\right)}{\frac{1}{2}} = \dfrac{93}{16}$

Exercise 4

1 Write down the 5th term and the nth term of the following geometric series.

a $2, 4, 8, \dots$

b $2, 1, \dfrac{1}{2}, \dots$

c $3, -6, 12, \dots$

d   first term 8, common ratio $-\dfrac{1}{2}$

e   first term 3, last term $\dfrac{1}{81}$, 6 terms

2 Find the sum, to the number of terms given, of the following geometric series.

a $3 + 6 + \dots$, 6 terms

b $3 - 6 + \dots$, 8 terms

c $1 + \dfrac{1}{2} + \dfrac{1}{4} + \dots$, 20 terms

d   first term 5, common ratio $\dfrac{1}{5}$, 5 terms

e   first term $\dfrac{1}{2}$, common ratio $-\dfrac{1}{2}$, 10 terms

f   first term 1, common ratio -1, 2001 terms.

3 The 6th term of a geometric series is 16 and the 3rd term is 2. Find the first term and the common ratio.

4 Find the common ratio, given that it is negative, of a geometric series when $a_1 = 8$ and $a_5 = \dfrac{1}{2}$.

5 The nth term of a geometric series is $\left(-\dfrac{1}{2}\right)^n$. Write down the first term and the 10th term.

6 Evaluate $\displaystyle\sum_{n=1}^{10} (1.05)^n$

7 Find the sum of the first n terms of the following series.

a $x + x^2 + x^3 + \dots$

b $x + 1 + \dfrac{1}{x} + \dots$

c $1 - y + y^2 - \dots$

d $x + \dfrac{x^2}{2} + \dfrac{x^3}{4} + \dfrac{x^4}{8} + \dots$

e $1 - 2x + 4x^2 - 8x^3 + \dots$

8 The sum of the first 3 terms of a geometric series is 14. The first term is 2.

Find the possible values of the sum of the first 5 terms.

9 Evaluate $\sum_{n=1}^{10} 3\left(\frac{3}{4}\right)^n$.

Convergence of series

When a piece of string, of length l, is cut up by first cutting it in half and keeping one piece, then cutting the remainder in half and keeping one piece, and so on, the sum of the lengths that are kept is

$$\frac{l}{2} + \frac{l}{4} + \frac{l}{8} + \frac{l}{16} + \ldots$$

This process can (in theory) be carried on indefinitely so the series formed above is infinite.

After several cuts have been made the part of the string that is left will be very small indeed, so the sum of the cut lengths will be very nearly equal to the total length, l, of the original piece of string. The more cuts that are made the closer to l this sum becomes.

Therefore, if after n cuts, the sum of the cut lengths is $\frac{l}{2} + \frac{l}{2^2} + \frac{l}{2^3} + \ldots + \frac{l}{2^n}$

then, as $n \to \infty$, $\frac{l}{2} + \frac{l}{2^2} + \ldots + \frac{l}{2^n} \to l$ or $\lim_{n \to \infty}\left[\frac{l}{2} + \frac{l}{2^2} + \ldots + \frac{l}{2^n}\right] = l$

l is called the *sum to infinity* of this series.

When S_n is the sum of the first n terms of any series and if $\lim_{n \to \infty}\left[S_n\right]$ exists and is finite then the series is **convergent**.

The sum to infinity of a convergent series, S_∞, is given by $S_\infty = \lim_{n \to \infty}\left[S_n\right]$.

The series $\frac{l}{2} + \frac{l}{2^2} + \frac{l}{2^3} + \ldots$ for example, is convergent as its sum to infinity is l.

However, for the series $1 + 2 + 3 + \ldots + n$, $S_n = \frac{1}{2}n(n+1)$.

As $n \to \infty$, $S_n \to \infty$ so this series does not converge, it is divergent.

For all arithmetic series $S_n = \frac{1}{2}n\left[2a + (n-1)d\right]$, and this always approaches infinity as $n \to \infty$.

Therefore all arithmetic series are divergent.

The sum to infinity of geometric series

For any geometric series, $a + ar + ar^2 + \ldots$, $S_n = \frac{a(1-r^n)}{1-r}$

when $|r| < 1$, $\lim_{n \to \infty} r^n = 0$.

($|r|$ means the positive value of r whether r is positive or negative, so $|r| < 1$ is a shorter way of writing $-1 < r < 1$.)

So $\lim_{n \to \infty} S_n = \lim_{n \to \infty}\left[\frac{a(1-r^n)}{1-r}\right] = \frac{a}{1-r}$

When $|r| \geq 1$, $\lim_{n \to \infty} = \infty$ so the series does not converge.

Therefore, provided that $|r| < 1$, a geometric series converges to a sum of $\dfrac{a}{1-r}$.

S_∞ is called the sum to infinity where for a geometric series

$S_\infty = \dfrac{a}{1-r}$ provided that $|r| < 1$.

Example 10

Find out whether each series converges. If it does, give its sum to infinity.

a $3 + 5 + 7 + \dots$

b $1 - \dfrac{1}{4} + \dfrac{1}{16} - \dfrac{1}{64} + \dots$

c $3 + \dfrac{9}{2} + \dfrac{27}{4} + \dots$

a $3 + 5 + 7 + \dots$ is an arithmetic series ($d = 2$) and so does not converge.

b $1 - \dfrac{1}{4} + \dfrac{1}{16} - \dfrac{1}{64} + \dots = 1 + \left(-\dfrac{1}{4}\right) + \left(-\dfrac{1}{4}\right)^2 + \left(-\dfrac{1}{4}\right)^3 + \dots$

which is a geometric series where $r = -\dfrac{1}{4}$, so $|r| < 1$.

Therefore this series converges and $S_\infty = \dfrac{a}{1-r} = \dfrac{1}{1 - \left(-\dfrac{1}{4}\right)} = \dfrac{4}{5}$.

c $3 + \dfrac{9}{2} + \dfrac{27}{4} + \dots = 3 + 3\left(\dfrac{3}{2}\right) + 3\left(\dfrac{9}{4}\right) + \dots = 3 + 3\left(\dfrac{3}{2}\right) + 3\left(\dfrac{3}{2}\right)^2 + \dots$

This series is a geometric series where $r = \dfrac{3}{2}$ and, because $|r| > 1$, the series does not converge.

Example 11

Find the condition satisfied by x so that $\displaystyle\sum_{n=1}^{\infty} \dfrac{(x-1)^n}{2^n}$ converges. Find S_∞ when $x = 1.5$.

$\displaystyle\sum_{n=1}^{\infty} \dfrac{(x-1)^n}{2^n} = \dfrac{x-1}{2} + \left(\dfrac{x-1}{2}\right)^2 + \dots$

This series is a geometric series with common ratio $\dfrac{x-1}{2}$ and so converges if $\left|\dfrac{x-1}{2}\right| < 1$, that is if $-1 < \dfrac{x-1}{2} < 1$

$\Rightarrow\ -1 < x < 3$.

When $x = 1.5$, the series converges

and $\displaystyle\sum_{n=1}^{\infty} \dfrac{(x-1)^n}{2^n} = \sum_{n=1}^{\infty} \left(\dfrac{1}{4}\right)^n = \dfrac{1}{4} + \left(\dfrac{1}{4}\right)^2 + \left(\dfrac{1}{4}\right)^3 + \dots$

using $S_\infty = \dfrac{a}{1-r}$ with $r = \dfrac{1}{4}$ and $a = \dfrac{1}{4}$ gives $S_\infty = \dfrac{\dfrac{1}{4}}{1 - \dfrac{1}{4}} = \dfrac{1}{3}$.

Example 12

Question

The 3rd term of a convergent geometric series is half the sum of the 1st and 2nd terms.

a Find the common ratio.

b The first term is 1. Find the sum to infinity.

Answer

a Using $a + ar + ar^2 + ar^3 + \ldots$

then $\qquad ar^2 = \dfrac{1}{2}(a + ar)$

$a \neq 0$, so $\qquad 2r^2 - r - 1 = 0 \quad \Rightarrow \quad (2r+1)(r-1) = 0$

so $\qquad r = -\dfrac{1}{2}$ or $r = 1$

The series is convergent so the common ratio is $-\dfrac{1}{2}$.

b When $r = -\dfrac{1}{2}$ and $a = 1$, $S_\infty = \dfrac{1}{1 + \dfrac{1}{2}} = \dfrac{2}{3}$.

Exercise 5

1 Decide if each of the series given below converges.

a $4 + \dfrac{4}{3} + \dfrac{4}{3^2} + \ldots$
b $9 + 7 + 5 + 3 + \ldots$

c $20 - 10 + 5 - 2.5 + \ldots$
d $\dfrac{5}{10} + \dfrac{5}{100} + \dfrac{5}{1000} + \ldots$

e $p + 2p + 3p + \ldots$
f $3 - 1 + \dfrac{1}{3} - \dfrac{1}{9} + \ldots$

2 Find the range of value of x for which the following series converge.

a $1 + x + x^2 + x^3 + \ldots$
b $x + 1 + \dfrac{1}{x} + \dfrac{1}{x^2} + \ldots$

c $1 + 2x + 4x^2 + 8x^3 + \ldots$
d $1 - (1-x) + (1-x)^2 - (1-x)^3 + \cdots$

e $(a+x) + (a+x)^2 + (a+x)^3 + \ldots$
f $(a+x) + 1 + \dfrac{1}{a+x} + \dfrac{1}{(a+x)^2} + \ldots$

3 Find the sum to infinity of those series in Question 1 that are convergent.

4 The sum to infinity of a geometric series is twice the first term. Find the common ratio.

5 The sum to infinity of a geometric series is 16 and the sum of the first 4 terms is 15. Find the first four terms.

6 a, b and c are the first three terms of a geometric series. Prove that $\sqrt{a} + \sqrt{b} + \sqrt{c}$ is another geometric series.

8.5 The Binomial theorem

$(1 + x)^{20}$ could be expanded using Pascal's triangle but it takes time to get down to row 20.

We sometimes need to expand expressions such as $(a + b)^4$ but the multiplication is tedious when the power is three or more.

We now describe a far quicker way of obtaining such expansions.
Consider the following expansions,

$$(a+b)^1 = a + b$$
$$(a+b)^2 = a^2 + 2ab + b^2$$
$$(a+b)^3 = a^3 + 3a^2b + 3ab^2 + b^3$$
$$(a+b)^4 = a^4 + 4a^3b + 6a^2b^2 + 4ab^3 + b^4$$

The first thing to notice is that the powers of a and b in the terms of each expansion form a pattern. Looking at the expansion of $(a+b)^4$ we see that the first term is a^4 and then the power of a decreases by 1 in each succeeding term while the power of b increases by 1. For all the terms, the sum of the powers of a and b is 4 and the expansion ends with b^4. There is a similar pattern in the other expansions.

Now consider just the coefficients of the terms. Writing these as a triangular array gives diagram 1.

This array is called *Pascal's Triangle* and it has a pattern. Each row starts and ends with 1 and each other number is the sum of the two numbers in the row above it, as shown. When the pattern is known, Pascal's triangle can be written down to as many rows as needed. Using Pascal's triangle to expand $(a+b)^6$, for example, we go as far as row 6, shown in diagram 2.

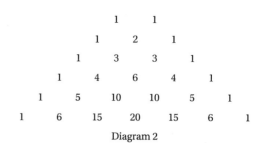

Diagram 1

```
      1       1
   1      2      1
1      3      3      1
1   4      6      4   1
```

```
         1       1
      1      2      1
   1      3      3      1
 1      4      6      4      1
1    5     10      10     5    1
1   6    15     20     15    6    1
```

Diagram 2

We then use our knowledge of the pattern of the powers, together with row 6 of the array, to fill in the coefficients,

so $(a+b)^6 = a^6 + 6a^5b + 15\,a^4b^2 + 20a^3b^3 + 15a^2b^4 + 6ab^5 + b^6$

The following worked examples show how expansions of other brackets can be found.

Example 13

Question

Expand $(x+5)^3$.

Answer

From Pascal's triangle $(a+b)^3 = a^3 + 3a^2b + 3ab^2 + b^3$

Replacing a by x and b by 5 gives $(x+5)^3 = x^3 + 3x^2(5) + 3x(5)^2 + (5)^3$
$$= x^3 + 15x^2 + 75x + 125$$

Example 14

Question

Expand $(2x-3)^4$.

Answer

From Pascal's triangle, $(a+b)^4 = a^4 + 4a^3b + 6a^2b^2 + 4ab^3 + b^4$

Replacing a by $2x$ and b by -3 gives

$(2x-3)^4 = (2x)^4 + 4(2x)^3(-3) + 6(2x)^2(-3)^2 + 4(2x)(-3)^3 + (-3)^4$
$$= 16x^4 - 96x^3 + 216x^3 - 216x + 81$$

A general method uses the binomial theorem which states that, if n is a positive integer,

$$(1+x)^n = 1 + nx + \frac{n(n-1)}{(2)(1)}x^2 + \frac{n(n-1)(n-2)}{(3)(2)(1)}x^3 + \frac{n(n-1)(n-2)(n-3)}{(4)(3)(2)(1)}x^4 + \ldots + nx^{n-1} + x^n.$$

The right-hand side of this expression is called the expansion of $(1+x)^n$.

The coefficients of the powers of x are called **binomial coefficients**.

The denominators of these coefficients involve the products of all the positive integers from the power of x in that term down to 1 and we can write these in a shorter form using factorial notation.

The product $(4)(3)(2)(1)$ is called '4 **factorial**' and is written 4! and 8! means the product of all the positive integers from 8 down to 1.

> When r is a positive integer,
> $r!$ means the product of all the positive integers from r down to 1.

A term somewhere in the middle of the series, involving x^r where r is an integer between 1 and n and the pattern of the binomial coefficients, shows that the coefficient of x^r is $\dfrac{n(n-1)(n-2)\ldots(n-r+1)}{r!}$ and this is denoted by $\binom{n}{r}$.

So

$$\binom{n}{r} = \frac{n(n-1)(n-2)\ldots(n-r+1)}{r!}$$

Note

$\dfrac{n(n-1)(n-1)(n-r-1)}{r!}$ simplifies to $\dfrac{n!}{r!(n-r)!}$

> To summarise, the binomial theorem states that, if n is a positive integer,

$$(1+x)^n = 1 + nx + \frac{n(n-1)}{2!}x^2 + \frac{n(n-1)(n-2)}{3!}x^3$$

$$+ \frac{n(n-1)(n-2)(n-3)}{4!}x^4 + \ldots + nx^{n-1} + x^n.$$

$$= 1 + \binom{n}{1}x + \binom{n}{2}x^2 + \binom{n}{3}x^3 + \ldots + \binom{n}{r}x^r + \ldots + x^n.$$

This shows that

1. the expansion of $(1+x)^n$ is a finite series with $n+1$ terms
2. the coefficient of x^r, $\dfrac{n(n-1)(n-2)\ldots(n-r+1)}{r!}$, has r factors in the numerator
3. the term containing x^2 is the third term, the term containing x^3 is the fourth term, and so on.

Look at $(a+b)^n$, where n is a positive integer.

$$(a+b)^n \equiv \left(a\left(1+\frac{b}{a}\right)\right)n = a^n\left(1+\frac{b}{a}\right)^n$$

Replacing x by $\dfrac{b}{a}$ in the binomial series gives

$$(a+b)^n = a^n\left[1 + \binom{n}{1}\left(\frac{b}{a}\right) + \binom{n}{2}\left(\frac{b}{a}\right)^2 + \ldots + \binom{n}{r}\left(\frac{b}{a}\right)^r + \ldots + \binom{n}{n}\left(\frac{b}{a}\right)^n\right]$$

$$= a^n + \binom{n}{1}a^{n-1}b + \binom{n}{2}a^{n-2}b^2 + \ldots + \binom{n}{r}a^{n-r}b^r + \ldots + \binom{n}{n}b^n$$

$$= a^n + na^{n-1}b + \frac{n(n-1)}{2!}a^{n-2}b^2 + \ldots + b^n$$

For example

$$(a+b)^6 = a^6 + 6a^5b + \frac{(6)(5)}{2!}a^4b^2 + \frac{(6)(5)(4)}{3!}a^3b^3 + \frac{(6)(5)(4)(3)}{4!}a^2b^4 + \frac{(6)(5)(4)(3)(2)}{5!}ab^5 + b^6$$

$$= a^6 + 6a^5b + 15a^4b^2 + 20a^3b^3 + 15a^2b^4 + 6ab^5 + b^6$$

Example 15

Question

Write down the first three terms in the expansion in ascending powers of x of

a $(1+x)^8$

b $(3-2x)^8$

Answer

a Using $(1+x)^n = 1 + nx + \frac{n(n-1)}{2!}x^2 + \ldots$ and replacing n by 8 gives

$$(1+x)^8 = 1 + 8x + \frac{8(8-1)}{2!}x^2 + \ldots$$

$$= 1 + 8x + 28x^2 + \ldots$$

b $(3-2x)^8 = 3^8\left(1 - \frac{2}{3}x\right)^8$.

Replacing x by $-\frac{2}{3}x$ and n by 8 in the expansion of $(1+x)^n$ gives

$$(3-2x)^8 = 3^8\left(1 + 8\left(-\frac{2}{3}x\right) + \left(\frac{8 \times 7}{2}\right)\left(-\frac{2}{3}x\right)^2 + \ldots\right)$$

Therefore the first three terms of this series are $3^8 - (16)(3^7)x + (112)(3^6)x^2$.

Example 16

Question

Find the 4th term in the expansion of $(a-2b)^{20}$ as a series in ascending powers of b.

Answer

$$(a-2b)^{20} = a^{20}\left(1 - \frac{2b}{a}\right)^{20}$$

The 4th term in the expansion of $(1+x)^n$ is $\binom{n}{3}x^3$. Replace x by $-\frac{2b}{a}$ and n by 20.

The 4th term is $a^{20}\binom{20}{3}\left(-\frac{2b}{a}\right)^3 = -\frac{(20)(19)(18)}{3!}(a)^{17}(8)(b)^3 = -9120a^{17}b^3$.

Example 17

Question

Write down the first three terms in the binomial expansion of $(1-2x)\left(1+\dfrac{1}{2}x\right)^{10}$.

Answer

The third term in the binomial expansion contains x^2, so start by expanding $\left(1+\frac{1}{2}x\right)^{10}$ as far as the term in x^2.

$$\left(1+\frac{1}{2}x\right)^{10} = 1+(10)\left(\frac{1}{2}x\right)+\frac{(10)(9)}{2!}\left(\frac{1}{2}x\right)^2+\dots$$

$$=1+5x+\frac{45}{4}x^2+\dots$$

Therefore $(1-2x)\left(1+\dfrac{1}{2}x\right)^{10} = (1-2x)\left(1+5x+\dfrac{45}{4}x^2+\dots\right)$

$$=1+5x+\frac{45}{4}x^2+\dots-2x-10x^2+\dots$$

$$=1+3x+\frac{5}{4}x^2+\dots$$

Note

The product of $-2x$ and $\dfrac{45}{4}x^2$ is not written down as terms in x^3 are not required.

Exercise 6

1   Write down the first four terms in the binomial expansion of

a $(1+3x)^{12}$

b $(1-2x)^9$

c $(2+x)^{10}$

d $\left(1-\dfrac{x}{3}\right)^{20}$

e $\left(2-\dfrac{3}{2}x\right)^7$

f $\left(\dfrac{3}{2}+2x\right)^9$

2   Write down the term given in the binomial expansion of each of the following functions.

a $(1-4x)^7$, 3rd term

b $\left(1-\dfrac{x}{2}\right)^{20}$, 2nd term

c $(2-x)^{15}$, 12th term

d $(p-2q)^{10}$, 5th term

e $(3a+2b)^8$, 2nd term

f $(1-2x)^{12}$, the term in x^4

g $\left(2+\dfrac{x}{2}\right)^9$, the term in x^5

h $(a+b)^8$, the term in a^3

3   Write down the binomial expansion of each function as a series of ascending powers of x as far as, and including, the term in x^2.

a $(1+x)(1-x)^9$

b $(1-x)(1+2x)^{10}$

c $(2+x)\left(1-\dfrac{x}{2}\right)^{20}$

d $(1+x)^2(1-5x)^{14}$

4   **a** Expand $(x+y)^5$ as a series of ascending powers of y.

   b By replacing x by 2 and y by 0.01, show that
   $(2.01)^5 = a + b(0.01) + c(0.01)^2 + d(0.01)^3 + e(0.01)^4 + f(0.01)^5$
   giving the values of the constants.

   c Hence calculate the exact value of $(2.01)^5$.

Summary

A sequence is a set of numbers in order, for example 1, 3, 5, 7, ...

A sequence can be generated when the nth term is given as a function of n, that is when $u_n = f(n)$ or when a term is known and a relationship between successive terms is given.

A sequence whose nth term approaches a finite value as n approaches infinity is convergent.

A series is the sum of the terms in a sequence, for example $1 + 3 + 5 + 7 + ...$

In an arithmetic series any term differs from the term before it by the same constant. This constant is called the *common difference*.

An arithmetic series with n terms can be written as
$a + (a + d) + (a + 2d) + ... + [a + (n-1)d]$.

The sum of the first n terms of an arithmetic series is either
$S_n = \frac{1}{2}n(a + l)$ or $S_n = \frac{1}{2}n[2a + (n-1)d]$.

The sum of the first n natural numbers is $\frac{1}{2}n(n+1)$.

A geometric series is a series where each term is the same constant multiple of the term before it. This constant multiplying factor is called the *common ratio*.

The sum of the first n terms of a geometric series is $S_n = \frac{a(1-r^n)}{1-r}$ where a is the first term and r is the common ratio.

The sum to infinity of a geometric series is $S_\infty = \frac{a}{1-r}$ provided that $|r| < 1$.

The binomial theorem states that, if n is a positive integer,

$$(1+x)^n = 1 + nx + \frac{n(n-1)}{2!}x^2 + \frac{n(n-1)(n-2)}{3!}x^3 + \frac{n(n-1)(n-2)(n-3)}{4!}x^4 + ... + nx^{n-1} + x^n$$

$$= 1 + \binom{n}{1}x + \binom{n}{2}x^2 + \binom{n}{3}x^3 + ... + \binom{n}{r}x^r + ... + x^n$$

Review

In Questions 1 to 3, write down the first six terms of each sequence and describe the behaviour of the terms as the sequence series.

1 $u_r = \frac{r}{r^2 + 1}$

2 $u_{r+1} = (u_r)^2 - u_r$

    a when $u_1 = 1$ **b** when $u_1 = 0.5$

3 $u_{r+1} = \frac{1}{1 - u_r}$ when

    a $u_1 = 2$ **b** $u_1 = 1$

In Questions 4 to 8, find the sum of each series.

4 $1 - \frac{1}{2} + \frac{1}{4} - \frac{1}{8} + ...$

5 $2 - (2)(3) + (2)(3)^2 - (2)(3)^3 + ... + (2)(3)^{10}$

6 $\displaystyle\sum_{r=2}^{n} ab^{2r}$

7 $\displaystyle\sum_{r=5}^{n} 4r$

8 $\displaystyle\sum_{r=1}^{\infty} \frac{1}{2^r}$

9 $u_1 = u_2$ and $u_{r+1} = \frac{1}{3}(2u_r^2 - 5)$. Find the possible values of u_1.

10 The sum of the first n terms of a series is n^3. Write down the first four terms and the nth term of the series.

11 The fourth term of an arithmetic series is 8 and the sum of the first 10 terms is 40. Find the first term and the 10th term.

12 Find the value of x for which the numbers $x+1$, $x+3$, $x+7$, are in geometric series.

13 The second term of a geometric series is $\frac{1}{2}$ and the sum to infinity of the series is 4. Find the first term and the common ratio of the series.

14 Find the first three terms and the last term in the expansion of $(1+2x)^9$ as a series of ascending powers of x.

15 Find the first four terms in the expansion of $(2-3x)^9$.

Assessment

1 a Expand $(1+3x)^5$ as a series of ascending powers of x.

   b Hence write down the expansion of $(1-3x)^5$.

   c Hence show that $(1+3\sqrt{3})^5 + (1-3\sqrt{3})^5 = 7832$.

2 The second term of a geometric series is 6 and the sixth term of the series is 96.

   a Find the common ratio of the series.

   b Find the first term of the series.

   c The nth term of the series is u_n. Show that $\displaystyle\sum_{n=1}^{30} u_n = 3(2^{30}-1)$.

3 The first term of an arithmetic series is 6. and the common difference is 1.

   a The nth term of the series is u_n. Show that $\displaystyle\sum_{n=1}^{n} u_n = An^2 + Bn$ giving the values of A and B.

   b Find the value of n for which $\displaystyle\sum_{n=1}^{n} u_n = n$.

4 A sequence of numbers is defined by $u_1 = 4$ and $u_{n+1} = 10 - 2u_n$.

   a Write down the first four terms of the sequence.

   b Show that $\displaystyle\sum_{n=1}^{4} u_n - \sum_{n-1}^{6} u_n = u_1$.

5 **a** Show that the first four terms in the expansion of $(1 - 2x)^n$ are $a + bx + cx^2 + dx^3$ giving the values of a, b, c and d in terms of n.

    b Use your answer to part **a** with the substitution $x = 0.01$ and a suitable value of n to find an approximate value for 0.98^6.

6 The nth term of a sequence is u_n. The sequence is defined by $u_{n+1} = pu_n + q$ where p and q are constants.

    The first two terms of the sequence are given by $u_1 = 60$ and $u_2 = 48$.

    The limit of u_n as n tends to infinity is 12.

    a Show that $p = \dfrac{3}{4}$ and find the value of q.

    b Find the value of u_3.

AQA MPC2 June 2011

7 **a** **i** Using the binomial expansion, or otherwise, express $(2 + y)^3$ in the form $a + by + cy^2 + y^3$, where a, b and c are integers.

      ii Hence show that $(2 + x^2)^3 + (2 - x^2)^3$ can be expressed in the form $p + qx^4$, where p and q are integers.

    b **i** Hence find $\displaystyle\int \left[(2 + x^{-2})^3 + (2 - x^{-2})^3\right] dx$

      ii Hence find the value of $\displaystyle\int_1^2 \left[(2 + x^{-2})^3 + (2 - x^{-2})^3\right] dx$

AQA MPC2 June 2013

Coordinate Geometry and Circles

Introduction

A point P can be anywhere in the xy plane. When the positions of P are kept to a line (curved or straight), the relationship between x and y which applies only to this line is called the equation of the line.

This chapter looks at a variety of methods for dealing with coordinate geometry problems involving the equations of circles.

Recap

You need to remember how to...

► Find the midpoint and the distance between two points in the xy plane.
► Solve a quadratic equation.
► Solve a pair of simultaneous equations when one equation is quadratic.
► Understand the meaning of a translation including when it is defined using vector notation.
► Find the equation of a straight line given a point on the line and its gradient.
► Remember the meaning of a normal to a curve.

Objectives

By the end of this chapter, you should know how to...

► Find the equation of a circle.
► Find the effect of a translation on the equation of a circle.
► Use the properties of tangents to circles to solve problems.

9.1 The equation of a circle

When a point P is always at the same distance r from a fixed point C, the path that P follows is a circle whose centre is C and whose radius is r.

A point $P(x, y)$ is at a constant distance, r, from the point $C(a, b)$.

Therefore P is on the circle if and only if $CP = r$, so $CP^2 = r^2$.

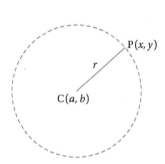

$$CP^2 = (x - a)^2 + (y - b)^2$$

$P(x, y)$ is on the circle $\iff$ $(x - a)^2 + (y - b)^2 = r^2$.
Therefore $(x - a)^2 + (y - b)^2 = r^2$ is the equation of a circle with centre (a, b) and radius r.

For example, the equation of a circle with centre $(-2, 3)$ and radius 1 is

$$[x - (-2)]^2 + [y - 3]^2 = 1 \implies (x + 2)^2 + (y - 3)^2 = 1 \implies x^2 + y^2 + 4x - 6y + 12 = 0$$

Recognising the equation of a circle

As well as being able to write down the equation of a circle given its centre and radius, it is also important to be able to recognise an equation as that of a circle.

Expanding and simplifying the equation $(x-a)^2+(y-b)^2=r^2$ gives

$$x^2+y^2-2ax-2by+(a^2+b^2-r^2)=0$$

which can be expressed as $x^2+y^2+2gx+2fy+c=0$ where g, f and c are constants.

Comparing the constants gives $r^2=f^2+g^2-c \Rightarrow r=\sqrt{f^2+g^2-c}$ and for r to have a real value, $g^2+f^2-c>0$.

> So an equation of the form $x^2+y^2+2gx+2fy+c=0$ is the equation of a circle provided that $g^2+f^2-c>0$.

Note

Notice that the coefficients of x^2 and y^2 are equal and that there is no xy term.

Example 1

Question

Find the centre and radius of the circle whose equation is

$$x^2+y^2+8x-2y+13=0$$

Answer

To compare the given equation with $(x-a)^2+(y-b)^2=r^2$, first rearrange the given equation as $x^2+8x+y^2-2y=-13$, then complete the square on x^2+8x and on y^2-2y.

$$x^2+y^2+8x-2y+13=0 \quad\Rightarrow\quad x^2+8x+16+y^2-2y+1=16+1-13$$
$$\Rightarrow\quad (x+4)^2+(y-1)^2=4$$

Therefore the centre of the circle is $(-4, 1)$ and the radius is 2.

Example 2

Question

a Show that $2x^2+2y^2-6x+10y=1$ is the equation of a circle and find its centre and radius.

b Find the y coordinates of the points where the circle cuts the y-axis.

Answer

a Divide both sides of the equation by 2 so that it can be compared with $(x-a)^2+(y-b)^2=r^2$.

$$2x^2+2y^2-6x+10y=1 \quad\Rightarrow\quad x^2+y^2-3x+5y=\frac{1}{2}$$

Completing the square on x^2-3x and on y^2+5y gives

$$\left(x-\frac{3}{2}\right)^2+\left(y+\frac{5}{2}\right)^2=\frac{1}{2}+\frac{9}{4}+\frac{25}{4}=9$$

Therefore the equation does represent a circle.

The centre is $\left(\frac{3}{2}, -\frac{5}{2}\right)$ and the radius is 3.

b The circle cuts the y-axis where $x=0$, and that is where $2y^2+10y=1$.

Solving this equation gives $2y^2+10y-1=0$

$$\Rightarrow\quad y=\frac{-10\pm\sqrt{100+8}}{4}=-\frac{5}{2}\pm3\sqrt{3}.$$

Therefore the circle cuts the y-axis where $y=-\frac{5}{2}+3\sqrt{3}$ and where $y=-\frac{5}{2}-3\sqrt{3}.$

Exercise 1

1 Write down the equation of the circle with

    a   centre $(1, 2)$, radius 3   **b**   centre $(0, 4)$, radius 1

    c   centre $(-3, -7)$, radius 2   **d**   centre $(4, 5)$, radius 3.

2 Find the centre and radius of the circle whose equation is

    a $x^2 + y^2 + 8x - 2y - 8 = 0$ **b** $x^2 + y^2 + x + 3y - 2 = 0$

    c $x^2 + y^2 + 6x - 5 = 0$ **d** $2x^2 + 2y^2 - 3x + 2y + 1 = 0$

    e $x^2 + y^2 = 4$ **f** $(x - 2)^2 + (y + 3)^2 = 9$

    g $2x + 6y - x^2 - y^2 = 1$ **h** $3x^2 + 3y^2 + 6x - 3y - 2 = 0$

3 Which of the following equations represent circles?

    a $x^2 + y^2 = 8$ **b** $2x^2 + y^2 + 3x - 4 = 0$

    c $x^2 - y^2 = 8$ **d** $x^2 + y^2 + 4x - 2y + 20 = 0$

    e $x^2 + y^2 + 8 = 0$ **f** $x^2 + y^2 + 4x - 2y - 20 = 0$

4 Find the x coordinates of the points where the circle $(x - 2)^2 + (y + 3)^2 = 25$ cuts the x-axis.

5 The diagram shows a circle whose centre C is the point $(3, 2)$ and whose radius is 3. The diameter parallel to the x-axis meets the circle at the point A.

    a   Find the coordinates of the point A.

    b   The line through A whose gradient is $\dfrac{1}{2}$ cuts the circle again at the point B. Find the coordinates of B.

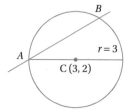

The effect of a translation on the equation of a circle

When the circle $(x - a)^2 + (y - b)^2 = r^2$ is translated by the vector $\begin{bmatrix} p \\ q \end{bmatrix}$, the centre moves but the radius of the circle does not change.

The centre is moved p units parallel to the x-axis and q units up the y-axis. Therefore the centre of the translated circle is the point $(a + p, b + q)$.

So the equation of this circle is

$$\left(x - (a + p)\right)^2 + \left(y - (b + q)\right)^2 = r^2 \implies (x - a - p)^2 + (y - b - q)^2 = r^2$$

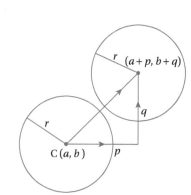

Example 3

The circle whose equation is $x^2 + y^2 - 4x + 6y + 7 = 0$ has centre C and is translated by a vector to the circle with centre D.

The equation of the translated circle is $x^2 + y^2 + 4y - 2 = 0$. Find the vector that describes the translation.

First express each equation in the form $(x - a)^2 + (y - b)^2 = r^2$ by completing the square.

$x^2 + y^2 - 4x + 6y + 7 = 0 \implies (x - 2)^2 + (y + 3)^2 = 4 + 9 - 7 = 6$

$x^2 + y^2 + 4y - 2 = 0 \implies x^2 + (y + 2)^2 = 6$

Therefore C is the point $(2, -3)$ and D is the point $(0, -2)$.

The diagram shows that C is translated to D by the vector $\begin{bmatrix} -2 \\ 1 \end{bmatrix}$.

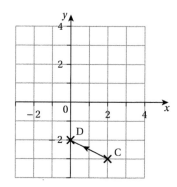

Exercise 2

1 A circle C has a radius of 10 and its centre is the point $(5, -2)$. C is translated by the vector $\begin{bmatrix} -4 \\ 6 \end{bmatrix}$ to the circle D.

Find the equation of circle D in the form $(x-a)^2 + (y-b)^2 = r^2$.

2 The equation of a circle C is $x^2 + y^2 - 8x - 10y + 5 = 0$. C is translated by the vector $\begin{bmatrix} 2 \\ -3 \end{bmatrix}$ to the circle D.

Find the equation of the circle D in the form $x^2 + y^2 + px + qy + w = 0$.

3 The circle C with equation $x^2 + y^2 - 2y - 7 = 0$ is translated to the circle $x^2 + y^2 + 6x - 2y + 1 = 0$.

Find the vector that describes the translation.

9.2 Geometric properties of circles

These important facts about circles are useful in solving problems.

The perpendicular bisector of a chord of a circle goes through the centre of the circle.

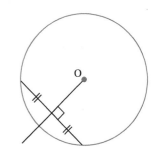

An angle in a semicircle is 90°. This is the angle subtended by a diameter on the circle.

Example 4

Question

A circle of radius 2 units which has its centre at the origin, cuts the x-axis at the points A and B and cuts the y-axis at the point C. Prove that angle ACB = 90°.

Answer

The diagram shows the information given in the question and properties known about the figure.

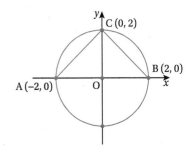

From the diagram, the gradient of AC is $\dfrac{2-0}{0-(-2)} = 1$

and the gradient of BC is $\dfrac{2-0}{0-2} = -1$.

Therefore (gradient of AC) × (gradient of BC) = −1

so AC is perpendicular to BC $\Rightarrow$ angle ACB = 90°.

Example 5

Question

A circle contains a triangle whose vertices are at the points A(0, 4), B(2, 3) and C(−2, −1). Find the centre of the circle.

When a circle circumscribes a figure, it passes through the vertices of the figure. The centre of the circle lies at the point of intersection of the perpendicular bisectors of two chords.

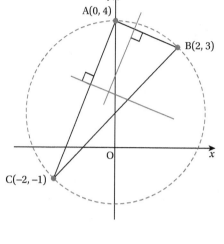

The midpoint of AC is $\left[\dfrac{0-2}{2}, \dfrac{4-1}{2}\right] \Rightarrow \left(-1, \dfrac{3}{2}\right)$.

The gradient of AC $= \dfrac{5}{2}$, so the gradient of the perpendicular bisector of AC is $-\dfrac{2}{5}$

and its equation is $y = -\dfrac{2}{5}x + \dfrac{11}{10} \Rightarrow 4x + 10y - 11 = 0$ [1]

The midpoint of AB is $\left(1, \dfrac{7}{2}\right)$ and its gradient is $-\dfrac{1}{2}$.

Therefore the gradient of the perpendicular bisector of AB is 2

and its equation is $y = 2x + \dfrac{3}{2} \Rightarrow 4x - 2y + 3 = 0$ [2]

Solving equations [1] and [2] simultaneously gives

$12y - 14 = 0 \Rightarrow y = \dfrac{7}{6}$ and $x = -\dfrac{1}{6}$.

Therefore the centre of the circle is the point $\left(-\dfrac{1}{6}, \dfrac{7}{6}\right)$.

Exercise 3

1 C(5, 3) is the centre of a circle of radius 5 units.

  a Find the equation of the circle in the form $(x-a)^2 + (y-b)^2 = r^2$.

  b Show that this circle cuts the x-axis at A(1, 0) and B(9, 0).

  c Prove that the radius that is perpendicular to AB goes through the midpoint of AB.

2 A triangle has its vertices at the points A(1, 3), B(5, 1) and C(7, 5). Prove that triangle ABC is right-angled and hence find the coordinates of the centre of the circle that circumscribes the triangle ABC.

3 The line joining A(5, 3) and B(4, −2) is a diameter of a circle. P(a, b) is a point on the circumference. Find a relationship between a and b.

4 A triangle has its vertices at the points A(1, 3), B(−2, 5) and C(4, −2).

  Find   **a** the coordinates of the centre of the circle that circumscribes triangle ABC

         b the radius of the circle that circumscribes triangle ABC. Give your answer correct to 3 significant figures.

9.3 Tangents to circles

When a line and a circle are drawn there are three possible positions of the line in relation to the circle. The line can miss the circle, or it can cut the circle in two distinct points, or it can touch the circle at one point.

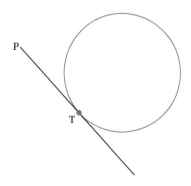

A line that touches a circle at one point is called a **tangent**.

The point at which it touches the circle is called the point of contact.

The length of a tangent drawn from a point to a circle is the distance from that point to the point of contact.

T is the point of contact.

PT is the length of the tangent from P.

A tangent to a circle is perpendicular to the radius drawn from the point of contact so AB is perpendicular to OT.

As the radius is perpendicular to the tangent at T, OT is the normal to the circle at T.

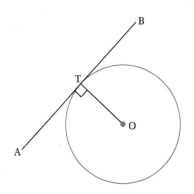

Example 6

Question

The centre of a circle is the point C(2, 5) and the radius of the circle is 3 units.

The equation of a line, L, is $x + y - 2 = 0$.

a Find the equation of the line J that goes through C and is perpendicular to L.

b Find the distance of C from L and hence determine whether L is a tangent to the circle.

Answer

a The line J is perpendicular to $x + y - 2 = 0$

so its equation is $x - y + k = 0$.

The point (2, 5) lies on J therefore $2 - 5 + k = 0 \implies k = 3$

so the equation of J is $x - y + 3 = 0$.

b The coordinates of A, the point of intersection of L and J, are needed to find the distance of C from L.

Solving the equations of L and J simultaneously gives $2x + 1 = 0$

$\implies \quad x = -\dfrac{1}{2}$ and $y = \dfrac{5}{2}$

so A is the point $\left(-\dfrac{1}{2}, \dfrac{5}{2}\right)$.

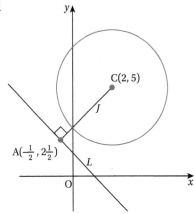

Therefore CA $= \sqrt{\left\{2 - \left(-\dfrac{1}{2}\right)\right\}^2 + \left\{5 - \dfrac{5}{2}\right\}^2} = 3.54$ to 3 significant figures.

For the line to be a tangent, CA would have to be 3 units exactly (equal to the radius).

CA > 3, therefore L is not a tangent.

Example 7

Question

Find the equation of the tangent at the point (3, 1) on the circle with equation $x^2 + y^2 - 4x + 10y - 8 = 0$.

Answer

$x^2 + y^2 - 4x + 10y - 8 = 0 \implies (x - 2)^2 + (y + 5)^2 = 37$

The centre of the circle is C(2, −5).

The tangent at A is perpendicular to the radius CA.

The gradient of CA is $\dfrac{1 - (-5)}{3 - 2} = 6$.

Therefore the gradient of the tangent at A is $-\dfrac{1}{6}$ and the tangent goes through A(3, 1).

So its equation is $y - 1 = -\dfrac{1}{6}(x - 3) \implies 6y + x = 9$.

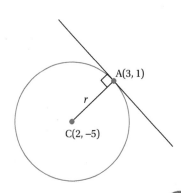

Example 8

Find the equation of the circle whose diameter is the line joining the points A(1, 5) and B(−2, 3).

We can use the fact that the angle in a semicircle is 90°.

$P(x, y)$ is a point on the circle if and only if (gradient AP) × (gradient BP) = −1.

The gradient of AP is $\dfrac{y-5}{x-1}$ and the gradient of PB is $\dfrac{y-3}{x+2}$

∴ $P(x, y)$ is on the circle $\Leftrightarrow \left(\dfrac{y-5}{x-1}\right)\left(\dfrac{y-3}{x+2}\right) = -1$

⇒ $(y-5)(y-3) = -(x-1)(x+2)$.

Therefore the equation of the circle is $x^2 + y^2 + x - 8y + 13 = 0$.

Alternatively the points A and B can be used to find the centre and radius of the circle.

The centre of the circle is the midpoint of AB, that is the point $(-\frac{1}{2}, 4)$.

The radius is the distance between the centre and A (or B),

so $r^2 = (1+\frac{1}{2})^2 + (5-4)^2 = \dfrac{13}{4}$.

Therefore the equation of the circle is $(x+\frac{1}{2})^2 + (y-4)^2 = \dfrac{13}{4}$

⇒ $x^2 + y^2 + x - 8y + 13 = 0$.

Exercise 4

1. The line $x - 2y + 4 = 0$ is a tangent to the circle whose centre is the point C(−1, 2).

   a. Find the equation of the line through C that is perpendicular to the line $x - 2y + 4 = 0$.

   b. Hence find the coordinates of the point of contact of the tangent and the circle.

2. The point A(6, 8) is on the circumference of a circle whose centre is the point C(3, 5). Find the equation of the tangent that touches the circle at A.

3. Write down the equation of the tangent to each circle at the given point.

   a. $x^2 + y^2 - 2x + 4y - 20 = 0$; (5, 1)

   b. $x^2 + y^2 - 10x - 22y + 129 = 0$; (6, 7)

   c. $x^2 + y^2 - 8y + 3 = 0$; (−2, 7)

4. Find the equation of the tangent at the origin to the circle

   $$x^2 + y^2 + 2x + 4y = 0$$

5. The line $y = 3x - 4$ is a tangent to the circle whose centre is the point (5, 2). Find the equation of the normal to the circle at the point of contact of the tangent.

6. The centre of a circle is the point C(5, 3). The gradient of the tangent to the circle at the point A is $-\dfrac{1}{2}$.

   Find the equation of the line through C and A.

Summary

The perpendicular bisector of a chord of a circle goes through the centre of the circle.

An angle in a semicircle is 90°.

The equation of a circle with centre (a, b) and radius r is $(x - a)^2 + (y - b)^2 = r^2$.

An equation of the form $ax^2 + by^2 + gx + fy + c = 0$ is the equation of a circle provided that $g^2 + f^2 - c > 0$.

Review

1. The equation $ax^2 + by^2 + fx + gy + c = 0$ is the equation of a circle with radius 3 and its centre at the origin. Find the values of f, g and c.

2. Which of the following equations is the equation of a circle? Give reasons for your answers.

   a $x^2 + 2y^2 = 1$ **b** $(x - 2)^2 - (y - 1)^2 = 4$

   c $x^2 + y^2 + 2x - 2y + 1 = 0$ **d** $x^2 + y^2 + 4 = 0$

3. The points A(5, 5) and B(−3, −1) are the ends of a diameter of the circle C.

   a Write down the coordinates of the centre of the circle.

   b Find the equation of the circle in the form $(x - a)^2 + (y - b)^2 = r^2$.

4. A circle C has equation $(x - 2)^2 + (y - 4)^2 = 20$.

   a Write down the coordinates of the centre of the circle and its radius.

   b Show that the point (−1, 3) lies inside the circle.

5. The circle C has equation $x^2 + y^2 + 2x - 6y - 3 = 0$.

   a Find the radius and the coordinates of the centre of C.

   b Show that the point (1, 6) lies on the circle.

6. Find the equations of the following circles.

   a A circle has its centre on the line $x + y = 1$ and passes through the origin and the point (4, 2).

   b The line joining (2, 1) to (6, 5) is a diameter of a circle.

   c A circle with centre (2, 7) passes through the point (−3, −5).

   d A circle intersects the y-axis at the origin and at the point (0, 6) and also touches the x-axis.

7. The equation of a circle is $(x - 4)^2 + (y - 2)^2 = 25$.

   a Find the coordinates of the points where the circle crosses the x-axis.

   b Find the equation of the tangent at the point where the circle cuts the positive x-axis.

8. The centre of a circle is the point C(−1, 3). The gradient of the tangent to the circle at the point A is $\dfrac{2}{3}$. Find the equation of the line through C and A.

Assessment

1 A circle has centre C(3, −1) and radius 10.

   a Give the equation of the circle in the form $(x-a)^2 + (y-b)^2 = c$.

   b Find the y-coordinates of the points where the circle cuts the y-axis.

   c The tangent to the circle at the point A has gradient $\frac{2}{3}$. Find an equation of the line CA.

2 A circle with centre C has equation $x^2 + y^2 - 2x - 6y - 7 = 0$.

   a Find the coordinates of C and the radius of the circle.

   b Find the y-coordinates of the points where the circle cuts the y-axis.

3 The circle with centre C(5, 3) touches the y-axis at the point A.

   a Give the equation of the circle in the form $(x-a)^2 + (y-b)^2 = c$.

   b Show that the point A(8, −1) lies on the circle.

   c Find the equation of the tangent to the circle at the point A.

4 A circle with centre C has equation $x^2 + y^2 - 4x + 6y = 12$.

   a Find the coordinates of C and the radius of the circle.

   b The point A (8, 4) lies outside the circle. Find the gradient of the line through C and A.

   c The line CA cuts the circle at P. Find the gradient of the tangent to the circle at P.

5 The points A and B have coordinates (2, 5) and (4, 12).

   a Find the coordinates of the midpoint of AB.

   b The line AB is a diameter of a circle. Find the radius of the circle.

   c Write down the equation of the circle in the form $(x-a)^2 + (y-b)^2 = c$.

   d Find the equation of the tangent to the circle at the point B.

6 A circle with centre C has equation $x^2 + y^2 + 2x - 6y - 40 = 0$.

 a Express this equation in the form $(x-a)^2 + (y-b)^2 = d$.

 b **i** State the coordinates of C.

   ii Find the radius of the circle, giving your answer in the form $n\sqrt{2}$.

 c The point P with coordinates $(4, k)$ lies on the circle. Find the possible values of k.

 d The points Q and R also lie on the circle, and the length of the chord QR is 2.

   Calculate the shortest distance from C to the chord QR.

AQA MPC1 June 2015

7 A circle has centre C(3, −8) and radius 10.

 a Express the equation of the circle in the form $(x-a)^2 + (y-b)^2 = k$.

 b Find the x-coordinates of the points where the circle crosses the x-axis.

 c The tangent to the circle at the point A has gradient $\dfrac{5}{2}$. Find an equation of the line CA, giving your answer in the form $rx + sy + t = 0$, where r, s and t are integers.

 d The line with equation $y = 2x + 1$ intersects the circle.

   i Show that the x-coordinates of the points of intersection satisfy the equation $x^2 + 6x - 2 = 0$

   ii Hence show that the x-coordinates of the points of intersection are of the form $m \pm \sqrt{n}$, where m and n are integers.

AQA MPC1 June 2011

10 Trigonometry

Introduction

Triangles are involved in many practical measurements so it is important to be able to make calculations from limited information about a triangle. This chapter shows how trigonometry is used to calculate sides, angles and areas of triangles.

Objectives

By the end of this chapter, you should know how to...

▶ Use trigonometry to find sides, angles and areas of triangles.

Recap

You need to remember how to...

▶ Use the sum of the interior angles of a triangle to help with calculations.
▶ Use Pythagoras' Theorem.
▶ Work with surds.

Applications

Triangulation is a process used in surveying which uses triangles to work out distance between places. The length between two points is measured and the angles to another distant point are measured from each end of the line. The triangle formed is used to work out the distances of the third point from each end of the line.

10.1 Trigonometric ratios of acute angles

The sine, cosine and tangent of an angle are called trigonometric ratios.

The sine, cosine and tangent of an acute angle A in a right-angled triangle are defined in terms of the sides of the triangle as

$$\sin A = \frac{\text{opposite}}{\text{hypotenuse}}, \quad \cos A = \frac{\text{adjacent}}{\text{hypotenuse}}, \quad \tan A = \frac{\text{opposite}}{\text{adjacent}}$$

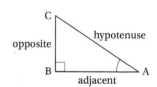

When any of these trigonometric ratios is given as a fraction, the lengths of two of the sides of the right-angled triangle can be marked. The third side can then be found using Pythagoras' theorem.

Example 1

Given that $\sin A = \dfrac{3}{5}$ find $\cos A$ and $\tan A$.

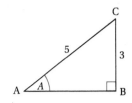

Answer

$\sin A = \dfrac{\text{opp}}{\text{hyp}}$, so draw a right-angled triangle with the side opposite angle A
3 units long and a hypotenuse 5 units long.

Using Pythagoras' theorem in triangle ABC gives

$(AB)^2 + 3^2 = 5^2 \quad \Rightarrow \quad AB = 4.$

Then $\quad \cos A = \dfrac{\text{adj}}{\text{hyp}} = \dfrac{4}{5} \quad$ and $\quad \tan A = \dfrac{\text{opp}}{\text{adj}} = \dfrac{3}{4}.$

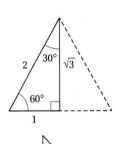

Exact values of trigonometric ratios

From the diagram (which is half an equilateral triangle of side 2 units)

$\sin 60° = \dfrac{\sqrt{3}}{2}, \cos 60° = \dfrac{1}{2}, \tan 60° = \sqrt{3}$

$\sin 30° = \dfrac{1}{2}, \cos 30° = \dfrac{\sqrt{3}}{2}, \tan 30° = \dfrac{1}{\sqrt{3}}$

From the diagram (which is an isosceles right angled triangle)

$\sin 45° = \cos 45° = \dfrac{1}{\sqrt{2}}, \tan 45° = 1$

These values (or how to derive them) should be learnt.

Trigonometric ratios of obtuse angles

The cosine of an obtuse angle

Using a calculator to find the cosines of angles from 0 to 180° gives the values
shown in the table.

θ	0	30°	45°	60°	90°	120°	135°	150°	180°
$\cos \theta$ (to 2 decimal places)	1	0.87	0.71	0.50	0	−0.50	−0.71	−0.87	−1

Plotting a graph of these values gives a shape called a cosine curve.

(The symbol θ is the most commonly used symbol for a variable angle.)

The values in the table and the graph show that an acute angle has a positive
cosine and an obtuse angle has a negative cosine. These values also show that

$\qquad \cos 60° = 0.5 \quad$ and $\quad \cos 120° = -0.5$

so $\quad \cos 120° = -\cos 60° \quad (120° + 60° = 180°)$

and $\quad \cos 45° = 0.71 \quad$ and $\quad \cos 135° = -0.71$

so $\quad \cos 135° = -\cos 45° \quad (135° + 45° = 180°)$

An accurate graph also shows that the cosine of an angle is equal to minus the
cosine of the supplementary angle, so

$\qquad \cos \theta = -\cos(180° - \theta)$

The sine of an obtuse angle

Using a calculator to find the sines of angles from 0 to 180° gives the values
shown in the table.

θ	0	30°	45°	60°	90°	120°	135°	150°	180°
$\sin \theta$ (to 2 decimal places)	0	0.5	0.71	0.87	1	0.87	0.71	0.5	0

Plotting these values gives this graph which is called a sine curve.

Again relationships can be seen between the sines of pairs of supplementary angles, for example

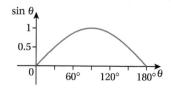

$$\sin 30° = 0.5 \text{ and } \sin 150° = 0.5$$

so $\quad \sin 150° = \sin 30° \quad (150° + 30° = 180°)$

also $\quad \sin 60° = 0.87 \text{ and } \sin 120° = 0.87$

so $\quad \sin 120° = \sin 60° \quad (120° + 60° = 180°)$

Therefore $\quad \sin \theta = \sin(180° - \theta)$.

An accurate graph also shows that $\quad \tan \theta = -\tan(180° - \theta)$.

Example 2

Question

Given $\sin \theta = \dfrac{1}{5}$ find two possible values for θ.

Answer

Using a calculator, the angle with a sine of 0.2 is 11.5°.

But $\sin \theta = \sin(180° - \theta)$ so $\sin 11.5° = \sin(180° - 11.5°)$,

therefore when $\sin \theta = \dfrac{1}{5}$, two possible values of θ are 11.5° and 168.5° correct to 1 decimal place.

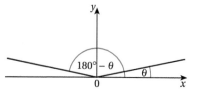

Exercise 1

In Questions 1 to 7, leave answers in surd form where necessary.

1 $\tan A = \dfrac{12}{5}$. Find $\sin A$ and $\cos A$.

2 $\cos X = \dfrac{4}{5}$. Find $\tan X$ and $\sin X$.

3 $\sin P = \dfrac{40}{41}$. Find $\cos P$ and $\tan P$.

4 $\tan A = 1$. Find $\sin A$ and $\cos A$.

5 $\cos Y = \dfrac{2}{3}$. Find $\sin Y$ and $\tan Y$.

6 $\sin A = \dfrac{1}{2}$.

    a Find $\cos A$.

    b Use your calculator to find the size of angle A correct to 1 decimal place.

7 $\sin X = \dfrac{a}{25}$ and $\tan X = \dfrac{7}{a}$.   Find $\cos X$.

In Questions 8 to 15, find X where X is an angle from 0 to 180°.

8 $\sin X = \sin 80°$ **9** $\cos X = -\cos 75°$

10 $\sin X = \sin 128°$ **11** $\cos 30° = -\cos X$

12 $\sin X = \sin 81°$ **13** $-\cos 123° = \cos X$

14 $\sin 90° = \sin X$ **15** $\cos 91° = -\cos X$

The unknown angles in Question 16 to 18 are in the range 0 to 180°.

16. $\cos X = -\dfrac{12}{13}$. Find $\sin X$.

17. $\sin \theta = \dfrac{4}{5}$. Find, to the nearest degree, two possible values of θ.

18. Given that $\sin A = 0.5$ and $\cos A = -0.8660$, find angle A correct to 1 decimal place.

19. Is there an angle X where
    - **a** $\cos X = 0$ and $\sin X = 1$
    - **b** $\sin X = 0$ and $\cos X = 1$
    - **c** $\cos X = 0$ and $\sin X = -1$?

20. Given $\cos A = -\cos B$, state a relationship between angle A and angle B.

21. An angle A has $\tan A = 1$. Write down the exact value of $\cos A$.

22. An angle X has $\sin X = \dfrac{\sqrt{2}}{2}$. Give two possible values of X that lie between $0°$ and $180°$.

10.2 The sine rule and cosine rule

A triangle can be defined without knowing every side and angle in the triangle. When enough information is known about a triangle, the remaining sides and angles can be calculated using a formula. The two formulae that are used most frequently are the sine rule and the cosine rule.

The notation used in a triangle ABC is that the side opposite to angle A is denoted by a, the side opposite to angle B by b and so on.

The sine rule

Look at the triangles both labeled ABC in which there is no right angle.

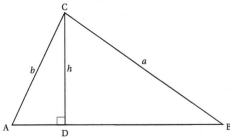 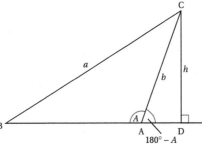

A line drawn from C perpendicular to AB, extended if necessary, divides each triangle ABC into two right-angled triangles, CDA and CDB.

In triangle CDA $\quad \sin A = \dfrac{h}{b} \quad \Rightarrow \quad h = b \sin A \quad (\sin(180 - A) = \sin A)$.

In triangle CDB $\quad \sin B = \dfrac{h}{a} \quad \Rightarrow \quad h = a \sin B$.

Therefore $\quad a \sin B = b \sin A \quad \Rightarrow \quad \dfrac{a}{\sin A} = \dfrac{b}{\sin b}$.

The triangles ABC can also be divided into two right-angled triangles by drawing the perpendicular line from A to BC (or from B to AC). This gives the result $\dfrac{b}{\sin B} = \dfrac{c}{\sin C}$.

Combining the two results gives the sine rule.

In any triangle ABC, $\dfrac{a}{\sin A} = \dfrac{b}{\sin B} = \dfrac{c}{\sin C}$ (use in this form to find a side)

or $\dfrac{\sin A}{a} = \dfrac{\sin B}{b} = \dfrac{\sin C}{c}$ (use in this form to find an angle)

The sine rule is made up of three separate fractions and only two of them can be used at a time. Therefore select the two fractions that contain three known quantities and only one unknown quantity.

To use the sine rule, the information needed is

either   two sides and the angle opposite one of them

or   two angles and a side.

When the sine rule is used to find an unknown angle, it can be used in the form $\dfrac{\sin A}{a} = \dfrac{\sin B}{b} = \dfrac{\sin C}{c}$.

Example 3

Question

In triangle ABC, BC = 5 cm, angle $A = 43°$ and angle $B = 61°$. Find the length of AC.

Answer

Angles A and B and side a are known. Side b is needed, so the two fractions to use from the sine rule are $\dfrac{a}{\sin A} = \dfrac{b}{\sin B}$.

$$\frac{5}{\sin 43°} = \frac{b}{\sin 61°}$$

$$\Rightarrow \quad b = \frac{5\sin 61°}{\sin 43°} = 6.412\ldots$$

Therefore AC = 6.41 cm correct to 3 significant figures.

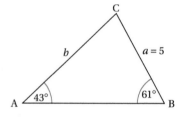

Example 4

Question

In ABC, AC = 17cm, angle $A = 105°$ and angle $B = 33°$. Find AB.

Answer

The two sides involved are b and c, so angle C must be found.

The sum of the angles in a triangle is 180° $\Rightarrow$ angle C = 42°.

Using the sine rule $\dfrac{b}{\sin B} = \dfrac{c}{\sin C}$

$$\Rightarrow \frac{17}{\sin 33°} = \frac{c}{\sin 42°} \quad \Rightarrow \quad c = \frac{17 \times \sin 42°}{\sin 33°} = 20.88\ldots$$

Therefore AB = 20.9cm correct to 3 significant figures.

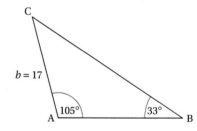

The ambiguous case

Two sides and one angle in a triangle are known.

When the angle is between the two sides there is only one possible triangle.

When the angle is not between the two given sides it is sometimes possible to draw two triangles from the given information.

Look at a triangle ABC in which angle $A = 20°$, $b = 10$ and $a = 8$.

The two triangles with these measurements are shown in the diagram; in one of them B is an acute angle, and in the other one B is obtuse.

There are *not* always two possible triangles.

For example when angle $A = 20°$, $b = 6$ and $a = 8$, there is only one triangle that fits the information.

Therefore when a known angle is not between the two known sides in a triangle, always check whether the obtuse angle is possible.

Example 5

In the triangle ABC, find C given that AB = 5 cm, BC = 3 cm and angle $A = 35°$.

Sides a and c and angle A are known so the sine rule can be used to find angle C.

$$\frac{\sin A}{a} = \frac{\sin C}{c} \Rightarrow \frac{\sin 35°}{3} = \frac{\sin C}{5} \Rightarrow \sin C = \frac{5 \times \sin 35°}{3} = 0.955...$$

$73°$ (to the nearest degree) is one angle whose sine is 0.955…

but $107°$ (to the nearest degree) is also an obtuse angle with the same sine.

If angle $C = 107°$, then   angle A + angle $C = 107° + 35° = 142°$

$\Rightarrow$ $B = 180° - 142° = 38°$.

So there is a triangle in which angle $C = 107°$ and there are two possible triangles.

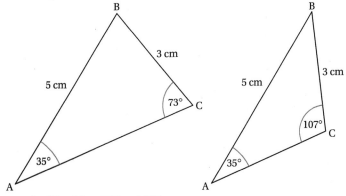

Therefore angle C is either $73°$ or $107°$.

Example 6

In the triangle XYZ, angle $Y = 41°$, XZ = 11 cm and YZ = 8 cm. Find angle X.

Use the part of the sine rule that involves x, y, angle X and angle Y

$$\frac{\sin X}{x} = \frac{\sin Y}{y} \Rightarrow \frac{\sin X}{8} = \frac{\sin 41°}{11} \Rightarrow \sin X = \frac{8 \times \sin 41°}{11} = 0.4771...$$

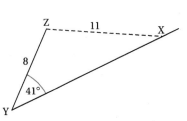

The two angles with a sine of 0.4771 … are $28°$ and $152°$ (to the nearest degree).

Checking whether $152°$ is a possible value for angle X shows that

angle X + angle $Y = 152° + 41° = 193°$.

(continued)

(*continued*)

This is greater than 180°, so angle X cannot be equal 152°.

In this case then, there is only one possible triangle containing the given data: the triangle in which angle X = 28°.

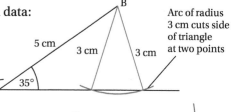

The diagrams show the two different situations in Examples 1 and 2.

When AB = 5 cm, BC = 3 cm and angle A = 35°, constructing the triangle gives

When XZ = 11cm, YZ = 8cm and angle Y = 41°, constructing the triangle gives

Exercise 2

Give answers correct to 3 significant figures.

1. In triangle ABC, AB = 9 cm, angle A = 51° and angle C = 39°. Find BC.

2. In triangle PQR, angle R = 52°, angle Q = 79° and PR = 12.7 cm. Find PQ.

3. In triangle DEF, DE = 174 cm, angle D = 48° and angle F = 56°. Find EF.

4. In triangle XYZ, angle X = 130°, angle Y = 21° and XZ = 53 cm. Find YZ.

5. In triangle PQR, angle Q = 37°, angle R = 101° and PR = 4.3 cm. Find PQ.

6. In triangle XYZ, XY = 92 cm, angle X = 59° and angle Y = 81°. Find XZ.

7. In triangle PQR, angle P = 78°, angle R = 38° and PR = 15 cm. Find QR.

8. In triangle ABC, AB = 10 cm, BC = 9.1 cm and AC = 17 cm. Can you use the sine rule to find angle A? If you answer YES, write down the two parts of the sine rule that you would use. If you answer NO, give your reason.

In Questions 9 to 14, find the angle shown by a question mark. Give two values when there are two possible triangles.

Give angles correct to the nearest degree.

	AB	BC	CA	A	B	C
9		2.9 cm	6.1 cm	?	40°	
10	5.7 cm		2.3 cm		20°	?
11	21 cm	36 cm		29.5°		?
12		2.7 cm	3.8 cm	?	54°	
13	4.6 cm		7.1 cm		?	33°
14	9 cm	7 cm		?		40°

The cosine rule

The sine rule cannot be used when only three sides of a triangle are known, but the cosine rule can be used.

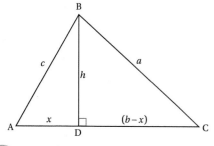
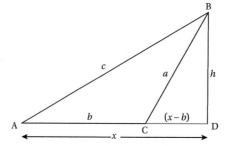

Triangle ABC has no right angle. BD is drawn perpendicular to AC.

Labelling AD as x, the length of CD is $(b - x)$ or $(x - b)$. Labelling BD as h, and using Pythagoras' theorem in each of the right-angled triangles BDA and BDC gives

$$h^2 = c^2 - x^2 \quad \text{and} \quad h^2 = a^2 - (b - x)^2 \quad \text{or} \quad (h^2 = a^2 - (x - b)^2)$$

Therefore $c^2 - x^2 = a^2 - (b - x)^2$

$$\Rightarrow \quad c^2 - x^2 = a^2 - b^2 + 2bx - x^2 \quad \Rightarrow \quad a^2 = b^2 + c^2 - 2bx.$$

In triangle DAB, $x = c \cos A$.

Therefore $a^2 = b^2 + c^2 - 2bc \cos A$.

When the height is drawn from A or from C similar expressions for the other sides of a triangle are obtained, giving

$$a^2 = b^2 + c^2 - 2bc \cos A, \ b^2 = c^2 + a^2 - 2ca \cos B \quad \text{and} \quad c^2 = a^2 + b^2 - 2ab \cos C$$

These formulae are valid when the angles are acute or obtuse.

Example 7

Question

In triangle ABC, BC = 7 cm, AC = 9 cm and $C = 61°$. Find AB.

Answer

Using the cosine rule, starting with c^2, we have

$$c^2 = a^2 + b^2 - 2ab \cos C$$

$$\Rightarrow \quad c^2 = 7^2 + 9^2 - (2)(7)(9) \cos 61°$$

$$\Rightarrow \quad c = 8.301\ldots$$

Therefore AB = 8.30 cm correct to 3 significant figures.

Example 8

Question

XYZ is triangle in which angle Y = 121°, XY = 14 cm and YZ = 26.9 cm. Find XZ.

Answer

Using $y^2 = z^2 + x^2 - 2zx \cos Y$ gives

$$y^2 = (14)^2 + (26.9)^2 - (2)(14)(26.9) \cos 121°$$

Hence $y^2 = 1307.53\ldots \quad \Rightarrow \quad y = 36.15\ldots$

Therefore XZ = 36.2 cm correct to 3 significant figures.

Using the cosine rule to find an angle

The cosine formula can be rearranged to find an unknown angle.

For example $c^2 = a^2 + b^2 - 2ab \cos C$ can be rearranged as $\cos C = \dfrac{a^2 + b^2 - c^2}{2ab}$

with similar expressions for $\cos A$ and $\cos B$.

Example 9

In triangle ABC, $a = 9$, $b = 16$ and $c = 11$. Find, to the nearest degree, the largest angle in the triangle.

The largest angle in a triangle is opposite to the longest side, so angle B is needed in this question.

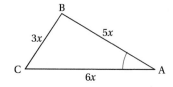

$$\cos B = \frac{c^2 + a^2 - b^2}{2ca}$$

$$= \frac{121 + 81 - 256}{(2)(11)(9)}$$

$$= -0.2727\ldots$$

The negative sign shows that angle B is obtuse.

Therefore angle $B = 106°$ to the nearest degree.

Example 10

The sides a, b, c of a triangle ABC are in the ratio $3 : 6 : 5$. Find the smallest angle in the triangle correct to 3 significant figures.

The lengths of the sides are not necessarily 3, 6 and 5 units so represent them by $3x$, $6x$ and $5x$.

The smallest angle is A (opposite to the smallest side).

$$\cos A = \frac{b^2 + c^2 - a^2}{2bc}$$

$$= \frac{36x^2 + 25x^2 - 9x^2}{60x^2} = \frac{52}{60} = 0.8666\ldots$$

Therefore the smallest angle in triangle ABC is $29.9°$ correct to 3 significant figures.

Exercise 3

In Questions 1 to 8, use the information given for triangle PQR to find the length of the third side. Give answers correct to 3 significant figures.

	PQ	QR	RP	P	Q	R
1		8 cm	4.6 cm			39°
2	11.7 cm		9.2 cm	75°		
3	29 cm	37 cm			109°	
4		2.1 cm	3.2 cm			97°
5	135 cm		98 cm	48°		
6	4.7 cm	8.1 cm			138°	
7		44 cm	62 cm			72°
8	19.4 cm		12.6 cm	167°		

In Questions 9 to 15, give angles correct to the nearest degree.

9 In triangle XYZ, XY = 34 cm, YZ = 29 cm and ZX = 21 cm. Find the smallest angle in the triangle.

10 In triangle PQR, PQ = 1.3 cm, QR = 1.8 cm and RP = 1.5 cm. Find angle Q.

11 In triangle ABC, AB = 51 cm, BC = 37 cm and CA = 44 cm. Find angle A.

12 Find the largest angle in triangle XYZ given that $x = 91$, $y = 77$ and $z = 43$.

13 In triangle BC, $a = 13$, $b = 18$ and $c = 7$. What is the size of

   a the smallest angle

   b the largest angle?

14 In triangle PQR the sides PQ, QR and RP are in the ratio 2 : 1 : 2. Find angle P.

15 ABCD is a quadrilateral in which AB = 5 cm, BC = 8 cm, CD = 11 cm, DA = 9 cm and angle ABC = 120°. Find

   a the length of AC correct to 3 significant figures

   b the size of the angle ADC.

General triangle calculations

A decision about whether to use the sine rule or the cosine rule to find sides or angles in a triangle depends on the facts that are known about the triangle.

The sine rule is easier to work with so use it whenever the given facts make this possible, that is whenever an angle and the opposite side are known. (Remember that if two angles are given, then the third angle is also known.)

The cosine rule is used only whenever the sine rule cannot be used.

For example, given the triangle PQR in the diagram, only one angle is known and the side opposite to it is not known.

Therefore the cosine rule must be used first to find the length of PR.

When q is found, angle Q is known so the sine rule can be used to find either of the remaining angles.

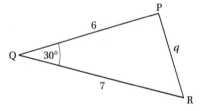

Exercise 4

Questions 1 to 6 refer to a triangle ABC. Fill in the unshaded spaces in the table. Give lengths correct to 3 significant figures and angles correct to 1 decimal place.

	A	B	C	a	b	c
1		80°	50°			68 cm
2			112°	15.7 cm	13 cm	
3	41°	69°		12.3 cm		
4	58°				131 cm	87 cm
5		49°	94°		206 cm	
6	115°		31°			21 cm

7 The three sides of a triangular field are bounded by straight fences of lengths 100 m, 80 m and 65 m. Find the angles between the boundary fences.

10.3 The area of a triangle

The simplest way to find the area of a triangle is to use the formula

$$\text{Area} = \frac{1}{2}\ \text{base} \times \text{perpendicular height}.$$

When the perpendicular height is not known, use the formula

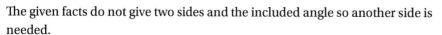

$$\text{area of triangle ABC} = \frac{1}{2} bc \sin A$$

with similar expressions for angles B and C, giving

$$\text{area} = \frac{1}{2} ab \sin C \qquad \text{and} \qquad \text{area} = \frac{1}{2} ac \sin B.$$

Each of these formulae can be remembered as

$$\text{area} = \frac{1}{2}\ \text{product of two sides} \times \text{sine of included angle}.$$

Example 11

Question

Find the area of triangle PQR, given that $P = 65°$, $Q = 79°$ and PQ = 30 cm.

Answer

The given facts do not give two sides and the included angle so another side is needed.

The sine rule can be used to find a side.

Angle $R = 180° - 65° - 79° = 36°$.

From the sine rule, $\dfrac{p}{\sin P} = \dfrac{r}{\sin R}$

$$\Rightarrow \qquad p = \frac{30 \times \sin 65°}{\sin 36°} = 46.25\ldots$$

Therefore area triangle $PQR = \dfrac{1}{2} pr \sin 79° = 681.1\ldots$

So the area of triangle PQR is 681 cm² (correct to 3 significant figures).

Exercise 5

Find the area of each triangle given in Questions 1 to 5. Give answers correct to 3 significant figures.

1. Triangle XYZ; XY = 180 cm, YZ = 145 cm, angle $Y = 70°$.
2. Triangle ABC; AB = 75 cm, AC = 66 cm, angle $A = 62°$.
3. Triangle PQR; QR = 69 cm, PR = 49 cm, angle $R = 85°$.
4. Triangle XYZ; $x = 30$, $y = 40$, angle $Z = 49°$.
5. Triangle PQR; $p = 9$, $r = 11$, angle $Q = 120°$.
6. In triangle ABC, AB = 6 cm, BC = 7 cm and CA = 9 cm. Find angle A and the area of the triangle.

7 In triangle PQR, angle $P = 60°$, angle $R = 50°$ and QR = 12 cm. Find PQ and the area of the triangle.

8 In triangle XYZ, XY = 150 cm, YZ = 185 cm and the area is 11 000 cm². Find angle Y and XZ.

9 The area of triangle ABC is 36.4 cm². Given that AC = 14 cm and angle $A = 98°$, find AB.

10 In triangle ABC, BD is perpendicular to AC. Using h as the length of BD, find an expression for h in triangle ABD. Hence prove that the area of triangle ABC is $\frac{1}{2} bc \sin A$.

Summary

The sine, cosine and tangent of an acute angle A in a right-angled triangle are defined as

$$\cos A = \frac{\text{adjacent}}{\text{hypotenuse}}, \ \sin A = \frac{\text{opposite}}{\text{hypotenuse}}, \ \tan A = \frac{\text{opposite}}{\text{adjacent}}$$

$$\sin 60° = \frac{\sqrt{3}}{2}, \cos 60° = \frac{1}{2}, \tan 60° = \sqrt{3}$$

$$\sin 30° = \frac{1}{2}, \cos 30° = \frac{\sqrt{3}}{2}, \tan 30° = \frac{1}{\sqrt{3}}$$

$$\sin 45° = \cos 45° = \frac{1}{\sqrt{2}}, \tan 45° = 1$$

$$\cos \theta = -\cos(180° - \theta)$$

$$\sin \theta = \sin(180° - \theta)$$

$$\tan \theta = -\tan(180° - \theta)$$

In any triangle ABC,

the sine rule is $\dfrac{a}{\sin A} = \dfrac{b}{\sin B} = \dfrac{c}{\sin C}$

the cosine rule is $a^2 = b^2 + c^2 - 2bc \cos A,$
$b^2 = c^2 + a^2 - 2ca \cos B$ and $c^2 = a^2 + b^2 - 2ab \cos C$
the area is $\frac{1}{2} bc \sin A = \frac{1}{2} ab \sin C = \frac{1}{2} ac \sin B.$

Review

1 Angle A is between 0° and 180°. Find the value of angle A when

   a $\cos A = -\cos 64°$ **b** $\sin 94° = \sin A$.

2 Angle X is acute and $\sin X = \dfrac{7}{25}$. Find $\cos(180° - X)$.

3 Given that $\sin A = \dfrac{5}{8}$, find $\tan A$ in surd form when

   a angle A is acute   **b** angle A is obtuse.

④ Use the information in the diagram to find, in surd form, $\sin\theta$ and $\cos\theta$.

a

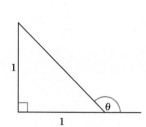

b

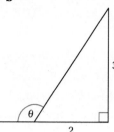

⑤ Given that $\sin X = \dfrac{12}{13}$ and X is obtuse, find $\cos X$.

⑥ In triangle ABC, BC = 11 cm, angle $B = 53°$ and angle $A = 76°$. Find AC correct to 3 significant figures.

⑦ In triangle PQR, $p = 3$, $q = 5$ and angle $R = 69°$. Find r correct to 3 significant figures.

⑧ In triangle XYZ, XY = 8 cm, YZ = 7 cm and ZX = 10 cm. Find angle Y correct to the nearest degree.

⑨ In triangle ABC, AB = 7 cm, BC = 6 cm and angle $A = 44°$. Find all possible values of angle C to the nearest degree.

⑩ Find the angles of a triangle whose sides are in the ratio $2 : 4 : 5$. Give answers correct to the nearest degree.

⑪ Use the cosine formula, $\cos A = \dfrac{b^2 + c^2 - a^2}{2bc}$, to show that

   a   angle A is acute if $a^2 < b^2 + c^2$ **b**   angle A is obtuse if $a^2 > b^2 + c^2$.

⑫ In triangle PQR, PQ = 11 cm, PR = 14 cm and angle QPR = 100°.

   Find the area of the triangle correct to 3 significant figures.

⑬ The area of ABC is 9 cm². AB = AC = 6 cm.

   a   Find $\sin A$.

   b   Are there two possible triangles? Give a reason for your answer.

Assessment

① Use the information in the diagram,

   a   to find angle ABC to the nearest degree

   b   to find the area of triangle ABC correct to 3 significant figures

   c   *hence* find the length of BD.

② In the triangle ABC, AB = 10 m and angle ABC = 150°.

   The area of the triangle is 16 m².

   a   Find the length of BC.

   b   Calculate the length of AC giving your answer correct to three significant figures.

   c   Find the size of the smallest angle in the triangle, giving your answer correct to one decimal place.

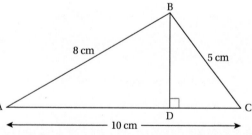

3 The line with equation $y = x$ and the x-axis form two sides of the triangle OAB.

The line with equation $y + 2x - 5 = 0$ intersects the line $y = x$ at A and the x-axis at B.

a Find the coordinates of A and B.

b State the size of the angle AOB.

c Find the area of triangle OAB.

4 The points P and Q are two points on land which are inaccessible. To find the distance PQ, a line AB of length 300 metres is marked out so that P and Q are on opposite sides of AB.

The directions of P and Q relative to the line AB are then measured and are shown in the diagram.

a Find the length of AP.

b Find the length of AQ.

c Hence calculate the length of PQ.

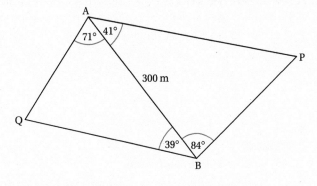

5 A surveyor stops at the point A and then walks 500 m due North along a road to a point B.

The points A and B are 500 m apart.

The point C is the base of a tower.

From A, the surveyor measures the angle between AB and AC as 62°.

From B the surveyor measures the angle between AB and BC as 84°.

a Find the distance AC.

b Find the area of triangle ABC.

c Hence find the perpendicular distance of the base of the tower from the road through A and B.

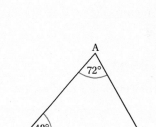

6 The diagram shows a triangle ABC.

The size of angle BAC is 72° and the size of angle ABC is 48°. The length of BC is 20 cm.

a Show that the length of AC is 15.6 cm, correct to three significant figures.

b The midpoint of BC is M. Calculate the length of AM, giving your answer, in cm, to three significant figures.

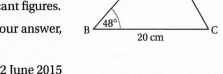

AQA MPC2 June 2015

7 The triangle ABC, shown in the diagram, is such that AB is 10 metres and angle BAC is 150°.

The area of triangle ABC is 40 m².

a Show that the length of AC is 16 metres.

b Calculate the length of BC, giving your answer, in metres, to two decimal places.

c Calculate the smallest angle of triangle ABC, giving your answer to the nearest 0.1°.

AQA MPC2 January 2012

11 Trigonometric Functions and Equations

Introduction

The angles used so far are angles that are measured in degrees and angles that can be found in a triangle, that is angles in the range 0° to 180°. This chapter introduces a different measure of angles and defines trigonometric functions for angles of any size.

Recap

You need to remember how to...

▶ Use the formulae for the circumference and radius of a circle.
▶ Use the exact values of the sine, cosine and tangent of 30°, 45° and 60°.
▶ Understand the meaning of rotational symmetry.
▶ Understand the effect on the equation of the curve $y = f(x)$ of a translation, a reflection and a one way stretch.

Objectives

By the end of this chapter, you should know how to...

▶ Define a radian as a measure of an angle.
▶ Find the length of an arc and the area of a sector of a circle.
▶ Define the trigonometric functions for $\sin x$, $\cos x$ and $\tan x$.
▶ Solve equations involving the sine, cosine and tangent of angles in a range of values of the angle.

11.1 Angle units

So far angles have been measured in degrees but there is another unit.

Before introducing that unit, here is a reminder of the names for parts of a circle.

Radians

Part of a circle is called an *arc*.

If the arc is less than half the circle it is called a *minor arc*;

if it is greater than half the circle it is called a *major arc*.

O is the centre of a circle and an arc PQ is drawn so that its length is equal to the radius of the circle.

The angle POQ is called a **radian**.

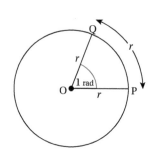

An arc equal in length to the radius of a circle subtends an angle of 1 radian at the centre of the circle.

The number of radians in a complete revolution is the number of times the radius divides into the circumference.

The circumference of a circle is $2\pi r$, so the number of radians in a revolution is $2\pi r \div r = 2\pi$.

> **Note**
>
> One radian is written 1 rad.

Therefore   **2π radians = 360°.**

So π radians $= 180°$ and $\dfrac{1}{2}\pi$ radians $= 90°$.

The radian symbol is not used when an angle is given in terms of π, so it is written as $180° = \pi$ (not $180° = \pi$ rad).

If an angle is a simple fraction of $180°$, it can be given in terms of π.

For example $60° = \dfrac{1}{3}$ of $180° = \dfrac{\pi}{3}$ and $135° = \dfrac{3}{4}$ of $180° = \dfrac{3\pi}{4}$.

Conversely, $\dfrac{7\pi}{6} = \dfrac{7}{6}$ of $180° = 210°$ and $\dfrac{2}{3}\pi = \dfrac{2}{3}$ of $180° = 120°$.

Angles that are not simple fractions of $180°$, or π, can be converted using the relationship $\pi = 180°$ and a calculator.

For example $73° = \dfrac{73}{180} \times \pi = 1.27$ rad (correct to 3 significant figures)

and 2.36 rad $= \dfrac{23.6}{\pi} \times 180° = 135°$ (correct to the nearest degree).

1 rad $= \dfrac{1}{\pi} \times 180° = 57°$ (correct to the nearest degree),

so 1 radian is a little less than $60°$.

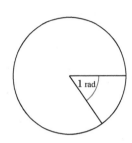

Example 1

Question

Express $75°$ in radians in terms of π.

Answer

$180° = \pi$ radians $\Rightarrow 1° = \dfrac{\pi}{180}$ radians;

so to convert degrees to radians, multiply by $\dfrac{\pi}{180}$.

$75° = 75 \times \dfrac{\pi}{180}$ radians $= \dfrac{5\pi}{12}$ radians

Example 2

Question

Express $\dfrac{1}{16}\pi$ radians in degrees.

Answer

π radians $= 180° \Rightarrow 1$ rad $= \dfrac{180°}{\pi}$; so to convert radians to degrees,

multiply by $\dfrac{180}{\pi}$.

$\dfrac{1}{16}\pi$ radians $= \dfrac{\pi}{16} \times \dfrac{180°}{\pi} = \dfrac{45°}{4} = 11\dfrac{1}{4}°$

Exercise 1

1 Express each of the following angles in radians as a fraction of π.

a $45°$	**b** $150°$	**c** $30°$	**d** $90°$
e $270°$	**f** $120°$	**g** $60°$	**h** $22.5°$
i $240°$	**j** $300°$	**k** $315°$	**l** $135°$
m $210°$	**n** $225°$		

2 Express each of the following angles in degrees.

a $\dfrac{1}{6}\pi$ **b** π **c** $\dfrac{1}{10}\pi$ **d** $\dfrac{\pi}{3}$

e $\dfrac{5}{6}\pi$ **f** $\dfrac{1}{12}\pi$ **g** $\dfrac{7\pi}{6}$ **h** $\dfrac{3}{4}\pi$

i $\dfrac{\pi}{9}$ **j** $\dfrac{3}{2}\pi$ **k** $\dfrac{4}{9}\pi$ **l** $\dfrac{1}{4}\pi$

m $\dfrac{3\pi}{5}$ **n** $\dfrac{1}{8}\pi$

3 Express each of the following angles in radians correct to 2 decimal places.

a $35°$ **b** $47.2°$ **c** $93°$ **d** $233°$

e $14.1°$ **f** $117°$ **g** $370°$

4 Express each of the following angles in degrees correct to 1 decimal place.

a 1.7 rad   **b** 3.32 rad   **c** 1 rad   **d** 2.09 rad

e 5 rad   **f** 6.28319 rad

5 Write down the value of

a $\sin\dfrac{1}{3}\pi$ **b** $\sin\dfrac{\pi}{6}$ **c** $\cos\dfrac{1}{2}\pi$ **d** $\cos\dfrac{\pi}{3}$

e $\tan\dfrac{1}{4}\pi$ **f** $\tan\pi$ **g** $\sin\dfrac{\pi}{2}$ **h** $\cos\pi$

i $\tan\dfrac{3\pi}{4}$ **j** $\cos\dfrac{2}{3}\pi$

6 Write down, as a fraction of π, the possible values of x in the range $0 \le x \le 2\pi$ for which

a $\cos x = 1$ **b** $\tan x = 1$ **c** $\sin x = \dfrac{1}{2}$

d $\cos x = \dfrac{1}{2}$ **e** $\sin x = -1$ **f** $\cos x = -1$

g $\sin x = 1$ **h** $\tan x = -1$ **i** $\sin x = -\dfrac{1}{2}$

j $\tan x = 0$ **k** $\cos x = 0$ **l** $\sin x = -\dfrac{1}{\sqrt{2}}$

For Questions 7 and 8 give answers correct to 3 significant figures.

7 Use your calculator to find

a $\sin 1.2$ rad   **b** $\cos 0.35$ rad

c $\tan 1.47$ rad   **d** $\cos 2.5$ rad

8 Use your calculator to find, in radians, the acute angle for which

a $\sin x = 0.28$ **b** $\tan x = 1.339$

c $\cos x = 0.7997$ **d** $\sin x = 0.0226$

> **Note**
>
> Make sure that the angle mode on your calculator is set to radians then $\sin\theta$ rad can be keyed in directly. Also with the mode in radians, a calculator will give the angle in radians for which for example, $\sin\theta = 0.7$.

11.2 The length of an arc

From the definition of a radian, the arc that subtends an angle of 1 radian at the centre of the circle is of length r.

Therefore when an arc subtends an angle of θ radians at the centre, the length of the arc is $r\theta$.

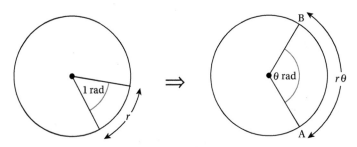

The length of arc AB = $r\theta$.

Example 3

Question

An arc subtends an angle of $\frac{\pi}{3}$ at the centre of a circle with radius 4.5 cm. Find the length of the arc in terms of π.

Answer

Length of arc = $r\theta = 4.5 \times \frac{\pi}{3}$ cm = 1.5π cm.

Example 4

Question

Find, in radians, the angle subtended at the centre of a circle of radius 4 cm by an arc of length 8 mm.

Answer

The units must be consistent when using any formula. In this question we use millimetres for both the length of the arc and the radius. Remember also that when using 'length of arc = $r\theta$' the angle must be measured in radians.

Length of arc = $r\theta$.

Therefore $8 = 40\theta \Rightarrow \theta = 0.2$ rad.

> **Note**
>
> When an angle is given or needed in degrees, use
> $$\frac{\text{length of arc}}{2\pi r} = \frac{\theta}{360}.$$

Exercise 2

1. The radius of a circle is 4 cm. Find, in terms of π, the length of the arc that subtends an angle of $\frac{1}{6}\pi$ radians at the centre of the circle.

2. An arc subtends an angle of $\frac{5}{4}\pi$ radians at the centre of a circle of radius 10 cm. Find, in terms of π, the length of the arc.

3. Find, in radians, the angle subtended at the centre of a circle of radius 5 cm by an arc of length 12 cm.

4. Find the size of the angle subtended at the centre of a circle of radius 65 mm by an arc of length 45 mm. Give your answer in radians correct to 3 significant figures.

5. Find in terms of π the radius of a circle in which an arc of length 15 cm subtends an angle of π radians at the centre.

6. An arc of length 20 cm subtends an angle of $\frac{4}{5}\pi$ radians at the centre of a circle. Find in terms of π the radius of the circle.

7. Find, in terms of π, the length of the arc that subtends an angle of $60°$ at the centre of a circle of radius 12 cm.

8. An arc of length 15 cm subtends an angle of $45°$ at the centre of a circle. Find, in terms of π, the radius of the circle.

9. Find, in degrees, the angle subtended at the centre of a circle of radius a cm by an arc of length $2a$ cm. Give your answer in terms of π.

10. Calculate, in degrees correct to 1 decimal place, the angle subtended at the centre of a circle of radius 2.7 cm by an arc of length 6.9 cm.

11.3 The area of a sector

The area enclosed by two radii and an arc is called a **sector**.

The area enclosed by a chord and an arc is called a **segment.**

If the segment is less than half a circle it is called a *minor segment*; if it is greater than half a circle it is called a *major segment*.

A sector contains an angle of θ radians at the centre of the circle.

The complete angle at the centre of the circle is 2π.

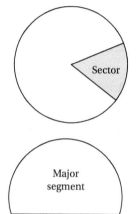

Therefore using $\dfrac{\text{area of sector}}{\text{area of circle}} = \dfrac{\text{angle contained in the sector}}{\text{complete angle at the centre}}$

gives $\dfrac{\text{area of sector}}{\text{area of circle}} = \dfrac{\theta}{2\pi}$.

The area of the circle is πr^2,

therefore area of sector $= \dfrac{\theta}{2\pi} \times \pi r^2 = \dfrac{1}{2} r^2 \theta$.

The area of sector AOB $= \dfrac{1}{2} r^2 \theta$.

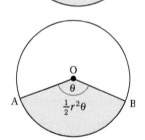

Example 5

Find, in terms of π, the area of the sector of a circle of radius 3 cm that contains an angle of $\dfrac{\pi}{5}$.

Area of sector $= \dfrac{1}{2} r^2 \theta = \dfrac{1}{2} (3)^2 \left(\dfrac{\pi}{5} \right)$ cm$^2 = \dfrac{9\pi}{10}$ cm^2.

Example 6

AB is a chord of a circle with centre O and radius 4 cm. AB is of length 4 cm and divides the circle into two segments. Find, correct to two decimal places, the area of the minor segment.

AOB is an equilateral triangle, so each angle is $60° = \dfrac{\pi}{3}$ rad.

To find the area of the minor segment, subtract the area of triangle AOB from the area of sector AOB.

Area of sector AOB $= \dfrac{1}{2}r^2\theta = \dfrac{1}{2}(4)^2\left(\dfrac{1}{3}\pi\right) = 8.3775\ldots$

Area of triangle AOB $= \dfrac{1}{2}r^2\sin\theta = \dfrac{1}{2}(4)(4)\left(\sin\dfrac{\pi}{3}\right) = 6.9282\ldots$

Area of minor segment = area of sector AOB − area of triangle AOB

$$= 8.3775 - 6.9282 = 1.4493$$

The area of the minor segment is 1.45 cm² correct to 3 significant figures.

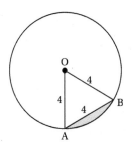

Exercise 3

Give answers that are not exact correct to 3 significant figures.

1 A sector of a circle of radius 4 cm contains an angle of 30°. Find the area of the sector.

2 A sector of a circle of radius 8 cm contains an angle of 135°. Find the area of the sector.

3 The area of a sector of a circle of radius 2 cm is π cm². Find the angle contained by the sector.

4 The area of a sector of a circle of radius 5 cm is 12 cm². Find the angle contained by the sector.

5 A sector of a circle of radius 10 cm contains an angle of $\dfrac{5}{6}\pi$. Find the area of the sector.

6 An arc of length 15 cm subtends an angle π at the centre of a circle. Find the radius of the circle and hence the area of the sector containing the angle π.

7 A sector of a circle has an area 3π cm² and contains an angle $\dfrac{1}{6}\pi$. Find the radius of the circle.

8 A sector of a circle has an area 67π cm² and contains an angle of 45°. Find the radius of the circle.

9 An arc of a circle is of length 5π cm and the sector it bounds has an area of 20π cm². Find the radius of the circle.

10 Calculate, in radians, the angle at the centre of a circle of radius 83 mm contained in a sector of area 974 mm².

11 A circle has centre O and radius 5 cm. AB is a chord of the circle of length 8 cm. Find

  a the area of triangle AOB **b** the area of the sector AOB.

12 A chord of length 10 mm divides a circle of radius 7 mm into two segments. Find the area of each segment.

13 A chord PQ, of length 12.6 cm, subtends an angle of $\dfrac{2}{3}\pi$ at the centre of a circle. Find

  a the length of the arc PQ **b** the area of the minor segment cut off by the chord PQ.

Note

To use the formula $A = \dfrac{1}{2}r^2\theta$ for the area of a sector, the angle must be in radians. When the angle is in degrees use $A = \pi r^2 \times \dfrac{\theta}{360}$.

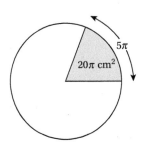

11.4 The trigonometric functions

The general definition of an angle

An angle is defined as a measure of rotation.

A line can rotate from its initial position OP_0 about the point O to any other position OP.

The amount of rotation is shown by the angle between OP_0 and OP, so

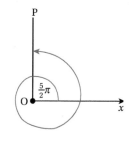

> **an angle is a measure of the rotation of a line about a fixed point.**

The anticlockwise sense of rotation is taken as positive and clockwise rotation is negative. Therefore an angle formed by the anticlockwise rotation of OP is a positive angle.

The rotation of OP is not limited to one revolution, so an angle can be as big as we want to make it.

If θ is any angle, then θ can be measured either in degrees or in radians and θ can take all real values.

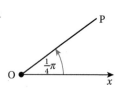
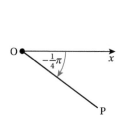
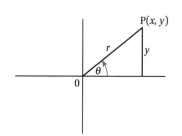

Definition of trigonometric ratios

The sine, cosine and the tangent of an angle are defined for angles of all values as follows.

OP is drawn on x- and y-axes as shown and, for all values of θ, the length of OP is r and the coordinates of P are (x, y), then

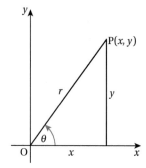

$$\sin \theta = \frac{y}{r}$$

$$\cos \theta = \frac{x}{r}$$

$$\tan \theta = \frac{y}{x}$$

The sine function

Measuring θ in radians, the definition of $f(\theta) = \sin \theta$ shows that:

For $0 \le \theta \le \frac{1}{2}\pi$, OP is in the first quadrant;

y is positive and increases in value from 0 to r as θ increases from 0 to $\frac{1}{2}\pi$.

r is always positive, so $\sin \theta$ increases from 0 to 1.

For $\frac{1}{2}\pi \le \theta \le \pi$, OP is in the second quadrant;

again y is positive but decreases in value from r to 0,

so $\sin \theta$ decreases from 1 to 0.

For $\pi \le \theta \le \frac{3}{2}\pi$, OP is in the third quadrant;

y is negative and decreases from 0 to $-r$,

so $\sin\theta$ decreases from 0 to -1.

For $\frac{3}{2}\pi \le \theta \le 2\pi$, OP is in the fourth quadrant;

y is still negative but increases from $-r$ to 0,

so $\sin\theta$ increases from -1 to 0.

For $\theta \ge 2\pi$, the cycle repeats itself as OP rotates round the quadrants again.

For negative values of θ, OP rotates clockwise round the quadrants in the order 4th, 3rd, 2nd, 1st, and so on.

So for negative values of θ, $\sin\theta$ decreases from 0 to -1, then increases to 0 and on to 1 before decreasing to zero and repeating the pattern.

This shows that $\sin\theta$ is positive for $0 < \theta < \pi$ and negative when $\pi < \theta < 2\pi$.

Also, $\sin\theta$ varies in value between -1 and 1 and the pattern repeats itself every revolution.

A plot of the graph of $f(\theta) = \sin\theta$ confirms these observations.

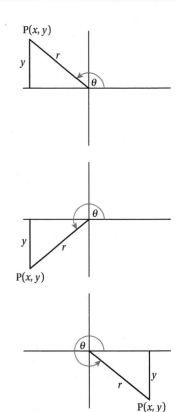

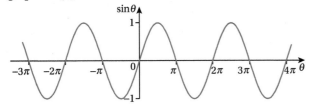

A graph of this shape is called a **sine wave** and shows the following properties of the sine function.

> The curve is continuous (it has no breaks).
> $-1 \le \sin\theta \le 1$

Note

Search online for 'animation of the sine graph' to see this graph develop from the definition.

The shape of the curve from $\theta = 0$ to $\theta = 2\pi$ is repeated for each complete revolution. Any function with a repetitive pattern is called **periodic** or **cyclic**. The width of the repeating pattern, as measured on the horizontal scale, is called the **period**.

> The period of the sine function is 2π.

Other properties of the sine function shown by the graph are as follows.

$$\sin\theta = 0 \text{ when } \theta = n\pi \text{ where } n \text{ is an integer.}$$

The curve has rotational symmetry about the origin so, for any angle α.

$$\sin(-\alpha) = -\sin\alpha, \text{ for example } \sin(-30°) = -\sin 30° = -\frac{1}{2}$$

An enlarged section of the graph for $0 \le \theta \le 2\pi$ shows further relationships.

The curve is symmetrical about the line $\theta = \frac{1}{2}\pi$, so

$$\sin(\pi - \alpha) = \sin\alpha, \text{ for example } \sin 130° = \sin(180° - 130°) = \sin 50°$$

The curve has rotational symmetry about $\theta = \pi$, so

$$\sin(\pi + \alpha) = -\sin\alpha \text{ and } \sin(2\pi - \alpha) = -\sin\alpha$$

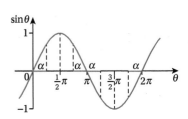

Example 7

Question

Find the exact value of $\sin\dfrac{4}{3}\pi$.

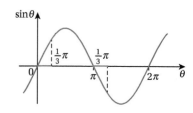

Answer

$$\sin\dfrac{4}{3}\pi = \sin\left(\pi+\dfrac{1}{3}\pi\right) = -\sin\dfrac{1}{3}\pi = -\dfrac{\sqrt{3}}{2}$$

Example 8

Question

Sketch the graph of $y = \sin\left(\theta-\dfrac{1}{4}\pi\right)$ for values of 6 between θ and 2π.

Answer

Use the fact that the curve $y = f(x-a)$ is a translation of the curve $y = f(x)$ by the vector $\begin{bmatrix} a \\ 0 \end{bmatrix}$.

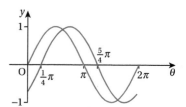

Exercise 4

For Questions 1 to 4, find the exact value of

1 $\sin 120°$ **2** $\sin -2\pi$

3 $\sin 300°$ **4** $\sin -210°$

5 Write down all the values of θ between 0 and 6π for which $\sin\theta = 1$.

6 Write down all the values of θ between 0 and -4π for which $\sin\theta = -1$.

For Questions 7 to 10 express in terms of the sine of an acute angle

7 $\sin 125°$ **8** $\sin 290°$

9 $\sin -120°$ **10** $\sin\dfrac{7}{6}\pi$

For Questions 11 to16 sketch the curve for values of θ in the range $0 \le \theta \le 3\pi$.

11 $y = \sin\left(\theta+\dfrac{1}{3}\pi\right)$ **12** $y = -\sin\theta$ **13** $y = \sin(-\theta)$

14 $y = 1 - \sin\theta$ **15** $y = \sin(\pi-\theta)$ **16** $y = \sin\left(\dfrac{1}{2}\pi-\theta\right)$

17 On the same set of axes draw the curves $y = \sin\theta$ and $y = \sin 3\theta$. What can you deduce about the relationship between the two curves?

18 Sketch the curves

    a $y = \sin 4\theta$ **b** $y = 4\sin\theta$

The cosine function

For any position of P, $\cos\theta = \dfrac{x}{r}$.

When P is in the first quadrant,

x decreases from r to 0 as θ increases, so $\cos\theta$ decreases from 1 to 0.

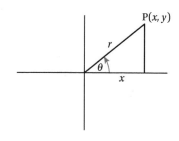

When P is in the second quadrant,
x decreases from 0 to −r,
so cos θ decreases from 0 to −1.

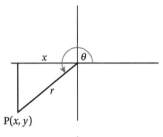

When P is in the third quadrant,
cos θ increases from −1 to 0.

When P is in the fourth quadrant,
cos θ increases from 0 to 1.

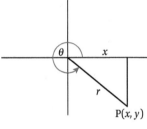

The cycle then repeats itself, giving this graph of f(θ) = cosθ.

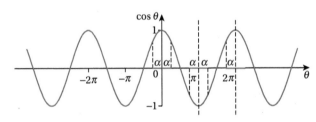

The graph shows the following properties of the cosine function.

The curve is continuous.

$-1 \leq \cos\theta \leq 1$

It is periodic with a period of 2π.

It is the same shape as the sine wave but is translated a distance $\frac{1}{2}\pi$ to the left.

$$\cos\theta = 0 \text{ when } \theta = \ldots -\frac{1}{2}\pi, \frac{1}{2}\pi, \frac{3}{2}\pi, \frac{5}{2}\pi, \ldots$$

The curve is symmetric about $\theta = 0$, so cos$-\alpha$ = cos α.

The curve has rotational symmetry about $\theta = \frac{1}{2}\pi$, so

$$\cos(\pi - \alpha) = -\cos\alpha$$

Symmetry also shows that

$$\cos(\pi + \alpha) = -\cos\alpha \quad \text{and} \quad \cos(2\pi - \alpha) = \cos\alpha$$

Exercise 5

1 Write in terms of the cosine of an acute angle

   a cos 123°   **b** cos 250°

   c cos(−20°)   **d** cos(−154°)

2 Find the exact value of

   a cos 150°   **b** $\cos\frac{3}{2}\pi$

   c $\cos\frac{5}{4}\pi$ **d** cos 6π

3 Sketch each of the following curves

    a $y = \cos(\theta + \pi)$ **b** $y = \cos\left(\theta - \frac{1}{3}\pi\right)$ **c** $y = \cos(-\theta)$

4 Sketch the graph of $y = \cos\left(\theta - \frac{1}{2}\pi\right)$. Describe the relationship this shows between $\sin\theta$ and $\cos\left(\theta - \frac{1}{2}\pi\right)$.

5 Sketch the curve $y = \cos\left(\theta - \frac{1}{4}\pi\right)$ for values of θ between $-\pi$ and π. Use the sketch to find the values of θ in this range for which

    a $\cos\left(\theta - \frac{1}{4}\pi\right) = 1$ **b** $\cos\left(\theta - \frac{1}{4}\pi\right) = -1$ **c** $\cos\left(\theta - \frac{1}{4}\pi\right) = 0$

6 On the same set of axes, sketch the graphs $y = \cos\theta$ and $y = 3\cos\theta$.

7 On the same set of axes, sketch the graphs $y = \cos\theta$ and $y = \cos3\theta$.

8 Sketch the graph of $f(\theta) = \cos4\theta$ for $0 \leq \theta \leq \pi$. Hence state the values of θ in this range for which $f(\theta) = 0$.

The tangent function

For any position of P, $\tan\theta = \dfrac{y}{x}$.

As OP rotates through the first quadrant, x decreases from r to 0, while y increases from 0 to r.

This means that the fraction $\dfrac{y}{x}$ increases from 0 to very large values therefore as $\theta \to \frac{1}{2}\pi$, $\tan\theta \to \infty$.

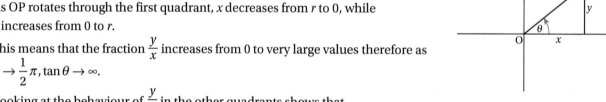

Looking at the behaviour of $\dfrac{y}{x}$ in the other quadrants shows that

    in the second quadrant, $\tan\theta$ is negative and increases from $-\infty$ to 0,

    in the third quadrant, $\tan\theta$ is positive and increases from 0 to ∞,

    in the fourth quadrant, $\tan\theta$ is negative and increases from $-\infty$ to 0.

The cycle then repeats itself giving this plot of the graph of $f(\theta) = \tan\theta$.

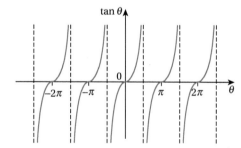

The graph shows that the properties of the tangent function are different from those of the sine and cosine functions.

    It is not continuous, as it is *undefined* when $\theta = \ldots -\frac{1}{2}\pi, \frac{1}{2}\pi, \frac{3}{2}\pi, \ldots$

    The range of values of $\tan\theta$ is unlimited.

    It is periodic with a period of π (not 2π as for $\sin\theta$ and $\cos\theta$).

The graph has rotational symmetry about $\theta = 0$, so

$$\tan(-\alpha) = -\tan\alpha$$

The graph has rotational symmetry about $\theta = \frac{1}{2}\pi$, giving

$$\tan(\pi - \alpha) = -\tan\alpha$$

The cycle repeats itself from $\theta = \pi$ to 2π, so

$$\tan(\pi + \alpha) = \tan\alpha \quad \text{and} \quad \tan(2\pi - \alpha) = -\tan\alpha$$

Example 9

Question

Express $\tan\dfrac{11}{4}\pi$ as the tangent of an acute angle.

Answer

$$\tan\left(\frac{11}{4}\pi\right) = \tan\left(2\pi + \frac{3}{4}\pi\right) = \tan\left(\frac{3}{4}\pi\right) = \tan\left(\pi - \frac{1}{4}\pi\right) = -\tan\frac{1}{4}\pi$$

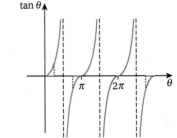

Exercise 6

1 Find the exact value of

   a $\tan\dfrac{9}{4}\pi$ **b** $\tan 120°$ **c** $\tan -\dfrac{2}{3}\pi$ **d** $\tan\dfrac{7}{4}\pi$

2 Write in terms of the tangent of an acute angle

   a $\tan 220°$ **b** $\tan\dfrac{12}{7}\pi$ **c** $\tan 310°$ **d** $\tan -\dfrac{7}{5}\pi$

3 Sketch the graph of $y = \tan\theta$ for values of θ in the range 0 to 2π. From this sketch find the values of θ in this range for which

   a $\tan\theta = 1$ **b** $\tan\theta = -1$ **c** $\tan\theta = 0$ **d** $\tan\theta = \infty$

11.5 Solving trigonometric equations

The definitions of $\sin\theta$, $\cos\theta$, and $\tan\theta$ are $\sin\theta = \dfrac{y}{r}$, $\cos\theta = \dfrac{x}{r}$, $\tan\theta = \dfrac{y}{x}$

where x, y, and r are the sides of a right-angled triangle.

So $\dfrac{\sin\theta}{\cos\theta} = \dfrac{y}{r} \div \dfrac{x}{r} = \dfrac{y}{x} = \tan\theta$

and $\sin^2\theta + \cos^2\theta = \dfrac{y^2}{r^2} + \dfrac{x^2}{r^2} = \dfrac{y^2 + x^2}{r^2}$, and from Pythagoras' theorem,

$y^2 + x^2 = r^2$, therefore $\sin^2\theta + \cos^2\theta = 1$.

> **Therefore for all values of θ, $\tan\theta = \dfrac{\sin\theta}{\cos\theta}$ and $\sin^2\theta + \cos^2\theta = 1$.**

These relationships, together with sketches of the graphs of the trigonometric functions can help to solve a number of trigonometric equations and you need to learn them.

Example 10

Question

Find the values of x between 0 and 360° for which $\sin x = -0.3$.

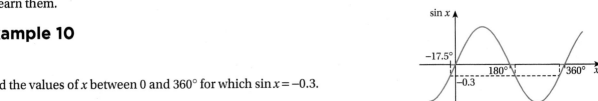

The value given for x by a calculator is $-17.5°$.

The graph shows that when $\sin x = -0.3$, the values of x in the given range are $180° + 17.5°$ and $360° - 17.5°$.

When $\sin x = -0.3$, $x = 197.5°$ and $342.5°$.

Example 11

Question

Find the smallest positive value of θ for which $\cos\theta = 0.7$ and $\tan\theta$ is negative.

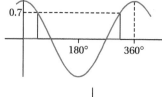

Answer

When $\cos\theta = 0.7$, the possible values of θ are $45.6°$, $314.4°$, ...
$\tan\theta$ is positive when θ is in the first quadrant and negative when θ is in the fourth quadrant. Therefore the value of θ is $314.4°$.

> **Note**
> When the range of values is given in degrees, the answer should also be given in degrees and the same is true for radians.

Example 12

Question

Solve the equation $2\cos^2\theta - \sin\theta = 1$ for values of θ in the range 0 to 2π.

Answer

The equation is quadratic, but it involves the sine and the cosine of θ, so use $\cos^2\theta + \sin^2\theta = 1$ to express the equation in terms of $\sin\theta$ only.

$2\cos^2\theta - \sin\theta = 1$
$\Rightarrow 2(1 - \sin^2\theta) - \sin\theta = 1$
$2\sin^2\theta + \sin\theta - 1 = 0$
$\Rightarrow (2\sin\theta - 1)(\sin\theta + 1) = 0 \Rightarrow \sin\theta = \dfrac{1}{2}$ or -1

When $\sin\theta = \dfrac{1}{2}$, $\theta = \dfrac{1}{6}\pi, \dfrac{5}{6}\pi$.

When $\sin\theta = -1$, $\theta = \dfrac{3}{2}\pi$.

Therefore the solution of the equation is $\theta = \dfrac{1}{6}\pi, \dfrac{5}{6}\pi, \dfrac{3}{2}\pi$.

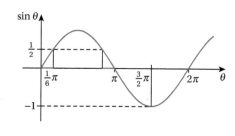

Example 13

Question

Solve the equation $\dfrac{\cos x}{\sin x} = \sin x$ for values of x from 0 to $360°$.

Answer

Both sides of an equation can be multiplied by any number *except* zero. The equation can be simplified by multiplying by $\sin x$ provided that $\sin x \neq 0$. So any values of x for which $\sin x = 0$ must be excluded from the solutions.

$\dfrac{\cos x}{\sin x} = \sin x \Rightarrow \cos x = \sin^2 x \Rightarrow \cos x = 1 - \cos^2 x \Rightarrow \cos^2 x + \cos x - 1 = 0$

This equation does not factorise so use the formula

$\Rightarrow \cos x = \dfrac{1}{2}(-1 \pm \sqrt{5}) \Rightarrow \cos x = 0.618$ or $\cos x = -1.618$

and there is no value of x for which $\cos x = -1.618$.

$\Rightarrow x = 51.8°$ or $308.2°$ ($\sin x \neq 0$ for either value of x.)

> **Note**
> Use $\cos^2 x + \sin^2 x = 1$.

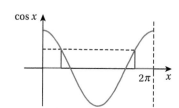

Exercise 7

Give answers that are not exact to 1 decimal place for degrees and to 3 significant figures for radians.

1 Within the range $-2\pi \le \theta \le 2\pi$, give all the values of θ for which

    a $\sin\theta = 0.4$ **b** $\cos\theta = -0.5$ **c** $\tan\theta = 1.2$

2 Within the range $0 \le \theta \le 720°$, give all the values of θ for which

    a $\tan\theta = -0.8$ **b** $\sin\theta = -0.2$ **c** $\cos\theta = 0.1$

3 Find the smallest angle in radians (positive or negative) for which

    a $\cos\theta = 0.8$ and $\sin\theta \ge 0$ **b** $\sin\theta = -0.6$ and $\tan\theta \le 0$

    c $\tan\theta = \sin\dfrac{1}{6}\pi$

4 Using $\tan\theta \equiv \dfrac{\sin\theta}{\cos\theta}$ show that the equation $\tan\theta = \sin\theta$ can be written as $\sin\theta\,(\cos\theta - 1) = 0$, provided that $\cos\theta \neq 0$.

    Hence find the values of θ between 0 and 2π for which $\tan\theta = \sin\theta$.

Solve the following equations for angles in the range $0 \le \theta \le 360°$.

5 $\sin\theta = \dfrac{\sqrt{3}}{2}$ **6** $\cos\theta = 0$ **7** $\tan\theta = -\sqrt{3}$

8 $\sin\theta = -\dfrac{1}{4}$ **9** $\cos\theta = -\dfrac{1}{2}$ **10** $\tan\theta = 1$

11 $\sin^2\theta = \dfrac{1}{4}$ **12** $4\cos^2\theta + 5\sin\theta = 3$

Solve the following equations for angles in the range $-\pi \le \theta \le \pi$.

13 $5\cos\theta - 4\sin^2\theta = 2$ **14** $2\cos\theta - 4\sin^2\theta + 2 = 0$

15 $2\sin\theta\cos\theta + \sin\theta = 0$ **16** $\sqrt{3}\tan\theta = 2\sin\theta$

Equations involving compound angles

Many trigonometric equations involve a compound angle.

For example $\sin\left(\theta - \dfrac{1}{2}\pi\right) = 1.$

Equations of this type can be solved by finding first the values of the compound angle and then, by solving a linear equation, the corresponding values of θ.

Example 14

Find the values of θ in the interval $-\pi \le \theta \le \pi$, for which $\cos\left(\theta - \dfrac{\pi}{3}\right) = \dfrac{1}{2}$.

Using $\left(\theta - \dfrac{\pi}{3}\right) = \phi$ gives $\cos\phi = \dfrac{1}{2}$.

Values of θ are to be found in the interval $-\pi \le \theta \le \pi$, so values of $\phi = \left(\theta - \dfrac{\pi}{3}\right)$

must be found in the interval $-\pi - \dfrac{\pi}{3} \le \phi \le \pi - \dfrac{\pi}{3}$.

In the range $-\pi - \dfrac{\pi}{3} \le \phi \le \pi - \dfrac{\pi}{3}$ the solutions of $\cos\phi = \dfrac{1}{2}$ are $-\dfrac{1}{3}\pi, \dfrac{1}{3}\pi$.

But $\phi = \left(\theta - \dfrac{\pi}{3}\right) \Rightarrow \theta = \phi + \dfrac{\pi}{3}$, therefore $\theta = -\dfrac{1}{3}\pi + \dfrac{1}{3}\pi, \dfrac{1}{3}\pi + \dfrac{1}{3}\pi = 0, \dfrac{2}{3}\pi$

$\Rightarrow \quad \theta = 0, \dfrac{2}{3}\pi.$

Exercise 8

In Questions 1 to 6, find the solution of the equation, for values of θ in the range $0 \le \theta \le 180°$.

1 $\tan(\theta - 30°) = 1$

2 $\cos(\theta - 45°) = -0.5$

3 $\sin(\theta + 15°) = -\dfrac{\sqrt{2}}{2}$

4 $\cos(\theta - 45°) = 0$

5 $\sin(\theta + 30°) = -1$

6 $\tan(\theta - 60°) = 0$

In Questions 7 to 11 solve the equations for values of θ in the range $0 \le \theta \le \pi$.

7 $\tan\left(\theta - \dfrac{\pi}{3}\right) = -\sqrt{3}$

8 $\cos(\theta - \pi) = \dfrac{1}{2}$

9 $\tan\left(\theta + \dfrac{\pi}{2}\right) = -1$

10 $\cos\left(\theta + \dfrac{1}{4}\pi\right) = \dfrac{1}{2}$

11 $\tan\left(\theta - \dfrac{1}{3}\pi\right) = -1$

Summary

One radian (1 rad) is the size of the angle subtended at the centre of a circle by an arc equal in length to the radius of the circle.

The length of arc AB is $r\theta$.

The area of sector AOB is $\dfrac{1}{2}r^2\theta$.

The sine function, $f(x) = \sin x$, is defined for all values of x,

is periodic with a period 2π,

has a maximum value of 1 when $x = \left(2n + \dfrac{1}{2}\right)\pi$, a minimum value of -1 when $x = \left(2n + \dfrac{3}{2}\right)\pi$, is zero when $x = n\pi$.

Sine function

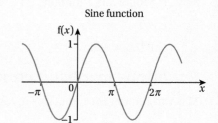

The cosine function, $f(x) = \cos x$, is defined for all values of x,

is periodic with a period 2π,

has a maximum value of 1 when $x = 2n\pi$, a minimum value of -1 when $x = (2n+1)\pi$, is zero when $x = \dfrac{1}{2}(2n+1)\pi$.

Cosine function

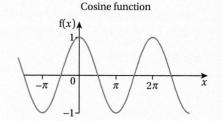

The tangent function, $y = \tan x$, is undefined for all odd multiples of $\dfrac{1}{2}\pi$, is periodic with period π.

$\tan\theta = \dfrac{\sin\theta}{\cos\theta}$ and $\sin^2\theta + \cos^2\theta = 1$

Tangent function

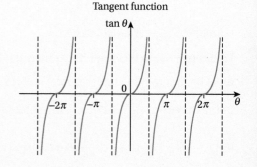

Review

In Questions 1 to 3 write down the letter giving the correct answer.

1 120° in radians is

   a $\dfrac{\pi}{2}$ **b** 2   **c** 2π **d** $\dfrac{\pi}{3}$ **e** $\dfrac{2\pi}{3}$

2 An angle of 1 radian is equivalent to:

   a 90°   **b** 60°   **c** 67.3°   **d** 57.3°   **e** 45°

3 An arc PQ subtends an angle of 60° at the centre of a circle of radius 1 cm. The length of the arc PQ is

   a 60 cm   **b** 30 cm   **c** $\dfrac{1}{6}\pi$ cm

   d $\dfrac{1}{3}\pi$ cm   **e** $\dfrac{1}{18}\pi^2$ cm

In Questions 4 to 9, give angles that are not exact correct to 1 decimal place.

4 When $\cos\beta = 0.5$, find possible values for $\sin\beta$ and $\tan\beta$, giving your answers in exact form.

5 Find the values of θ for which $\tan\left(\theta - \dfrac{1}{3}\pi\right) = 1$ in the interval $-\pi \le \theta \le \pi$.

6 Find the solution of the equation $\tan\theta = 3\sin\theta$ for values of θ in the range $-180° \le \theta \le 180°$.

7 Solve the equation $\cos\theta = \dfrac{1}{2}\sqrt{3}$ giving values of θ from 0 to 180°.

8 Find, in the range $-180° \le \theta \le 180°$, the values of θ that satisfy the equation $2\cos^2\theta - \sin\theta = 1$.

Assessment

1 **a** Solve the equation $\sin x = \dfrac{1}{2}$ in the interval $0° < x < 360°$.

   b Show that $\sin^2\theta - \sin\theta\cos\theta = 12$ can be expressed as $\tan^2\theta - \tan\theta - 12 = 0$

   c Hence solve the equation $\sin^2\theta - \sin\theta\cos\theta = 12$ in the interval $0° < x < 360°$ giving your answer to the nearest degree.

2 The diagram shows an arc ADC of a circle with centre O and radius 4 cm. The angle AOC $= \dfrac{2\pi}{3}$.

   a Find the length of the arc ADC.

   b Find the area of the sector OADC.

   c The line AB and AC are tangents to the circle at A and B.

      Find the length of OB.

   d Hence find the area enclosed by the arc ADC and the lines AB and BC.

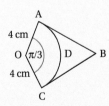

3 The diagram shows a sector of a circle whose centre is O and whose radius is r m.

   The angle AOB is θ radians and the perimeter of the sector AOB is 20 m.

   a Show that $\theta = \dfrac{20 - 2r}{r}$.

   b The area of the sector is A m². Find an expression for A in terms of r.

4 **a** Solve the equation $2\sin^2 x + 3\cos x = 0$ for values of x in the interval $0° < x < 180°$.

    b Hence solve the equation $2\sin^2\left(x - \dfrac{1}{3}\pi\right) + 3\cos\left(x - \dfrac{1}{3}\pi\right) = 0$ for values of x in the interval $0° < x < \pi$.

5 **a** Find the values of $\sin x$ for which $5\cos^2 x = 1 - \sin x$.

    b Hence solve the equation $5\cos^2 x = 1 - \sin x$ for values of x in the interval $0° < x < 180°$.

6 The diagram shows a sector OAB of a circle with centre O and radius 5 cm.

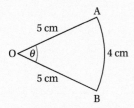

The angle between the radii OA and OB is θ radians.

The length of the arc AB is 4 cm.

    a Find the value of θ.

    b Find the area of the sector OAB.

AQA MPC2 January 2011

7 **a** Given that $\dfrac{\cos^2 x + 4\sin^2 x}{1 - \sin^2 x} = 7$, show that $\tan^2 x = \dfrac{3}{2}$

    b Hence solve the equation $\dfrac{\cos^2 2\theta + 4\sin^2 2\theta}{1 - \sin^2 2\theta} = 7$ in the interval $0° < \theta < 180°$, giving your values of θ to the nearest degree.

AQA MPC2 June 2014

12 Exponentials and Logarithms

Introduction

Exponentials and logarithms are closely related.

This chapter explains what an exponent is and what a logarithm is and how to work with them.

Recap

You need to remember how to...

► Describe the effect on the equation of the curve $y = f(x)$ of a translation, a reflection and a one way stretch.
► Use the laws of indices.
► Use the factor theorem to find factors of cubic polynomials.

Objectives

By the end of this chapter, you should know how to...

► Understand exponential functions and their graphs.
► Express a power of a number as a logarithm.
► Find and use the laws of logarithms.
► Use logarithms to solve equations such as $2^x = 1.3$.

Applications

Exponential equations often appear in financial calculations and logarithms can help with the solution of exponential equations. For example, the formula for finding how much an investment is worth after n years at a fixed rate of interest involves an expression to the power n. The only way to find out when the investment is worth more than a given amount, is to use an equation to find the value of n and this can be done with logarithms.

12.1 Exponential functions

Exponent is another word for index or power.

An exponential function is one where the variable is in the index.

For example, 2^x, 3^{-x}, 10^{x+1} are exponential functions of x.

The shape of the graph of f(x) = a^x

The table gives corresponding values of x and 2^x.

x	$-\infty \leftarrow \ldots$	-10	-1	$-\dfrac{1}{10}$	0	$\dfrac{1}{10}$	1	10	$\ldots \rightarrow \infty$
$f(x)$	$0 \leftarrow \ldots$	$\dfrac{1}{1024}$	$\dfrac{1}{2}$	0.93	1	1.07	2	1024	$\ldots \rightarrow \infty$

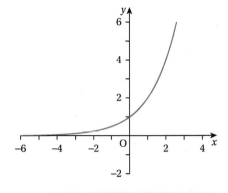

The table shows that

▶ 2^x has real values for all real values of x
▶ 2^x is positive for all values of x
▶ as $x \to -\infty$, $f(x) \to 0$
▶ as x increases, $f(x)$ increases at a rapidly increasing rate.

A plot of the curve $y = 2^x$ confirms these observations and looks like this.

The graph shows that

▶ the curve crosses the crosses y-axis at $(0, 1)$, so $f(0) = 1$
▶ as $x \to -\infty$, the curve gets closer and closer to the x-axis
▶ the curve never crosses the x-axis
▶ the x-axis is an **asymptote**.

Any function of the form $f(x) = a^x$, where $a > 1$, crosses the y-axis at $(0, 1)$ because $a^0 = 1$.

The curve $f(x) = a^x$ is similar in shape to that of $f(x) = 2^x$.

> **Note**
>
> Any line that a curve gets closer and closer to but never crosses is called an asymptote.

Example 1

Sketch the graph of   **a** $y = 3^{x-1}$ **b** $y = \dfrac{1}{3^x}$

a The curve $y = 3^{x-1}$ is the translation of the curve $y = 3^x$ by 1 unit in the positive direction of the x-axis.

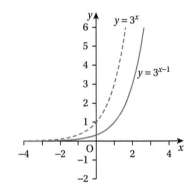

b $y = \dfrac{1}{3^x} = 3^{-x}$ which is the reflection of the curve $y = 3^x$ in the y-axis.

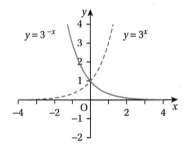

Exercise 1

1 a Write down the values of $f(x) = \left(\dfrac{1}{2}\right)^x$ corresponding to $x = -4, -3, -2, -1, 0, 1, 2, 3$ and 4. Use these values to deduce the behavior of $f(x)$ as $x \to \pm \infty$ and hence sketch the graph of the function.

b Explain the relationship between the curve $y = 2^x$ and the curve $y = \left(\dfrac{1}{2}\right)^x$.

2 Sketch the graph of

a $y = 4^x$ **b** $y = 4^{x+2}$ **c** $y = 4^{2x}$ **d** $y = \dfrac{1}{4^x}$

12.2 Logarithms

The statement $10^2 = 100$ can be read as 'the base 10 raised to the power 2 is equal to 100'.

This relationship can be rearranged to give the same information, but in a different order:

2 is the power to which the base 10 must be raised to equal 100.

In this form the power is called a **logarithm** (log).

The whole relationship can then be shortened to read

2 is the logarithm to the base 10 of 100   or $2 = \log_{10} 100$

$10^2 = 100$ and $2 = \log_{10} 100$ are equivalent.

In the same way, $2^3 = 8$ $\Leftrightarrow$ $3 = \log_2 8$

and $3^4 = 81$ $\Leftrightarrow$ $4 = \log_3 81$

Similarly $\log_5 25 = 2$ $\Leftrightarrow$ $25 = 5^2$

and $\log_9 3 = \dfrac{1}{2}$ $\Leftrightarrow$ $3 = 9^{\frac{1}{2}}$

The base of a logarithm can be any positive number, including an unspecified positive number represented by a letter, for example

$$b = a^c \Leftrightarrow \log_a b = c$$

The power of a positive number always gives a positive result, for example

$$4^2 = 16, \; 4^{-2} = \frac{1}{64}.$$

Therefore when $\log_a b = c$, $b = a^c$, so b must be positive, therefore logarithms of positive numbers exist, but

the logarithm of a negative number does not exist.

Example 2

Question

a   Write $\log_2 64 = 6$ in index form.

b   Write $5^3 = 125$ in logarithmic form.

c   Complete the statement $2^{-3} = ?$ and then write it in logarithmic form.

Answer

a $\log_2 64 = 6$ means that the base is 2, the number is 64 and the power is 6.

    $\log_2 64 = 6$ $\Rightarrow$ $64 = 2^6$

b $5^3 = 125$ means that the power is 3, the base is 5 and the number is 125.

    $5^3 = 125$ $\Rightarrow$ $3 = \log_5 125$

c $2^{-3} = \dfrac{1}{8}$

The base is 2, the power is -3 and the number is $\dfrac{1}{8}$.

$2^{-3} = \dfrac{1}{8}$ $\Rightarrow$ $-3 = \log_2 \left(\dfrac{1}{8} \right)$

Evaluating logarithms

A simple equation in index form is usually easier to solve than an equation in logarithmic form. Therefore change a logarithm to index form to find the value of the logarithm without using a calculator.

For example to find the value of $\log_{49} 7$

write $\qquad x = \log_{49} 7 \quad \Rightarrow \quad 49^x = 7$ so $x = \dfrac{1}{2}$

therefore $\quad \log_{49} 7 = \dfrac{1}{2}$.

In particular, for any base b,

when $\qquad x = \log_b 1$ then $b^x = 1 \quad \Rightarrow \quad x = 0$

therefore   **the logarithm of 1 to any base is zero.**

Using a calculator

A scientific calculator can be used to find the values of logarithms with a base of 10.

The button marked **[log]** gives the value of a logarithm with a base of 10.

Exercise 2

Convert each of the following facts to logarithmic form.

1 $10^3 = 1000$

2 $2^4 = 16$

3 $10^4 = 10000$

4 $3^2 = 9$

5 $4^2 = 16$

6 $5^2 = 25$

7 $10^{-2} = 0.01$

8 $9^{\frac{1}{2}} = 3$

9 $5^0 = 1$

10 $4^{\frac{1}{2}} = 2$

11 $12^0 = 1$

12 $8^{\frac{1}{3}} = 2$

13 $p = q^2$

14 $x^y = 2$

15 $p^q = r$

Convert each of the following facts to index form.

16 $\log_{10} 100000 = 5$

17 $\log_4 64 = 3$

18 $\log_{10} 10 = 1$

19 $\log_2 4 = 2$

20 $\log_2 32 = 5$

21 $\log_{10} 1000 = 3$

22 $\log_5 1 = 0$

23 $\log_3 9 = 2$

24 $\log_4 16 = 2$

25 $\log_3 27 = 3$

26 $\log_{36} 6 = \dfrac{1}{2}$

27 $\log_a 1 = 0$

28 $\log_x y = z$

29 $\log_a 5 = b$

30 $\log_p q = r$

In Questions 31 to 46, find the value of the logarithm without using a calculator.

31 $\log_2 4$

32 $\log_{10} 1000000$

33 $\log_2 64$

34 $\log_3 81$

35 $\log_8 64$

36 $\log_4 64$

37 $\log_9 3$

38 $\log_{\frac{1}{2}} 4$

39 $\log_{10} 0.1$

40 $\log_{121} 11$

41 $\log_5 1$

42 $\log_2 2$

43 $\log_{64} 4$

44 $\log_{99} 1$

45 $\log_{27} 3$

46 $\log_a a^3$

47 Use a calculator to evaluate, correct to 3 significant figures,

    a $\log_{10} 3$

    b $\log_{10} 2.4$

    c $\log_{10} 0.201$

    d $\log_{10} 17.3$

    e $\log_{10} 5.6$

    f $\log_{10} 250$

12.3 The laws of logarithms

Indices obey certain rules in the multiplication and division of numbers. Because logarithm is just another word for index or power, logarithms also obey laws.

When $\quad b = \log_a x \quad$ and $\quad c = \log_a y$

$\Rightarrow \quad\quad a^b = x \quad$ and $\quad a^c = y$

$\quad\quad\quad xy = (a^b)(a^c) \quad \Rightarrow \quad xy = a^{b+c}$

Therefore $\quad \log_a(xy) = b + c$

$\Rightarrow \quad\quad \log_a(xy) = \log_a x + \log_a y$

This is the first law of logarithms. As a can represent *any* base, this law applies to the log of *any product provided that the same base is used for all the logarithms in the formula.*

Using b and c again $\quad \dfrac{x}{y} = \dfrac{a^b}{a^c} \quad \Rightarrow \quad \dfrac{x}{y} = a^{b-c}$

Therefore $\quad\quad\quad \log_a\left(\dfrac{x}{y}\right) = b - c$

$\Rightarrow \quad\quad\quad\quad \log_a\left(\dfrac{x}{y}\right) = \log_a x - \log_a y$

This is the second law of logarithms and again this law applies to the log of *any product provided that the same base is used for all the logarithms in the formula.*

Using $\quad b = \log_a(x^k) \quad \Rightarrow \quad a^b = x^k$

$\Rightarrow \quad\quad a^{\frac{b}{k}} = x$

Therefore $\quad \dfrac{b}{k} = \log_a x \quad \Rightarrow \quad b = k \log_a x$

$\Rightarrow \quad\quad \log_a(x^k) = k \log_a x$

This is the third law of logarithms. Because all three laws are true for *any* base, the base does not need to be included.

In each of these laws the base of every logarithm must be the same.

$\log x + \log y = \log(xy)$

$\log x - \log y = \log\left(\dfrac{x}{y}\right)$

$k \log x = \log(x^k)$

Example 3

Express $\log(pq^2 \sqrt{r})$ in terms of $\log p$, $\log q$ and $\log r$.

$\log(pq^2 \sqrt{r}) = \log p + \log q^2 + \log \sqrt{r}$

$\quad\quad\quad\quad\quad = \log p + 2 \log q + \dfrac{1}{2} \log r$

Example 4

Question

Express $3 \log p + n \log q - 4 \log r$ as a single logarithm.

Answer

$$3 \log p + n \log q - 4 \log r = \log p^3 + \log q^n - \log r^4$$
$$= \log \left(\frac{p^3 q^n}{r^4} \right)$$

Example 5

Question

Express $\log \left(\frac{x+1}{x^2} \right)$ as separate logarithms.

Answer

$$\log \left(\frac{x+1}{x^2} \right) = \log (x+1) - \log x^2 = \log (x+1) - 2 \log x$$

Example 6

Question

Express $\log (x+1) + \log 4 - \frac{1}{2} \log x$ as a single logarithm.

Answer

$$\log (x+1) + \log 4 - \frac{1}{2} \log x = \log 4(x+1) - \log \sqrt{x}$$
$$= \log \left(\frac{4(x+1)}{\sqrt{x}} \right)$$

Exercise 3

Express in terms of $\log p$, $\log q$, and $\log r$

1 $\log (pq)$

2 $\log (pqr)$

3 $\log \left(\dfrac{p}{q} \right)$

4 $\log \left(\dfrac{pq}{r} \right)$

5 $\log \left(\dfrac{p}{qr} \right)$

6 $\log (p^2 q)$

7 $\log \left(\dfrac{q}{r^2} \right)$

8 $\log \left(p\sqrt{q} \right)$

9 $\log \left(\dfrac{p^2 q^3}{r} \right)$

10 $\log \left(\sqrt{\dfrac{q}{r}} \right)$

11 $\log q^n$

12 $\log(p^n q^m)$

Express as a single logarithm

13 $\log p + \log q$

14 $2 \log p + \log q$

15 $\log q - \log r$

16 $3 \log q + 4 \log p$

17 $n \log p - \log q$

18 $\log p + 2 \log q - 3 \log r$

Express as the sum or difference of the simplest possible logarithms.

19 $\log(5x)$ **20** $\log(5x^2)$ **21** $\log[3(x+1)]$

22 $\log\left(\dfrac{x}{x+1}\right)$ **23** $\log\left(\dfrac{2x}{x-1}\right)$ **24** $\log(xy^2)$

25 $\log[x(x+4)]$ **26** $\log(x^2-1)$ **27** $\log[x^2(x+y)]$

28 $\log[a^2x(x-b)]$

Express as a single logarithm.

29 $\log 2 + \log x$ **30** $\log 3 - \log x$ **31** $2\log x - \log 4$

32 $\log x - 2\log(1-x)$ **33** $2\log x - \dfrac{1}{2}\log(x-1)$ **34** $2\log x - 3\log y$

12.4 Equations containing logarithms or x as a power

When x forms part of an index, first look to see if the value of x is obvious.

For example, when $4^x = 16$, it is obvious that $x = 2$ because $4^2 = 16$.

Not so obvious is the equation $4^x = 32$, but 4 and 32 can both be expressed as powers of 2:

$$4^x = 32 \implies (2^2)^x = 2^5, \quad \text{so} \quad 2^{2x} = 2^5$$

Therefore $2x = 5 \implies x = 2.5$.

When the value of the unknown is not obvious, using logarithms will often transform the index into a factor.

For example, when $5^x = 10$, taking logarithms of both sides gives

$$x\log 5 = \log 10 \implies x = \frac{\log 10}{\log 5} = 1.43 \text{ correct to 3 s.f. (using a calculator)}$$

> **Note**
>
> $\dfrac{\log 10}{\log 5}$ is NOT equal to $\log\dfrac{10}{5}$.

When an equation contains logs involving x, first look to see if there is an obvious solution. For example, when $\log_a x = \log_a 4$, it is obvious that $x = 4$.

When the solution is not so obvious, express the log terms as a single logarithm and then remove the logarithm.

For example, when $2\log_2 x - \log_2 8 = 1$, using the laws of logs gives $\log_2\left(\dfrac{x^2}{8}\right) = 1$.

Expressing in index from gives $\dfrac{x^2}{8} = 2^1 \implies x^2 = 16 \implies x = 4$.

$x = -4$ is not a solution because $\log_2(-4)$ does not exist so it is essential that all roots are checked in the original equation, when solving equations of any form.

Exercise 4

Solve the equations in Questions 1 to 12. Give answers that are not exact correct to 3 significant figures.

1 $3^x = 9$ **2** $3^x = \dfrac{1}{9}$

3 $9^x = 27$ **4** $3^x = 6$

5 $2^{2x} = 5$ **6** $5^x = 4$

7 $3^{x-1} = 7$ **8** $4^{2x+1} = 8$

9 $\log_2 x = \log_2(2x-1)$ **10** $\log_4 x = 2$

11 $\log x = 2\log(x-2)$ **12** $\log 2 + 2\log x = \log(x+3)$

> **Note**
>
> When the base of a logarithm is not given, assume that the base is 10.

13 Express $\log_x 5 - 2\log_x 3$ as a single log term. Hence find the value of x when $\log_x 5 - 2\log_x 3 = 2$.

14 Solve the equation $\log 4 - 2\log(x+1) = \log x$.

15 Express $\log_3 y - 2\log_3 x$ as a single logarithm.

Hence express y in terms of x when $\log_3 y - 2\log_3 x = 1$.

16 Given that $y = 2^x$, express 2^{2x} in terms of y.

By substituting y for 2^x, solve the equation $2^{2x} - 2^x - 2 = 0$.

Summary

Exponentials

An exponential function is one where the variable is in the index, for example $f(x) = 2^x$.

The shape of the curve $y = a^x$ is

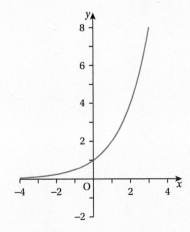

For all values of a, the curve crosses the y-axis at the point $(0, 1)$.

Logarithms

$b = a^c \Leftrightarrow \log_a b = c$

The logarithm of a negative number does not exist.

Laws of logarithms

$$\log x + \log y = \log(xy) \qquad \log x - \log y = \log\left(\frac{x}{y}\right) \qquad k\log x = \log x^k$$

(In each of these laws every logarithm must be to the same base.)

Review

1 Sketch the graph of $f(x) = \left(\frac{1}{3}\right)^x$.

2 Evaluate

  a $\log_2 128$ b $\log_{25} 5$ c $\log_{12} 1$ d $\log_{10} 10^5$

3 Express in terms of $\log a$, $\log b$ and $\log c$

a $\log\left(\dfrac{a^3}{(bc^2)}\right)$ **b** $\log\left(\dfrac{a^n}{b}\right)$

c $\log\left(\dfrac{ab}{c}\right)$ **d** $\log\left(a\sqrt{1+b}\right)$

4 Express as a single logarithm

a $3\log a - \log b$ **b** $\log\left(\dfrac{1}{a}\right) + \log 1$

5 Express as a single logarithm

a $\log x - \log y$ **b** $2 + \log(x+1)$

c $\log A + \log x$ **d** $\log x - \log xy + \log y^2$

Solve the equations in Questions 6 to 9.

6 $5^x = 125$ **7** $\log 2x = \log(x+2)$

8 $3^x = 10$ **9** $\log 2 - 2\log(x-1) = \log x$

Assessment

1 **a** Sketch the graph of $y = \dfrac{1}{4^x}$.

b Show that $\dfrac{1}{4^x} = \dfrac{5}{6}$ can be written as $4^x = 1.2$.

c Solve the equation $\dfrac{1}{4^x} = \dfrac{5}{6}$.

2 **a** Express $2\log 5 - \log 20 + \dfrac{1}{2}\log x$ as a single logarithm.

b Hence solve the equation $2\log 5 - \log 20 + \dfrac{1}{2}\log x = 5$.

3 **a** Express $\log_3 6 - 2\log_3 2 + \log_3 x$ as a single logarithm.

b Given that $\log_3 6 - 2\log_3 2 + \log_3 x = 2 + \log_3 y$ express y in terms of x in a form not including logarithms.

4 **a** Express $\log x^2 - \log xy - \log y^2$ as a single logarithm.

b Given that $\log_{10} x^2 - \log_{10} xy - \log_{10} y^2 = 0$ express x in terms of y in a form not including logarithms.

5 **a** Given that $y = 5x^2$ show that $\log_5 y = 1 + 2\log_5 x$.

b Hence solve the equation $1 + 2\log_5 x = \log_5(2x+3)$.

6 **a** Solve the equation $6^x = 12$ giving your answer correct to three significant figures.

b Sketch the graph of $y = 6^x - 12$ showing where the graph crosses the axes.

c Solve the equation $\log_6(2x+1) = -1$.

7 **a** Sketch the curve with equation $y = 4^x$, indicating the coordinates of any point where the curve intersects the coordinate axes.

b Describe the geometrical transformation that maps the graph of $y = 4^x$ onto the graph of $y = 4^x - 5$.

c **i** Use the substitution $Y = 2^x$ to show that the equation $4^x - 2^{x+2} - 5 = 0$ can be written as $Y^2 - 4Y - 5 = 0$.

ii Hence show that the equation $4^x - 2^{x+2} - 5 = 0$ has only one real solution.

Use logarithms to find this solution, giving your answer to three decimal places.

AQA MPC2 June 2011

8 **a** Given that $\log_a b = c$, express b in terms of a and c.

b By forming a quadratic equation, show that there is only one value of x which satisfies the equation $2 \log_2 (x + 7) - \log_2 (x + 5) = 3$.

AQA MPC2 June 2013

13 Probability

Introduction

The development of probability was stimulated by this enquiry to Blaise Pascal (1623–1662) from his friend the Chevalier de Mere:

Which is the more likely to occur, a throw of six at least once in 4 throws of a single die or a throw of double six at least once in 24 throws of a pair of dice?

It was the resulting correspondence between Pascal and Fermat (1601–1665) which laid the foundations for the serious study of probability.

Recap

In this chapter you will review many of the methods studied at GCSE, drawing them together formally in the probability laws. This will enable you to tackle complex probability problems.

Apps

Probability is used widely in everyday life, from assessment of risk to reliability of consumer products, from trading in financial markets to testing of new medicines; in the biological and physical sciences; in engineering; in business; in politics; the list is endless.

Objectives

By the end of this chapter, you should know how to...

► Assign probabilities to events using relative frequencies or equally likely outcomes.

► Apply the addition law of probability and adapt it for mutually exclusive events.

► Apply the multiplication law of probability and adapt it for independent events.

► Use tree diagrams.

. . . .

13.1 Introduction to probability

The **probability** of an event is a measure of the likelihood that it will happen and it is given on a numerical scale from 0 to 1. Probabilities can be written as a decimal, fraction or percentage.

The two extremes on the probability scale are impossibility at one end and certainty at the other end.

A probability of 0 indicates that the event is impossible.
A probability of 1 (or 100%) indicates that the event is certain to happen.
All other events have a probability between 0 and 1.

For example:

There is an 'even' chance of a fair coin showing heads when it is tossed; the probability is $\frac{1}{2} = 0.5 = 50\%$.

There is a one in four chance of a fair tetrahedral die marked 1, 2, 3, 4 landing on 3;

the probability is $\frac{1}{4} = 0.25 = 25\%$.

The weather forecaster may say that there is a 70% chance of rain;

the probability is $70\% = 0.7$.

If you select a tennis ball from a box of yellow tennis balls

the probability that you will select a yellow ball is 1 (certain)

the probability that you will select a blue ball is 0 (impossible).

These probabilities are shown on a probability scale below:

Practical probability

When a drawing pin is dropped it lands in one of two positions: 'point-up' or 'point-down'

point-up   point-down

What is the probability that the drawing pin will land 'point-up'?

You could estimate the probability by dropping the pin a number of times and working out the proportion, known as the relative frequency, that land 'point-up', where

$$\text{relative frequency} = \frac{\text{number of 'points-up'}}{\text{total number of times the pin is dropped}}$$

How many times should you drop the drawing pin to get a good estimate?

To investigate this, a drawing pin was dropped 200 times and the relative frequency of the number that landed 'point-up' was calculated after every 10 throws. Here is part of the table of results:

Number of times pin is dropped	10	50	100	150	200
Cumulative number of 'point-up'	3	28	55	86	122
Relative frequency (2 dp)	0.3	0.56	0.55	0.57	0.61

All the relative frequencies were plotted on a graph.

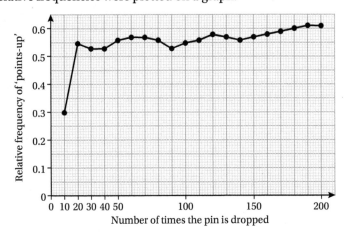

You can see that as the pin is dropped more and more times, the relative frequency appears to be settling to a value around 0.6. This **limiting value** is taken as an estimate of the probability that the pin will land 'point-up'.

In general, if an event occurs r times in n trials, an estimate of the probability is given by the **long-term relative frequency**, which is the limiting value of $\frac{r}{n}$. The reliability of the estimate increases as n increases.

Theoretical probability

Suppose you wanted the probability of a fair coin showing heads when it is tossed. You would give the answer $\frac{1}{2}$ straight away, without bothering to toss a coin a large number of times to work out the long-term relative frequency. Intuitively, you would use the definition of probability that applies when outcomes are equally likely:

$$\text{probability} = \frac{\text{number of successful outcomes}}{\text{total number of possible outcomes}}$$

When a fair coin is tossed, there are two possible outcomes, head or tail, only one of which (getting a head) is successful. Since the coin is unbiased, each outcome is equally likely, so the probability of a head is $\frac{1}{2}$.

Definitions and notation

Any statistical experiment or trial has a number of possible **outcomes**.

The set of all distinct possible outcomes is the **possibility space** S.

The number of outcomes in S is written $n(S)$.

An **event** A consists of one or more of the outcomes in the possibility space. The number of outcomes resulting in event A is written $n(A)$.

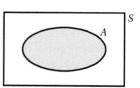

When outcomes are **equally likely**, the **probability** of event A is given by

$$P(A) = \frac{n(A)}{n(S)}$$ ← number of outcomes resulting in event A
← total number of outcomes in the possibility space S

The **complement** of A is A', where A' is the event 'A does not occur', so A' denotes the event 'not A'.

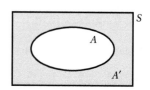

Since an event either occurs or does not occur,

$$P(A) + P(A') = 1$$

so $$P(A') = 1 - P(A)$$

Example 1

a   A **fair cubical die** is thrown. The 'score' is the number on the uppermost face. Find the probability that

   i   the score is less than 3   ii   the score is at least 3.

b   Two fair cubical dice are thrown. Find the probability that

   i   the sum of the scores is 6   ii   the sum of the scores is not 6.

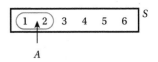

a Possibility space $S = \{1, 2, 3, 4, 5, 6\}$, so $n(S) = 6$.

If event A is 'the score is less than 3', then $A = \{1, 2\}$ and $n(A) = 2$.

i P(score is less than 3) = $P(A) = \dfrac{n(A)}{n(S)} = \dfrac{2}{6} = \dfrac{1}{3}$

ii P(score is at at least 3) = $P(A') = 1 - P(A) = 1 - \dfrac{1}{3} = \dfrac{2}{3}$

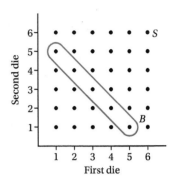

b When two fair dice are thrown, there are 36 possible outcomes, shown by dots on the possibility space diagram, so $n(S) = 36$.

If B is the event 'the sum of the scores on the dice is 6' then $B = \{(1, 5), (2, 4), (3, 3), (4, 2), (5, 1)\}$, shown ringed, so $n(B) = 5$.

i P(sum of scores is 6) = $P(B) = \dfrac{n(B)}{n(S)} = \dfrac{5}{36}$

ii P(sum of the scores is not 6) = $P(B') = 1 - \dfrac{5}{36} = \dfrac{31}{36}$

Example 2

A box contains 20 counters numbered 1, 2, 3, ... up to 20. A counter is picked at random from the box. Find the probability that the number on the counter is

a a multiple of 5   **b** not a multiple of 5   **c** higher than 7.

When you pick a counter at random from the box, there are 20 possible outcomes and each outcome is equally likely to occur.

a The multiples of 5 are {5, 10, 15, 20}, so there are 4 successful outcomes.

P(multiple of 5) = $\dfrac{4}{20} = 0.2$

b P(not a multiple of 5) = $1 - 0.2 = 0.8$

c The numbers higher than 7 are {8, 9, ...20}, so there are 13 successful outcomes.

P(higher than 7) = $\dfrac{13}{20} = 0.65$

Playing cards are often used to demonstrate probability ideas.

An ordinary pack of playing cards consists of 52 cards, split equally into four suits. Diamonds and hearts are red suits; clubs and spades are black suits.

Jack of hearts   2 of diamonds   Ace of spades   9 of clubs

In each suit there are 13 cards:

    Ace, 2, 3, 4, 5, 6, 7, 8, 9, 10, Jack, Queen, King

The Jack, Queen and King are called picture cards.

Example 3

A card is dealt from a well-shuffled ordinary pack of 52 playing cards.

a Find the probability that the card is

i the 4 of spades   **ii** a diamond or the 4 of spades.

b The first card dealt is placed face-up on the table. It is the 3 of diamonds. What is the probability that the second card is from a red suit?

a The card is equally likely to be any of the 52 cards.

  i $\text{P}(4 \text{ of spades}) = \dfrac{1}{52}$

  ii $\text{P}(\text{a diamond or the 4 of spades}) = \dfrac{14}{52} = \dfrac{7}{26}$

b Since the 3 of diamonds has been removed, there are 51 cards left in the pack. Of these, 12 are diamonds and 13 are hearts.

  $\text{P}(\text{from a red suit}) = \dfrac{25}{51}$

Example 4

A party bag contains a selection of balloons, as shown in the frequency table.

	Blue	Red	Green	Yellow	TOTAL
Long	17	20	10	18	65
Round	15	19	14	7	55
TOTAL	32	39	24	25	120

Yizi takes a balloon at random from the bag. Find the probability that the balloon is

a red

b a blue round balloon

c not yellow.

a $\text{P}(\text{red})$

  $= \dfrac{39}{120} = \dfrac{13}{40}$

a	B	R	G	Y	
Long					
Round					
		39			**120**

b $\text{P}(\text{a blue round balloon})$

  $= \dfrac{15}{120} = \dfrac{1}{8}$

b	B	R	G	Y	
Long					
Round	15				
					120

c $\text{P}(\text{not yellow})$

  $= \dfrac{32+39+24}{120} = \dfrac{19}{24}$

Alternatively:

$\text{P}(\text{not yellow})$

  $= 1 - \text{P}(\text{yellow})$

  $= 1 - \dfrac{25}{120} = \dfrac{19}{24}$

c	B	R	G	Y	
Long					
Round					
	32	39	24		**120**

Exercise 1

1 An ordinary fair cubical die is thrown. Find the probability that the score is

   a even   **b** lower than 7

   c a factor of 6   **d** at least 4

   e higher than 1.

2 In a box there are 10 red, 15 blue, 5 green and 10 yellow highlighters. One fifth of the highlighters have dried up and will not write. Che picks a highlighter at random from the box.

   Find the probability that the highlighter

   a is blue   **b** is neither green nor yellow

   c is purple   **d** will write.

3 An integer is picked at random from the integers from 1 to 20 inclusive. A is the event 'the integer is a multiple of 3' and B is the event 'the integer is a multiple of 4'. Find

   a $P(A)$ **b** $P(B')$.

4 A card is dealt from a well-shuffled ordinary pack of 52 playing cards.

   a Find the probability that the card dealt is

      i a Queen

      ii a heart or a diamond

      iii a picture card showing spades.

   b Two cards are dealt and put face-up on the table. They are the 4 of clubs and the 7 of diamonds. A third card is now dealt. What is the probability that it is a club or a 7?

5 Every work day Kusuma catches a bus to work. The bus is never early but it is sometimes late.

   Kusuma decided to record the number of minutes the bus is late over a period of 10 days. Here are his results.

     0, 3, 4, 1, 0, 0, 5, 4, 6, 0

   a Find the probability that on a randomly chosen day from the 10 days

      i the bus was on time

      ii the bus was more than the mean number of minutes late.

   b Jamie estimates that the probability that his bus will be late when he is on his way home from work is 0.75. What is the probability that his bus will not be late when he is on his way home from work?

6 A cubical die, with faces numbered 1 to 6, is weighted so that a 6 is twice as likely to occur as any other number.

   Find the probability that when the die is thrown the score is

   a 6   **b** odd.

7 A fair five-sided spinner has sides numbered 1, 1, 2, 3, 3.

The spinner is spun twice. Use the possibility space diagram to find the probability that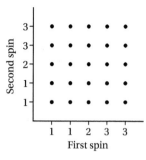

a the spinner stops at 1 at least once

b the sum of the numbers on the two spins is even.

8 At a Driving Test centre there are three examiners, I, II and III. The table shows the results of all the people who took their driving test during the first week of September.

	Examiner I	Examiner II	Examiner III
Pass	32	27	41
Fail	11	14	15

A person is selected at random from those who took their driving test that week. Find the probability that the person

a passed the driving test

b was assigned to Examiner II

c was assigned to Examiner III and passed the test

d was not assigned to Examiner III.

9 Two fair coins are tossed together. Find the probability that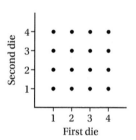

a exactly one tail is obtained

b at most one head is obtained.

10 A **tetrahedral die** has four faces and they are labelled 1, 2, 3, 4. When the die is thrown, the score is the number on which the die lands. Two fair tetrahedral dice are thrown. Find the probability that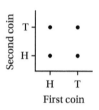

a the sum of the scores is divisible by 4

b the product of the scores is an even number

c the scores differ by at least 2.

11 Two ordinary fair cubical dice are thrown together and the scores are multiplied. P(N) denotes the probability that the number N is obtained.

a Find **i** P(9) **ii** P(4) **iii** P(14) **iv** P(37)

b If P(N) = $\frac{1}{9}$, find the possible values of N.

13.2 Combined events

The following definitions relate to two events in the possibility space and are illustrated using a Venn diagram.

A and B

The outcomes in event **A and B** are those in *both* A and B. This is the **intersection** of A and B and is written $A \cap B$.

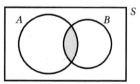

$A \cap B$ means A <u>and</u> B

P(A and B) = P($A \cap B$) = P(both A and B occur)

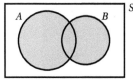

A or B

The outcomes in event **A or B** are those in A or B or both. This is the **union** of A and B and is written $A \cup B$.

$$P(A \text{ or } B) = P(A \cup B) = P(A \text{ occurs or } B \text{ occurs or both occur})$$

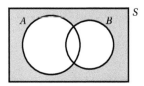

$A \cup B$ means A <u>or</u> B <u>or</u> both

Neither A nor B

The outcomes in event **neither A nor B** are those that are not in the union of A and B.

$$P(\text{neither } A \text{ nor } B) = 1 - P(A \text{ or } B)$$

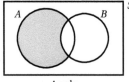

Neither A <u>nor</u> B

A only

The outcomes in event **A only** are outcomes that are in A but not in B. You can think of this in two ways, where

$$P(A \text{ only}) = P(A) - P(A \text{ and } B)$$
or $\quad P(A \text{ only}) = P(A \text{ or } B) - P(B)$

A only

The addition law of probability

The addition law gives the relationship between P(A or B) and P(A and B). It is illustrated below using equally likely outcomes, but holds for probabilities in general.

For two overlapping sets A and B, if you add together the number of outcomes in A and the number of outcomes in B, you will count the overlap twice.

So to find the number of outcomes in event $A \cup B$, you need to subtract the number in the intersection:

$$n(A \cup B) = n(A) + n(B) - n(A \cap B)$$

Dividing by $n(S)$ gives the **addition law of probability**:

$$P(A \cup B) = P(A) + P(B) - P(A \cap B)$$
so $\quad P(A \text{ or } B) = P(A) + P(B) - P(A \text{ and } B)$

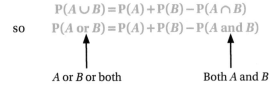

A or B or both   Both A and B

The addition law is stated in set notation in the *Formulae and Statistical Tables* booklet. Note that, although it will not be essential to use set notation in the examination, you are expected to understand it.

Example 5

Events A and B are such that $P(A) = 0.7$, $P(A \text{ or } B) = 0.8$ and $P(A \text{ and } B) = 0.25$. Find

a $P(B')$

b   P(only A occurs)

c   P(exactly one of A or B occurs).

a By the addition law:

$$P(A \text{ or } B) = P(A) + P(B) - P(A \text{ and } B)$$

$$0.8 = 0.7 + P(B) - 0.25$$

$$P(B) = 0.35$$

$$P(B') = 1 - P(B)$$

$$= 1 - 0.35$$

$$= 0.65$$

b $P(A \text{ only}) = P(A) - P(A \text{ and } B)$

$$= 0.7 - 0.25$$

$$= 0.45$$

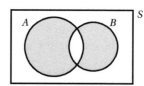

c $P(\text{exactly one of } A \text{ or } B) = P(A \text{ or } B) - P(A \text{ and } B)$

$$= 0.8 - 0.25$$

$$= 0.55$$

Alternatively, you could use

$$P(\text{exactly one of } A \text{ or } B) = P(A \text{ only}) + P(B \text{ only})$$

Example 6

All 100 pupils in a year group were asked whether they had read two particular comics, Whizz and Wham, during the past week. The results showed that 65 had read Whizz, 55 had read Wham, 30 had read both Whizz and Wham and some pupils had not read either comic.

A pupil was selected at random from the year group to answer more questions. Find the probability that the pupil had read

a Whizz or Wham

b neither of the comics

c Whizz but not Wham.

Let Z be the event 'the pupil had read Whizz' and M be the event 'the pupil had read Wham.'

a $P(Z) = \dfrac{65}{100} = 0.65$, $P(M) = \dfrac{55}{100} = 0.55$, $(Z \text{ and } M) = \dfrac{30}{100} = 0.3$

$$P(Z \text{ or } M) = P(Z) + P(M) - P(Z \text{ and } M)$$

$$= 0.65 + 0.55 - 0.3$$

$$= 0.9$$

$P(Z \cup M) = P(Z) + P(M) - P(Z \cap M)$

b $P(\text{neither } Z \text{ nor } M) = 1 - P(Z \text{ or } M)$

$$= 1 - 0.9$$

$$= 0.1$$

c $P(Z \text{ but not } M) = P(Z \text{ only})$

$$= P(Z) - P(Z \text{ and } M)$$

$$= 0.65 - 0.3$$

$$= 0.35$$

Alternatively, you could use a Venn diagram directly as follows:

Fill in the 30 who read both comics in the intersection (overlap).

Then fill in those who read Whizz only and those who read Wham only.

(continued)

(continued)

a $\text{P(Whizz or Wham)} = \dfrac{35+20+35}{100} = \dfrac{90}{100} = 0.9$

b $\text{P(neither comic)} = \dfrac{10}{100} = 0.1$

c $\text{P(Whizz but not Wham)} = \dfrac{35}{100} = 0.35$

Notice that it is easy to find other probabilities from the Venn diagram.

For example,

$\text{P(pupil had read only one of the comics)} = \dfrac{35+25}{100} = 0.6$

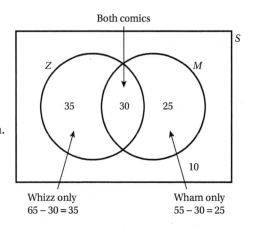

Both comics

S

Z 35   30   25 M

10

Whizz only
$65 - 30 = 35$

Wham only
$55 - 30 = 25$

Probability tables

Often actual frequencies are not as important as the relative frequencies between categories and it can be more convenient to work with proportions or percentages of the total.

So, in Example 6 above, since there were 100 pupils in the year group, you know that the proportion that read Whizz was 0.65, the proportion that read Wham was 0.55 and the proportion that read both was 0.3.

These proportions (relative frequencies) can be shown in a two-way table **probability table**, sometimes called a **relative frequency table**.

Step 1

From the given information you are able to fill in the following entries. Remember that the total proportion (probability) is 1.

	Z	**Z'**	Total
M	0.3		0.55
M'			
Total	0.65		1

Step 2

Now complete the missing entries by making the totals agree. These are shown in red.

	Z	**Z'**	Total
M	0.3	0.25	0.55
M'	0.35	0.1	0.45
Total	0.65	0.35	1

You can see directly from the table that

$\text{P}(Z \text{ or } M) = 0.3 + 0.25 + 0.35 = 0.9$ green, blue and yellow cells

$\text{P}(\text{neither } Z \text{ nor } M) = 0.1$ pink cell

$\text{P}(Z \text{ but not } M) = 0.35$ blue cell

$\text{P}(\text{only one of } Z \text{ or } M) = 0.35 + 0.25 = 0.6$ blue and yellow cells

A probability table is also useful when you are given the probabilities of events directly.

In general, for two events A and B, label the columns A and A' and the rows B and B'.

The column totals show P(A) and P(A') and the row totals show P(B) and P(B').

	A	A'	Total
B	P(A and B)	P(only B)	**P(B)**
B'	P(only A)	P(neither)	**P(B')**
Total	**P(A)**	**P(A')**	**1**

The entries in the four cells give the probabilities of the combined events shown.

The grand total shows the sum of all the probabilities, so it is always 1.

Total probability = 1

Example 7

Question

Mario always contacts two particular friends, Jack and Sakchai, by text message. The event that he contacts Jack is denoted by A and the event that he contacts Sakchai is B. On any one day, Mario may contact neither, or one, or both friends.

a Complete the table of probabilities for these events, where A' and B' denote the events not A and not B respectively.

	A	A'	Total
B			**0.55**
B'	0.4		
Total		**0.35**	

b Hence find the probability that on a particular day, Mario:
  i contacts both Jack and Sakchai
  ii contacts either Jack or Sakchai, but not both
  iii contacts Jack or Sakchai.

Answer

a Use the row totals, column totals and 'grand total' to complete missing entries. These are written in red in the table.

	A	A'	Total
B	0.25	0.3	**0.55**
B'	**0.4**	0.05	**0.45**
Total	**0.65**	**0.35**	**1**

b i P(A and B) = 0.25   green cell

  ii P(either A or B, but not both)

    = 0.4 + 0.3   blue and yellow cells

    = 0.7

  iii P(A or B)

    = 0.25 + 0.4 + 0.3   green, blue and yellow cells

    = 0.95

Note that if you are given P(A or B), you cannot enter it directly into the table, but you can use it to find P(neither A nor B), since P(neither A nor B) = 1 − P(A or B).

	A	A'
B	P(A and B)	P(only B)
B'	P(only A)	

The sum of these entries gives P(A or B).

	A	A'
B		
B'		P(neither)

This entry gives P(neither A nor B).

Example 8

Question

Two events, A and B, are such that

$P(A \text{ or } B) = 0.95$, $P(A') = 0.45$, $P(B) = 0.65$

a Find $P(\text{neither } A \text{ nor } B)$.

b Draw up a two-way probability table for events A and B.

c Find

    i $P(A \text{ and } B)$ **ii** $P(A)$ **iii** $P(A \text{ only})$

Answer

a $P(\text{neither } A \text{ nor } B) = 1 - P(A \text{ or } B)$

                              $= 1 - 0.95$

                              $= 0.05$

b Enter the known values into the table, along with the grand total of 1.
You can then fill in the missing values, shown below in red.

	A	A'	Total
B	0.25	0.4	0.65
B'	0.3	0.05	0.35
Total	0.55	0.45	1

c **i** $P(A \text{ and } B) = 0.25$

   ii $P(A) = 0.55$

   iii $P(A \text{ only}) = 0.3$ Remember that $P(A \text{ only}) = P(A) - P(A \text{ and } B)$.

Mutually exclusive events

Events are **mutually exclusive** if they cannot occur at the same time.

For example, when a die is thrown, the events 'the score is 3' and 'the score is 5' are mutually exclusive, since the score cannot be 3 and 5 at the same time. However, the events 'the score is an even number' and 'the score is a prime number' can occur at the same time (since 2 is both even and prime), so they are not mutually exclusive.

When A and B are mutually exclusive there is no overlap between them, so

    $P(A \text{ and } B) = 0$

Note that in a probability table, the entry for $P(A \text{ and } B)$ will be zero.

This leads to a special case of the addition law.

The **addition law for two mutually exclusive events** is

    $P(A \text{ or } B) = P(A) + P(B)$

    so $P(A \cup B) = P(A) + P(B)$

The law can be extended to **n mutually exclusive** events, where

    $P(A_1 \text{ or } A_2 \text{ or } A_3 \text{ or } \cdots \text{ or } A_n) = P(A_1) + P(A_2) + P(A_3) + \cdots + P(A_n)$

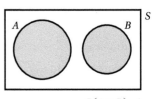

$P(A \cap B) = 0$

> **Note**
>
> This is the **'or'** rule for mutually exclusive events.

Example 9

Question

A card is dealt from an ordinary pack of 52 playing cards. Find the probability that the card is

a   a club (C) or a diamond (D)

b   a club (C) or a King (K).

Answer

a   A card cannot be a club and a diamond at the same time, so the events are mutually exclusive.

$P(C) = \dfrac{13}{52}$, $P(D) = \dfrac{13}{52}$

So $P(C \text{ or } D) = P(C) + P(D) = \dfrac{13}{52} + \dfrac{13}{52} = \dfrac{26}{52} = \dfrac{1}{2}$

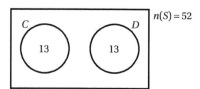

b   The events C and K are not mutually exclusive since the King of clubs is both a King and a club.

$P(C \text{ and } K) = \dfrac{1}{52}$

$P(C \text{ or } K) = P(C) + P(K) - P(C \text{ and } K)$

$= \dfrac{13}{52} + \dfrac{4}{52} - \dfrac{1}{52} = \dfrac{16}{52} = \dfrac{4}{13}$

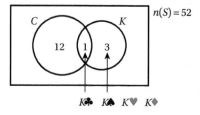

Example 10

Question

A fair cubical die is thrown. Events are defined as follows:

  A: the score is at most 3

  B: the score is at least 3

  C: the score is lower than 3

  D: the score is higher than 3

Identify pairs of events that are mutually exclusive.

Answer

List the outcomes in each event.

  $A = \{1, 2, 3\}$ $B = \{3, 4, 5, 6\}$ $C = \{1, 2\}$ $D = \{4, 5, 6\}$

Look for pairs where there is no overlap.

  $P(A \text{ and } D) = 0$, $P(B \text{ and } C) = 0$, $P(C \text{ and } D) = 0$

The following pairs of events are mutually exclusive:

  A and D, B and C, C and D

Example 11

Question

Jo, Paula and Maladee are playing a game in which there can be only one winner. The probability that Jo wins is 0.3, the probability that Paula wins is 0.2 and the probability that Maladee wins is 0.4. Find the probability that

a   Jo or Maladee wins

b   Jo or Paula or Maladee wins

c   someone else wins.

Since there can only be one winner in the game, the events are mutually
exclusive.

a P(Jo or Maladee wins) = 0.3 + 0.4 = 0.7

b P(Jo or Maladee or Paula wins) = 0.3 + 0.4 + 0.2 = 0.9

c P(someone else wins) = 1 − 0.9 = 0.1

Exercise 2

1 An ordinary fair cubical die is thrown.
Find the probability that the score on the die is

   a even   **b** prime

   c even or prime   **d** even and prime.

2 All the students in a class of 30 study at least one of the subjects, physics
and biology. Of these, 20 study physics and 21 study biology. Find the
probability that a student chosen at random studies

   a both physics and biology

   b only physics

   c only one of the subjects.

3 From an ordinary pack of 52 playing cards the seven of diamonds has been
lost. A card is dealt from the well-shuffled pack. Find the probability that
the card is

   a a diamond   **b** a Queen

   c a diamond or a Queen   **d** a diamond or a seven

   e a diamond and a seven.

4 In a quality control test, all the components produced by three machines
on a particular day are tested. The results are summarised in the table.

	Machine A	Machine B	Machine C
Faulty	2	3	1
Not faulty	80	72	42

A machine operator selects one of these components at random. Find the
probability that the component

   a is from machine A

   b is a faulty component from machine C

   c is not faulty or is from machine A.

5 A class consists of 9 boys and 11 girls. Of these, 4 boys and 3 girls are in the
athletics team. A pupil is chosen at random from the class to take part in
the 100 metres race on Sports Day.
Find the probability that the pupil is

   a in the athletics team   **b** a girl

   c a girl in the athletics team   **d** a girl or in the athletics team.

6. Two fair cubical dice are thrown. Using a possibility space diagram, find the probability that the sum of the scores is

   a   a multiple of 5   b   greater than 9

   c   a multiple of 5 or greater than 9   d   a multiple of 5 and greater than 9.

7. The probability that a randomly chosen boy in Class 2 is in the football team is 0.4, the probability that he is in the chess team is 0.5 and the probability that he is in both teams is 0.2.

   Find the probability that a boy chosen at random from the class

   a   is in the football team, but not in the chess team

   b   is in the football team or the chess team

   c   is not in either team   d   is in just one of the teams.

8. Events X and Y are such that $P(X) = 0.75$, $P(Y') = 0.45$ and $P(X \text{ and } Y) = 0.5$.

   Complete the probability table and use it to find

   a $P(X \text{ or } Y)$ b $P(\text{neither } X \text{ nor } Y)$.

	X	X'	Total
Y	0.5		
Y'			0.45
Total	0.75		

9. Events A and B are such that $P(A) = P(B)$, $P(A \text{ and } B) = 0.1$ and $P(A \text{ or } B) = 0.7$.

   Find

   a $P(A')$ b $P(B \text{ only})$.

10. Events A and B are such that

    $P(A \text{ occurs}) = 0.6$, $P(B \text{ occurs}) = 0.7$, $P(\text{neither } A \text{ nor } B \text{ occurs}) = 0.1$.

    Find

    a $P(\text{both } A \text{ and } B \text{ occur})$ b $P(\text{at least one of } A \text{ and } B \text{ occurs})$

    c $P(A \text{ occurs or } B \text{ occurs but not both } A \text{ and } B \text{ occur})$.

11. Events X and Y are mutually exclusive and $P(X) = 0.5$, $P(Y) = 0.25$.

    Find

    a $P(X \text{ or } Y)$ b $P(X \text{ and } Y)$.

12. Events A, B and C are mutually exclusive.

    $P(A \text{ or } B \text{ or } C) = 0.8$, $P(A') = 0.95$, $P(A \text{ or } B) = 0.45$.

    Find

    a $P(B)$ b $P(C)$ c $P(A \text{ or } C)$ d $P(B \text{ and } C)$.

13. Two fair cubical dice are thrown.

    Event A is the scores on the dice are the same.

    Event B is the product of the scores is a multiple of 3.

    Event C is the sum of the scores is 7.

    State, with a reason, whether the following pairs of events are mutually exclusive:

    a A and B b A and C c B and C.

13.3 Conditional events

Suppose you have a box of milk and plain chocolates. If you randomly select a chocolate, eat it, then randomly select another one, the probability that the second chocolate is plain depends on whether the first one was plain or milk. The events are **conditional**.

The **conditional probability** that B occurs, given that A has occurred, is $P(B$ given $A)$, written $P(B\,|\,A)$.

Since A has occurred, the possibility space is reduced to just A.

So, $P(B\,|\,A) = \dfrac{n(B \text{ and } A)}{n(A)}$.

Dividing the top and bottom of the fraction by $n(S)$ gives

$$P(B\,|\,A) = \dfrac{P(B \text{ and } A)}{P(A)}$$

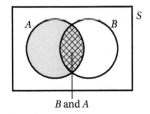

B and A

Example 12

Of the 120 first year students at a college, 36 study chemistry, 60 study biology and 10 study both chemistry and biology.

A first year student is selected at random to represent the college at a conference. Find the probability that the student studies

a   chemistry, given that the student studies biology

b   biology, given that the student studies chemistry.

With obvious notation, $P(C) = \dfrac{36}{120}, P(B) = \dfrac{60}{120}, P(C \text{ and } B) = \dfrac{10}{120}$

a $P(C\,|\,B) = \dfrac{P(C \text{ and } B)}{P(B)} = \dfrac{\frac{10}{120}}{\frac{60}{120}} = \dfrac{1}{6}$

b $P(B\,|\,C) = \dfrac{P(B \text{ and } C)}{P(C)} = \dfrac{\frac{10}{120}}{\frac{36}{120}} = \dfrac{10}{36} = \dfrac{5}{18}$

Alternatively, use a Venn diagram directly:
Fill in the 10 who study both by putting them in the intersection (overlap), then the number in C but not in the intersection (26) and the number in B but not in the intersection (50).

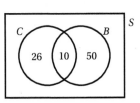

a   Look only at the reduced possibility space of B.

$$P(C\,|\,B) = \dfrac{n(C \text{ and } B)}{n(B)} = \dfrac{10}{60} = \dfrac{1}{6}$$

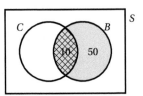

b   Look only at the reduced possibility space of C.

$$P(B\,|\,C) = \dfrac{n(B \text{ and } C)}{n(C)} = \dfrac{10}{36} = \dfrac{5}{18}$$

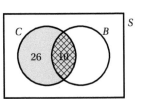

Example 13

Last month a consultant saw 60 men and 65 women suspected of having a particular eye condition. Tests were carried out and the results are shown in the table.

C is the event: the patient had the eye condition, C' is the event: the patient did not have the eye condition, M is the event: the patient is a man; W is the event: the patient is a woman.

	C	C'	Totals
M	25	35	60
W	20	45	65
Totals	45	80	125

One of the patients was selected at random to take part in a survey. Find the probability that the patient selected

a   was a woman, given that the patient had the eye condition

b   did not have the eye condition, given that the patient was a man.

Focus on the relevant numbers in the table.

a   P(woman, given eye condition)

$= P(W \mid C)$

$= \dfrac{20}{45}$

$= \dfrac{4}{9}$

	C	C'	
M			
W	20		
	45		

The possibility space has been reduced to the 45 patients in C.

b   P(did not have eye condition, given man)

$= P(C' \mid M)$

$= \dfrac{35}{60}$

$= \dfrac{7}{12}$

	C	C'	
M		35	60
W			

The possibility space has been reduced to the 60 men.

The multiplication law of probability

For the conditional probability $P(B \mid A)$

$$P(B \mid A) = \frac{P(B \text{ and } A)}{P(A)}$$

This can be rearranged to give the **multiplication law of probability**

$$P(B \text{ and } A) = P(A) \times P(B \mid A)$$

Similarly $P(A \text{ and } B) = P(B) \times P(A \mid B)$

In general, since $P(A \text{ and } B) = P(B \text{ and } A)$,

$$P(A \text{ and } B) = P(A) \times P(B \mid A) = P(B) \times P(A \mid B)$$

Example 14

Two events X and Y are such that $P(X) = 0.2$, $P(Y) = 0.25$ and $P(Y \mid X) = 0.4$. Find

a $P(X \text{ and } Y)$ **b** $P(X \mid Y)$ **c** $P(X \text{ or } Y)$.

a $P(X \text{ and } Y) = P(X) \times P(Y \mid X)$

$$= 0.2 \times 0.4$$

$$= 0.08$$

b $P(X \mid Y) = \dfrac{P(X \text{ and } Y)}{P(Y)}$

$$= \dfrac{0.08}{0.25}$$

$$= 0.32$$

c $P(X \text{ or } Y) = P(X) + P(Y) - P(X \text{ and } Y)$

$$= 0.2 + 0.25 - 0.08$$

$$= 0.37$$

Example 15

There are 5 red counters and 7 blue counters in a bag. Vukašin selects a counter at random from the bag and puts it on the table. He then selects another counter at random.

Find the probability that he selects

a a red counter then a blue counter

b two counters that are the same colour

c at least one blue counter.

> **Note**
>
> This is known as **sampling without replacement**, as the first counter is not put back into the bag before the second counter is taken out. The events are conditional.

a $P(R_1 \text{ and } B_2)$

$$= P(R_1) \times P(B_2 \mid R_1)$$

$$= \dfrac{5}{12} \times \dfrac{7}{11} \qquad \text{There are now only 11 counters in the bag, of which 7 are blue.}$$

$$= \dfrac{35}{132}$$

b $P(\text{same colour})$

$$= P(\text{both red}) + P(\text{both blue})$$

$$= P(R_1) \times P(R_2 \mid R_1) + P(B_1) \times P(B_2 \mid B_1)$$

$$= \left(\dfrac{5}{12} \times \dfrac{4}{11} \right) + \left(\dfrac{7}{12} \times \dfrac{6}{11} \right)$$

$$= \dfrac{5}{33} + \dfrac{7}{22}$$

$$= \dfrac{31}{66}$$

(continued)

(continued)

Answer

c P(at least one blue)

$$= 1 - \text{P(both red)} \qquad \text{\small from part b}$$

$$= 1 - \frac{5}{33}$$

$$= \frac{28}{33}$$

Alternatively, P(at least one blue) = P(both blue) + P(R_1 and B_2) + P(B_1 and R_2).

The results can be extended to more than two events.

Example 16

Question

There are three shelves in the classroom bookcase and the number and type of books on each shelf is shown in the table.

	Hardback	Paperback	Total
Shelf 1	21	9	30
Shelf 2	8	7	15
Shelf 3	15	12	27
Total	44	28	72

A student is asked to select three books at random from the bookcase.

Calculate, to three decimal places, the probability that the student selects

a a hardback, then a hardback, then a paperback

b two hardbacks and one paperback

c one book from each shelf.

Answer

> If you pick a hardback first, there are only 43 hardbacks left, so *reduce* the *numerator* on the *second* selection. There are, however, still 28 paperbacks.

a $P(H, H, P) = \dfrac{44}{72} \times \dfrac{43}{71} \times \dfrac{28}{70}$

> After the first book has been selected, there are only 71 books to choose from and after the second selection there are only 70, so *reduce* the *denominators* on the *second* and *third* selections.

$$= 0.1480\ldots = 0.148 \text{ (3 dp)}$$

b P(two hardbacks and one paperback)

$$= P(H, H, P) + P(H, P, H) + P(P, H, H)$$

$$= \left(\frac{44}{72} \times \frac{43}{71} \times \frac{28}{70} \right) + \left(\frac{44}{72} \times \frac{28}{71} \times \frac{43}{70} \right) + \left(\frac{28}{72} \times \frac{44}{71} \times \frac{43}{70} \right)$$

$$= 3 \times \frac{44}{72} \times \frac{43}{71} \times \frac{28}{70}$$

$$= 0.4441\ldots$$

$$= 0.444 \text{ (3 dp)}.$$

> Notice that the probability is the same for each arrangement, so you could just find $3 \times P(H, H, P)$.

(continued)

(continued)

c   There are 6 arrangements for the selection of the shelves:

   1 2 3,   1 3 2,   2 3 1,   2 1 3,   3 2 1,   3 1 2

The probability is the same for each arrangement, so find the probability of one arrangement and multiply by 6.

P(one book from each shelf)

$$= 6 \times P(\text{Shelf 1, Shelf 2, Shelf 3})$$

$$= 6 \times \frac{30}{72} \times \frac{15}{71} \times \frac{27}{70}$$

$$= 0.2037\ldots$$

$$= 0.204 \,(3\text{ dp})$$

Exercise 3

1 Each of the numbers 1, 2, 3, 4, 5, 6, 7, 8, 9 is written on a card and the nine cards are shuffled.

A card is then dealt.

Given that the card is a multiple of 3, find the probability that the card is

a   even

b   a multiple of 4.

2 In a large group of people it is known that 10% have a hot breakfast, 20% have a hot lunch and 25% have a hot breakfast or a hot lunch. Find the probability that a person chosen at random from this group

a   had a hot breakfast and a hot lunch

b   had a hot lunch, given that the person had a hot breakfast

c   had a hot breakfast, given that the person did not have a hot lunch.

3 In a group of 100 college students, 80 own a laptop computer, 65 own a desktop computer and 50 own both a laptop computer and a desktop computer. Find the probability that a student chosen at random from the group

a   owns a desktop computer, given that the student owns a laptop computer

b   does not own a laptop computer, given the student owns a desktop computer

c   does not own a laptop computer or a desktop computer.

4 Two fair tetrahedral dice, each with faces labelled 1, 2, 3, 4, are thrown and the number on which each lands is noted.

Find the probability that

a   the sum of the two numbers is even, given that at least one die lands on a 3

b   at least one die lands on a 3, given that the score is even.

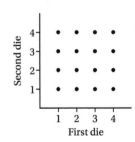

5 The table shows how pupils at a certain school travelled to school on the first day of term.

	Walk	Cycle	Bus	Car
Boy	60	22	48	70
Girl	72	15	52	71

Find the probability that a randomly selected pupil from the school

a   was a boy who had travelled to school by car

b   had cycled to school or was a boy

c   was a girl, given that the pupil travelled by bus

d   travelled by car, given a boy was selected

e   was a boy, given that the pupil did not walk or cycle.

6   Two events A and B are such that $P(A) = 0.3$, $P(B) = 0.45$ and $P(A \mid B) = 0.2$. Find

a $P(A$ and $B)$ b $P(B \mid A)$ c $P(B$ or $A)$.

7   Two events W and X are such that $P(W \mid X) = 0.5$, $P(X \mid W) = 0.4$ and $P(W) = 0.7$. Find

a $P(W$ and $X)$ b $P(X)$

c $P(W$ or $X)$ d $P($neither W nor $X)$.

8   A bag contains 20 identical sweets apart from the colour: 10 are pink, 7 are green and 3 are yellow. Jovian randomly selects two sweets from the bag, one after the other, and eats them. Find the probability that

a   she eats two pink sweets

b   the first sweet is green and the second sweet is yellow

c   she eats exactly one pink sweet

d   neither sweet is green.

9   A card is picked from an ordinary pack containing 52 playing cards. It is then replaced in the pack, the pack is shuffled and a second card is picked. Find the probability that

a   both cards are the seven of diamonds

b   the first card is a heart and the second card is a spade

c   at least one card is a Queen.

10 A large apartment block consists of apartments with either 1, 2, 3 or 4 bedrooms. They are either owner-occupied or let to tenants.

The number of each type is shown in the table.

	1 bedroom	2 bedrooms	3 bedrooms	4 bedrooms
Owner-occupied	8	27	45	11
Let to tenants	12	34	10	3

a   Two apartments are selected at random. Find the probability that

i   neither has 4 bedrooms

ii   one is let to tenants and the other is owner occupied.

b   Three apartments are selected at random.

Find the probability that one is owner-occupied and two are let to tenants.

Independent events

In general, for events A and B:

$$P(A \text{ and } B) = P(A) \times P(B \mid A)$$

However, if either event can occur without being affected by the other, the events are said to be **independent**.

Since it makes no difference that A has occurred,

$$P(B \mid A) = P(B)$$

In the same way $P(A \mid B) = P(A)$

This gives the **multiplication law for independent events**

$$P(A \text{ and } B) = P(A) \times P(B)$$

so $\qquad P(A \cap B) = P(A) \times P(B)$

The rule can be extended to **n independent events**

$$P(A_1 \text{ and } A_2 \text{ and } A_3 \text{ and } \cdots \text{ and } A_n) = P(A_1) \times P(A_2) \times P(A_3) \times \cdots \times P(A_n)$$

> **Note**
>
> This is the '**and**' rule for independent events.

Example 17

Question

a A fair cubical die is thrown twice. Find the probability that

  i the score is 3 on both throws

  ii the score is 3 only on one throw

  iii neither score is 3.

b The die is thrown four times. Find the probability that the score is even on all four throws.

Answer

a The scores are independent so, for each throw, $P(3) = \dfrac{1}{6}$, $P(\text{not } 3) = \dfrac{5}{6}$.

  i $P(3 \text{ on both}) = \dfrac{1}{6} \times \dfrac{1}{6} = \dfrac{1}{36}$

  ii $P(\text{only one } 3) = \left(\dfrac{1}{6} \times \dfrac{5}{6}\right) + \left(\dfrac{5}{6} \times \dfrac{1}{6}\right) = \dfrac{10}{36} = \dfrac{5}{18}$

  iii $P(\text{neither shows } 3) = \dfrac{5}{6} \times \dfrac{5}{6} = \dfrac{25}{36}$

The outcomes in parts i, ii and iii cover all the possible outcomes. They are also mutually exclusive, so the total probability is 1.

Check:

$$\dfrac{1}{36} + \dfrac{10}{36} + \dfrac{25}{36} = \dfrac{36}{36} = 1$$

b The scores are independent so, for each throw, $P(\text{even}) = \dfrac{1}{2}$.

$$P(\text{score is even on all four throws}) = \left(\dfrac{1}{2}\right)^4 = \dfrac{1}{16}.$$

Example 18

Question

There are 5 red counters and 7 blue counters in a bag. Eliza takes a counter from the bag, notes its colour and then puts it back into the bag. Obren then takes a counter from the bag. Find the probability that

a Eliza takes a red counter and Obren takes a blue counter

b Obren's counter is the same colour as Eliza's counter.

> **Note**
>
> This is known as **sampling with replacement,** as the first counter is put back into the bag before the second counter is taken out. The events are independent.

Answer

As Eliza's counter is put back into the bag, the events are independent so, for both Eliza and Obren,

$$P(R) = \dfrac{5}{12}, P(B) = \dfrac{7}{12}$$

a P(Eliza picks red and Obren picks blue)

$$= P(R_E) \times P(B_O)$$

$$= \dfrac{5}{12} \times \dfrac{7}{12}$$

There are still 12 counters in the bag, of which 7 are blue.

$$= \dfrac{35}{144}$$

(continued)

(continued)

b   P(same colour) = P(both red) + P(both blue)

$$= P(R_E) \times P(R_O) + P(B_E) \times P(B_O)$$

$$= \frac{5}{12} \times \frac{5}{12} + \frac{7}{12} \times \frac{7}{12}$$

$$= \frac{37}{72}$$

Example 19

Three coffee machines are installed in an office. The probability that this type of machine will break down on any day is 0.1. Find the probability that on Wednesday

a   all three machines break down

b   exactly one machine breaks down

c   at least one machine is working.

Let B be the event 'the machine breaks down' and W the event 'the machine is working'.

a   P(all three machines break down)

$$= P(B_1\, B_2\, B_3)$$

$$= 0.1 \times 0.1 \times 0.1$$

$$= 0.001$$

b   P(exactly one machine breaks down)

$$= P(B_1\, W_2\, W_3) + P(W_1\, B_2\, W_3) + P(W_1\, W_2\, B_3)$$

$$= (0.1 \times 0.9 \times 0.9) + (0.9 \times 0.1 \times 0.9) + (0.9 \times 0.9 \times 0.1)$$

$$= 0.243$$

> Notice that each probability is the same, so you could just calculate $3 \times 0.1 \times 0.9^2$.

c   P(at least one machine is working)

$$= 1 - P(\text{all three break down})$$

$$= 1 - 0.001 \quad \text{from part } \mathbf{a}$$

$$= 0.999$$

Example 20

A supermarket sells three types of milk: red top, green top and blue top.
The probability that a customer will buy a particular type of milk is shown in the table.

Milk	Probability
Red top (R)	0.3
Green top (G)	0.55
Blue top (B)	0.15

Three customers were selected at random from those who had bought one type of milk. Find the probability that

a   they all bought green top

b   exactly two bought red top,

c   one bought red top, one bought green top and one bought blue top.

The events R, G and B are independent.

a $\text{P(all bought green top)} = \text{P}(GGG)$

$$= 0.55 \times 0.55 \times 0.55$$
$$= 0.166375$$
$$= 0.166 \, (3 \, \text{sf})$$

b $\text{P}(R) = 0.3$, $\text{P}(R') = 0.7$

There are 3 arrangements of RRR' and each has the same probability, so

$\text{P(exactly two bought red top)} = 3 \times \text{P}(RRR')$

$$= 3 \times 0.3 \times 0.3 \times 0.7$$
$$= 0.189$$

c There are 6 arrangements of RGB, each with the same probability, so

$\text{P(one } R \text{, one } G \text{, one } B) = 6 \times \text{P}(RGB)$

$$= 6 \times 0.3 \times 0.55 \times 0.15$$
$$= 0.1485$$

Exercise 4

1 a A fair cubical die is thrown twice. Find the probability of obtaining

  i a score of 2 on both throws

  ii a score of 2 on just one of the two throws

  iii a score of 4 on at least one throw

  iv a score lower than 3 on both throws.

  b The die is thrown 3 times. Find the probability that the score is even on all three throws.

2 A manufacturer makes pens. When the process is going well, only 2.5% of the pens are defective. The supervisor selects two pens at random from the production line. Find the probability that

a both pens are defective

b exactly one pen is defective.

3 A coin is biased so that it is three times as likely to show heads than tails.

a The coin is tossed once.

  Find the probability of obtaining a head.

b The coin is tossed twice.

  Find the probability of obtaining at least one head.

c The coin is tossed three times.

  Find the probability of obtaining at most 2 heads.

4 Anzer, Barbara and Clefta are members of the skating club. The independent probabilities that Anzer, Barbara and Clefta will go to the club on a Friday evening are shown in the table.

Name	Probability
Anzer	0.4
Barbara	0.3
Clefta	0.5

Find the probability that, on a randomly selected Friday,

a all three girls go to the skating club

b only Barbara goes to the skating club.

5 In an experiment, a coloured ball is selected at random from a bag containing 6 red balls, 3 yellow balls and 1 blue ball. Its colour is noted and the ball is put back into the bag.

    a The process is performed twice. Find the probability that

      i two red balls are selected

      ii a yellow and a blue ball are selected

      iii neither ball is blue.

    b The process is performed 3 times. Find the probability that

      i one ball is red and two balls are yellow

      ii a ball of each colour is selected.

    c The process is performed 5 times. Find the probability that all the balls selected are red.

13.4 Tree diagrams

You may prefer to show the outcomes and probabilities on a **tree diagram**. Tree diagrams can be used both for conditional events and for independent events.

Steps:

▶ Show the *events* in *layers* on the tree.

▶ Show the *outcomes* of each event in the *branches*, writing the appropriate probability on each branch. Remember that the probabilities on a set of branches add up to 1.

▶ To find the *final outcomes*, follow each path.

▶ To work out the probability of each final outcome, *multiply* the probabilities along the path.

▶ When more than one final outcome satisfies what you want, *add* the probabilities of the final outcomes.

The method is illustrated in the following example.

> **Note**
>
> You may use tree diagrams in the examination, but questions will not specifically require you to use this method.

Example 21

Each day I travel to work by route *A* or route *B*. The probability that I choose route *A* is 0.2. The probability that I am late for work if I choose route *A* is 0.3 and the probability that I am late if I choose route *B* is 0.65.

a Find the probability that I am not late for work.

b Given that I am not late for work, find the probability that I chose route *B*. Give your answer to 3 decimal places.

Question

Event A: I go by route A	$P(A) = 0.2$
Event B: I go by route B	$P(B) = 1 - 0.2 = 0.8$
Event L: I am late for work	$P(L \mid A) = 0.3, P(L' \mid A) = 0.7$
	$P(L \mid B) = 0.65, P(L' \mid B) = 0.35$

First draw a set of branches to show the routes and write in the probabilities.

Then draw a set of branches from A to show whether or not I am late for work.

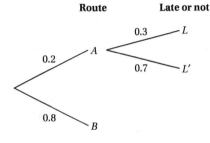

The events are conditional.

The probabilities on the branches meeting at a point must add up to 1.

Now complete the tree:

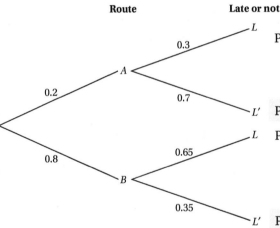

Multiply along the branches.

$P(A \text{ and } L) = P(A) \times P(L \mid A) = 0.2 \times 0.3 = 0.06$

$P(A \text{ and } L') = P(A) \times P(L' \mid A) = 0.2 \times 0.7 = 0.14$

$P(B \text{ and } L) = P(B) \times P(L \mid B) = 0.8 \times 0.65 = 0.52$

$P(B \text{ and } L') = P(B) \times P(L' \mid B) = 0.8 \times 0.35 = 0.28$ *

Notice that the sum of these probabilities is 1.

a P(not late for work)

$= P(L')$

$= P(A \text{ and } L') + P(B \text{ and } L')$ highlighted outcomes

$= 0.14 + 0.28$

$= 0.42$

b P(chose route B | not late for work)

$= \dfrac{P(B \text{ and } L')}{P(L')}$

marked *

from part **a**

$= \dfrac{0.28}{0.42}$

$= 0.6666...$

$= 0.667$ (3 dp)

> **Note**
>
> The possibility space has been reduced to the highlighted outcomes.

Example 22

A coin is biased so that the probability it shows heads is 0.8. The coin is tossed twice.

Find the probability that there is

a   exactly one head

b   at least one head

c   exactly one head, given there is at least one head.

For each toss, $P(H) = 0.8$, $P(T) = 0.2$.

First toss   **Second toss**   The events are independent.

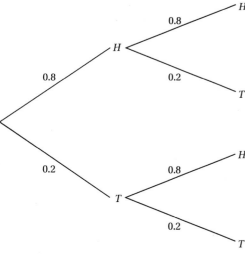

$P(H_1 \text{ and } H_2) = P(H_1) \times P(H_2) = 0.8 \times 0.8 = 0.64$

$P(H_1 \text{ and } T_2) = P(H_1) \times P(T_2) = 0.8 \times 0.2 = 0.16$ *

$P(T_1 \text{ and } H_2) = P(T_1) \times P(H_2) = 0.2 \times 0.8 = 0.16$ *

$P(T_1 \text{ and } T_2) = P(T_1) \times P(T_2) = 0.2 \times 0.2 = 0.04$

a   P(exactly one head)

   $= 0.16 + 0.16$ ← marked *

   $= 0.32$

b   P(at least one head)

   $= $ P(exactly one head) + P(two heads) ← highlighted outcomes

   $= 0.32 + 0.64$

   $= 0.96$

c   P(exactly one head | at least one head)

   $= \dfrac{0.32}{0.96}$ ← from part **a** / from part **b**

   $= \dfrac{1}{3}$

Example 23

Question

Dafydd, Eli and Fabio are members of an amateur cycling club that holds a time trial each Sunday during summer. The independent probabilities that Dafydd, Eli and Fabio takes part in any one of these trials are 0.6, 0.7 and 0.8.

Find the probability that, on a particular Sunday during summer:

a none of the three cyclists takes part;

b Fabio is the only one of the three cyclists to take part;

c exactly one of the three cyclists takes part;

d either one or two of the three cyclists takes part.

AQA MS1B January 2007

> **Note**
>
> A tree diagram is not essential, but it may help to clarify the probabilities.

Event D: Daffyd takes part $P(D) = 0.6$, $P(D') = 0.4$

Event E: Eli takes part $P(E) = 0.7$, $P(E') = 0.3$

Event F: Fabio takes part $P(F) = 0.8$, $P(F') = 0.2$

The events are D, E and F are independent.

Answer

a   P(none takes part)

    $= P(D'E'F')$

    $= 0.4 \times 0.3 \times 0.2$

    $= 0.024$

b   P(only F takes part)

    $= P(D'E'F)$

    $= 0.4 \times 0.3 \times 0.8$

    $= 0.096$

c   P(exactly one takes part)

    $= P(\text{only } D) + P(\text{only } E) + P(\text{only } F)$

    $= (0.6 \times 0.3 \times 0.2) + (0.4 \times 0.7 \times 0.2) + 0.096$

    $= 0.188$

d   P(either one or two)

    $= 1 - P(\text{none}) - P(\text{all three take part})$

    $= 1 - 0.024 - (0.6 \times 0.7 \times 0.8)$

    $= 0.64$

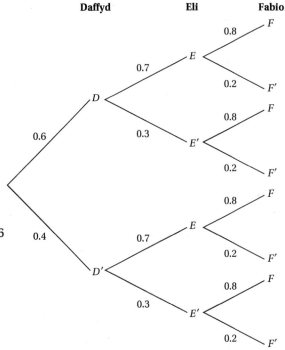

Exercise 5

Think carefully about whether events are independent or conditional.

1   A mother and her daughter are both entering a cake making competition at a show. From past experience they estimate that the probability that the mother will win a prize is $\frac{1}{6}$ and, independently of her mother's result, the probability that the daughter will win a prize is $\frac{2}{7}$.

Find the probability that

    a either the mother or the daughter, but not both, wins a prize,

    b the mother wins a prize, given that just one of them wins a prize,

    c at least one of them wins a prize.

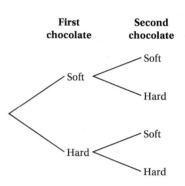

First chocolate   Second chocolate

2 A box contains 20 chocolates, of which 15 have soft centres and 5 have hard centres. Sadie takes two chocolates from the box and eats them.

    a Find the probability that just one of the chocolates has a soft centre.

    b Find the probability that the first chocolate has a soft centre, given that just one of the chocolates has a soft centre.

3 In a large batch of flower seeds 70% have been treated to improve germination. The treated seeds have a probability of 0.8 of germinating, whereas the untreated seeds have a probability of 0.3 of germinating. A seed is selected at random from the batch. By drawing a tree diagram, or otherwise, find the probability that

    a the seed will germinate

    b the seed has been treated, given that it has germinated.

4 A box contains 9 pens of which 6 are red and 3 are blue. Verlind is doing an experiment as part of his mathematics homework.

    a He takes a pen from the box at random, notes its colour and then puts it back in the box. He does this a second time, then a third time.

        Find the probability that he takes out

        i a red pen each time

        ii at least one blue pen.

    b Verlind now repeats the experiment, but this time he does not return the pen to the box each time.

        Find the probability that he takes out

        i a red pen each time

        ii at least one blue pen.

5 At a children's party each of the 12 guests is to receive a toy. Mohammed is hoping to get a torch and Rebekah wants a ball. The 12 toys, consisting of 4 balls, 3 torches and 5 pens, are placed in a bag. Mohammed receives the first toy drawn out of the bag and Rebekah receives the second toy drawn out of the bag. Assume that at each stage each toy has an equal chance of being drawn.

    a Find the probability that Rebekah will get a ball.

    b When Rebekah's parents arrive at the end of the party, Rebekah shows them the ball that she got in the draw. Find the probability that Mohammed got a torch.

6 Some students are answering multiple choice questions. In each question, there are four choices.

    a Amy does not know the answer to a particular question, so she guesses. What is the probability that Amy guesses the correct answer?

b Fahed guesses the answers to two of the multiple choice questions. Find the probability that

   i both answers are correct

   ii exactly one answer is correct.

c Imogen guesses the answers to three of the multiple choice questions. Find the probability that

   i all three are incorrect

   ii exactly two answers are correct

   iii at least two answers are correct

   iv fewer than two answers are correct.

d Caliti guesses the answers to four multiple choice questions. Find the probability that all his answers are correct.

7 Pippa is playing a game at a fund raiser event in which she randomly selects a coloured disc from a bag. The bag contains 10 discs of which 9 are blue and 1 is white. If she selects the white disc she will win a teddy bear. She is allowed three attempts and no disc is returned to the bag once it has been chosen.

a Find the probability that Pippa wins the teddy bear.

b Given that she wins the teddy bear, find the probability that she only needed one attempt.

8 Garuda is playing table tennis. Each time that she serves, the probability that she wins the point is 0.6, independent of the result of any preceding serves. At the start of a particular game she serves for the first five points.

a Find the probability that, for the first two points of the game,

   i she wins both points

   ii she wins exactly one of the points.

b Calculate the probability that she loses all five points.

9 When a particular firm needs to hire a taxi, the receptionist calls one of three firms, X, Y or Z.

40% of the calls are to X, 50% are to Y and 10% are to Z.

9% of the taxis hired from X are late, 6% of those hired from Y are late and 20% of those hired from Z are late.

Find the probability that the next taxi hired

a will be from X and will not arrive late

b will arrive late

c is from X, given that it arrives late.

10 Jodie and Kapil are playing a game in which they have two bags containing coloured cubes.

Bag A has 7 red cubes and 3 blue cubes.

Bag B has 4 red cubes and 6 blue cubes.

  a Jodie takes a cube at random from bag A and Kapil takes a cube at random from bag B. Find the probability that

    i both cubes are red

    ii just one of the cubes is red.

  b The cubes are returned to their correct bags. Jodie now takes a cube at random from bag A and after noting the colour she puts it in bag B.

    Kapil now takes a cube at random from bag B. Find the probability that it is red.

Showing whether events are mutually exclusive or independent

Mutually exclusive events occur in the context of several possible outcomes in one experiment.

▶ To prove that two events are **mutually exclusive**, you must give working to show that *either* of the following is satisfied

$$P(A \text{ or } B) = P(A) + P(B)$$

$$P(A \text{ and } B) = 0$$

Independent events occur in the context of two or more experiments taking place together or being repeated one or more times.

▶ To prove that two events are **independent**, you must give working to show that *any one* of the following is satisfied

$$P(A \text{ and } B) = P(A) \times P(B)$$

$$P(A \mid B) = P(A)$$

$$P(B \mid A) = P(B)$$

Example 24

When a cubical die is thrown, the number on the uppermost face is the score. Two fair dice are thrown, one red and one blue.

The outcomes resulting in events *A*, *B* and *C* are shown on the possibility space diagram, where

    A is the event: the score on the red die is more than 4;

    B is the event: the sum of scores is 7;

    C is the event: the sum of the scores is 8.

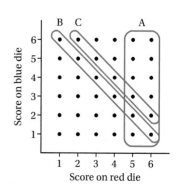

a Giving a reason for your answer:

  i state which two of the events *A*, *B* and *C* are mutually exclusive

  ii show that events *A* and *B* are independent.

b **i** Find $P(A \mid C)$.

  ii Are events *A* and *C* independent? Give a reason.

a i Since it is not possible for the sum to be 7 and 8,

P(B and C) = 0, so events B and C are mutually exclusive.

ii $P(A \text{ and } B) = \dfrac{2}{36} = \dfrac{1}{18}$

$P(A) = \dfrac{12}{36} = \dfrac{1}{3}$, $P(B) = \dfrac{6}{36} = \dfrac{1}{6}$,

so $P(A) \times P(B) = \dfrac{1}{3} \times \dfrac{1}{6} = \dfrac{1}{18}$

Since P(A and B) = P(A) × P(B), events A and B are independent.

b i $P(A \mid C) = \dfrac{2}{5}$

ii From part **a**, you know that $P(A) = \dfrac{1}{3}$

Therefore P(A | C) ≠ P(A), so events A and C are not independent.

Example 25

The table shows an analysis of the tickets purchased for a performance at a theatre. There are three types of tickets available: child (up to 15 years old); adult (16–59) and senior (60 or over) and the tickets are either for the balcony or the stalls.

	Child	Adult	Senior	Totals
Balcony	12	20	13	45
Stalls	8	15	7	30
Totals	20	35	20	75

a State, giving a reason, whether the event of selecting a balcony ticket and the event of selecting a child's ticket are independent.

b Give an event that is mutually exclusive to the event of selecting a ticket in the stalls.

a Let B be the event 'selecting a balcony ticket' and C the event 'selecting a child's ticket'

$P(B) = \dfrac{45}{75} = 0.6$, $P(B \mid C) = \dfrac{12}{20} = 0.6$

Since P(B | C) = P(B), events B and C are independent.

Alternatively, show P(C | B) = P(C) or P(B and C) = P(B) × P(C).

b The event 'selecting a balcony ticket' is mutually exclusive to 'selecting a ticket in the stalls'.

Example 26

Events A and B are such that P(A) = 0.3, P(B) = 0.6 and P(A or B) = 0.72.

State, giving a reason in each case, whether events A and B are

a mutually exclusive

b independent.

a $P(A \text{ or } B) = 0.72$

$P(A) + P(B) = 0.3 + 0.6 = 0.9$

Since $P(A) + P(B) \neq P(A \text{ or } B)$, events A and B are not mutually exclusive.

b   By the addition law

$P(A \text{ or } B) = P(A) + P(B) - P(A \text{ and } B)$

$0.72 = 0.3 + 0.6 - P(A \text{ and } B)$

so $P(A \text{ and } B) = 0.18$

Now, $P(A) \times P(B) = 0.3 \times 0.6 = 0.18$.

Since $P(A \text{ and } B) = P(A) \times P(B)$, the events A and B are independent.

Alternatively, you could draw up a probability table and make your conclusions from it:

- Fill in $P(A) = 0.3$ and $P(B) = 0.6$ and the grand total of 1.
- Use $P(A \text{ or } B) = 0.72$ to fill in $P(\text{neither } A \text{ nor } B) = 1 - 0.72 = 0.28$.
- Now fill in the remaining values by making totals agree.

	A	A'	
B	0.18	0.42	0.6
B'	0.12	0.28	0.4
	0.3	0.7	1

From the table:

a $P(A \text{ and } B) = 0.18$

Since $P(A \text{ and } B) \neq 0$, A and B are not mutually exclusive.

b $P(A) \times P(B) = 0.3 \times 0.6 = 0.18 = P(A \text{ and } B)$, so events A and B are independent.

> **Note**
>
> Remember to state your conclusion.

Exercise 6

1 The events A and B are such that $P(A \mid B) = 0.4$, $P(B \mid A) = 0.25$ and $P(A \text{ and } B) = 0.12$.

a   Are A and B independent? Give a reason for your answer.

b   Find $P(A \text{ or } B)$.

2 Events A and B are such that $P(A) = 0.45$, $P(B) = 0.35$ and $P(A \text{ or } B) = 0.7$.

a   Find $P(A \text{ and } B)$.

b   Show that events A and B are not independent.

c   Find $P(A \mid B)$.

3 Events A and B are such that $P(A) = 0.7$, $P(A \text{ only}) = 0.49$, $P(B) = 0.3$

	A	A'	
B			0.3
B'	0.49		
	0.7		1

a   Complete the probability table.

b   State, with a reason whether

i   events A and B are independent

ii   events A and B are mutually exclusive

iii events A' and B are independent.

④ Events X and Y are mutually exclusive where $P(X) = 0.3$ and $P(Y) = 0.4$.

Are events X' and Y' mutually exclusive? Give a reason for your answer.

⑤ A school has 100 teachers. In a survey on the use of the school car park, the teachers were asked whether they had driven a car to school on that day.

Of the 70 full-time teachers, 45 had driven a car to school and of the 30 part-time teachers, 12 had driven a car to school.

a Copy and complete the two-way table, where C denotes the event 'the teacher had driven a car to school that day'.

b Find the probability that a teacher chosen at random

   i is a part-time teacher who had driven a car to school

   ii is a full-time teacher who had not driven a car to school

   iii is a full-time teacher or had driven a car to school

   iv is a part-time teacher, given that the teacher had driven a car to school.

	C	C'	Total
Full-time teacher			
Part-time teacher			
Total			100

c Are the events 'the teacher had driven a car to school' and 'the teacher is full-time' independent? Give a reason for your answer.

d Describe two events that are mutually exclusive.

13.5 Further applications

Example 27

1 A basket in a stationery store contains a total of 400 marker and highlighter pens. Of the marker pens, some are permanent and the rest are non-permanent. The colours and types of pen are shown in the table.

Type	Colour			
	Black	Blue	Red	Green
Permanent marker	44	66	32	18
Non-permanent marker	36	53	21	10
Highlighter	0	41	37	42

A pen is selected at random from the basket. Calculate the probability that it is:

a   a blue pen;

b   a marker pen;

c   a blue pen or a marker pen;

d   a green pen, given that it is a highlighter pen;

e   a non-permanent marker pen, given that it is a red pen.

AQA MS1B June 2008

Work out the totals, checking that the row totals and column totals add up to 400:

	Colour				
Type	**Black**	**Blue**	**Red**	**Green**	
Permanent marker	44	66	32	18	160
Non-permanent marker	36	53	21	10	120
Highlighter	0	41	37	42	120
	80	160	90	70	**400**

a $P(\text{blue}) = \dfrac{160}{400} = \dfrac{2}{5}$

b $P(\text{marker pen}) = \dfrac{160+120}{400} = \dfrac{7}{10}$

First two rows.

c $P(\text{blue pen or marker pen}) = \dfrac{321}{400}$

Count all the marker pens and the blue highlighters.

d $P(\text{green, given highlighter pen}) = \dfrac{42}{120} = \dfrac{7}{20}$

The sample space is reduced to just highlighters.

e $P(\text{non-permanent marker, given red}) = \dfrac{21}{90} = \dfrac{7}{30}$

The sample space is reduced to just red pens.

Example 28

Roger is an active retired lecturer. Each day after breakfast, he decides whether the weather for that day is going to be fine (F), dull (D) or wet (W). He then decides on only one of four activities for the day: cycling (C), gardening (G), shopping (S) or relaxing (R). His decisions from day to day may be assumed to be independent.

The table shows Roger's probabilities for each combination of weather and activity.

		Weather		
		Fine (F)	**Dull (D)**	**Wet (W)**
Activity	**Cycling (C)**	0.30	0.10	0
	Gardening (G)	0.25	0.05	0
	Shopping (S)	0	0.10	0.05
	Relaxing (R)	0	0.05	0.10

a Find the probability that, on a particular day, Roger decided:

i that it was going to be fine and that he would go cycling;

ii on either gardening or shopping;

iii to go cycling, given that he had decided it was going to be fine;

iv that it was going to be fine, given that he did **not** go cycling.

(continued)

b Calculate the probability that, on a particular Saturday and Sunday, Roger decided that it was going to be fine and decided on the same activity for both days.

AQA MS1A January 2013

a i $P(F \text{ and } C) = 0.30$

ii Since G and S are mutually exclusive,

$$P(G \text{ or } S) = P(G) + P(S)$$
$$= (0.25 + 0.05) + (0.10 + 0.05)$$
$$= 0.45$$

iii $P(C \mid F) = \dfrac{P(C \text{ and } F)}{P(F)} = \dfrac{0.30}{0.55} = \dfrac{6}{11}$

iv $P(F \mid C') = \dfrac{P(F \text{ and } C')}{P(C')} = \dfrac{0.25}{0.6} = \dfrac{5}{12}$

b $P(F \text{ and } C) = 0.30$, so $P(F \text{ and } C \text{ on both days}) = 0.30^2$

$P(F \text{ and } G) = 0.25$, so $P(F \text{ and } G \text{ on both days}) = 0.25^2$

P(fine and same activity on both days)

$$= 0.30^2 + 0.25^2$$
$$= 0.1525$$

Example 29

Twins Alec and Eric are members of the same local cricket club and play for the club's under 18 team.

The probability that Alec is selected to play in any particular game is 0.85.

The probability that Eric is selected to play in any particular game is 0.60.

The probability that both Alec and Eric are selected to play in any particular game is 0.55.

a By using a table, or otherwise:

i show that the probability that neither twin is selected for a particular game is 0.10;

ii find the probability that at least one of the twins is selected for a particular game;

iii find the probability that exactly one of the twins is selected for a particular game.

b The probability that the twins' younger brother, Cedric, is selected for a particular game is:

0.30 given that both of the twins have been selected;

0.75 given that exactly one of the twins has been selected;

0.40 given that neither of the twins has been selected.

Calculate the probability that, for a particular game:

i all three brothers are selected;

ii at least two of the three brothers are selected.

AQA MS1A January 2012

a Draw up a probability table

	A	**A′**	Total
E	**0.55**	0.05	**0.60**
E′	0.30	0.10	0.40
Total	**0.85**	0.15	1

From the table

i P(neither) = P(A' and E') = 0.10

ii P(at least one) = 1 − P(neither)

$$= 1 - 0.10$$

$$= 0.90$$

iii P(exactly one) = P(A only) + P(E only)

$$= 0.30 + 0.05$$

$$= 0.35$$

b Draw a tree, with the first layer showing the three outcomes for Alec and Eric (both, only one, neither) and the second layer showing the conditional probabilities for Cedric (selected or not).

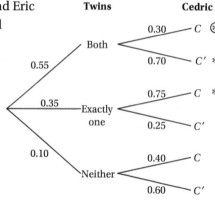

i P(all three are selected)

= P(both and C)   marked ⊛

$$= 0.55 \times 0.30$$

$$= 0.165$$

ii P(at least two)   marked *

= P(all three) + P(both and C′) + P(exactly one and C)

$$= 0.165 + (0.55 \times 0.70) + (0.35 \times 0.75)$$

$$= 0.165 + 0.385 + 0.2625$$

$$= 0.8125$$

Summary

The probability of an event is a measure of the likelihood it will happen.

A probability of 0 indicates that the event is impossible.

A probability of 1 indicates that the event is certain to happen.

All other events have a probability between 0 and 1.

Estimating probability using relative frequency

If event A occurs r times in n trials of an experiment, the probability of A is estimated using the long-term relative frequency, which is the limiting value of $\dfrac{r}{n}$. The reliability of the estimate increases as n increases.

Equally likely outcomes

$$P(A) = \dfrac{n(A)}{n(S)}$$
number of outcomes resulting in event A
total number of outcomes in the possibility space S

Complement

The complement of A is A', where A' is the event 'A does not occur'.

$$P(A') = 1 - P(A)$$

Addition law for two events

For events A and B,

$$P(A \text{ or } B) = P(A) + P(B) - P(A \text{ and } B) \qquad \text{A \textbf{or} B means A or B or both}$$

In set notation $\quad P(A \cup B) = P(A) + P(B) - P(A \cap B) \qquad \text{A \textbf{and} B means both A and B}$

Mutually exclusive events

For mutually exclusive events A and B,

$$P(A \text{ and } B) = 0,$$

so $\qquad P(A \text{ or } B) = P(A) + P(B) \qquad$ '**or**' rule for mutually exclusive events

In set notation $\quad P(A \cup B) = P(A) + P(B)$

For n mutually exclusive events

$$P(A_1 \text{ or } A_2 \text{ or } A_3 \text{ or ... or } A_n) = P(A_1) + P(A_2) + P(A_3) + \cdots + P(A_n)$$

Conditional probability

$$P(A, \text{given } B) = P(A \mid B) = \frac{P(A \text{ and } B)}{P(B)}$$

$$P(B, \text{given } A) = P(B \mid A) = \frac{P(B \text{ and } A)}{P(A)}$$

Multiplication law for two events

$$P(A \text{ and } B) = P(B) \times P(A \mid B)$$

Similarly $\qquad P(B \text{ and } A) = P(A) \times P(B \mid A) \qquad$ Remember that $P(A \text{ and } B) = P(B \text{ and } A)$

Independent events

For independent events A and B,

$$P(A \mid B) = P(A)$$

Also, $\qquad P(B \mid A) = P(B)$

$$P(A \text{ and } B) = P(A) \times P(B) \qquad \text{'\textbf{and}' rule for independent events}$$

In set notation $\quad P(A \cap B) = P(A) \times P(B)$

For n independent events:

$$P(A_1 \text{ and } A_2 \text{ and } A_3 \text{ and ... and } A_n) = P(A_1) \times P(A_2) \times P(A_3) \times \cdots \times P(A_n).$$

Probability (relative frequency) tables

	A	A'	Total
B	$P(A \text{ and } B)$	$P(\text{only } B)$	$P(B)$
B'	$P(\text{only } A)$	$P(\text{neither})$	$P(B')$
Total	$P(A)$	$P(A')$	1

When A and B are mutually exclusive, the entry for $P(A \text{ and } B)$ is zero.

Tree diagrams

For events A and B,

> multiply the probabilities along the branches

> add the final probabilities of the successful end outcomes.

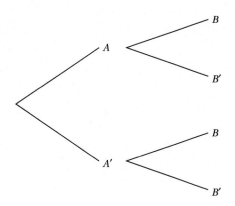

First event	Second event

B $P(A \text{ and } B) = P(A) \times P(B \mid A)$

A

B' $P(A \text{ and } B') = P(A) \times P(B' \mid A)$

B $P(A' \text{ and } B) = P(A') \times P(B \mid A')$

A'

B' $P(A' \text{ and } B') = P(A') \times P(B' \mid A')$

Review

1 Events C and D are such that $P(C) = \dfrac{19}{30}$, $P(D) = \dfrac{2}{5}$ and $P(C \text{ or } D) = \dfrac{4}{5}$. Find

    a $P(C \text{ and } D)$ **b** $P(D \text{ only})$ **c** $P(C \text{ or } D \text{ but not both})$.

2 In a survey of the members of an arts club, it was found that 73% had seen the most recent play, 49% had seen the most recent ballet and 15% had not seen the play or the ballet.

    Find the probability that a person chosen at random from the arts club had seen

    a   the play or the ballet   **b**   both the play and the ballet

    c   only the ballet.

3 Two fair coins are tossed.

    a   Events A and B are mutually exclusive.

       A is the event 'at least one head is obtained'. Define event B.

    b X is the event 'one head is obtained'. Define an event Y such that X and Y are not mutually exclusive.

4 A local greengrocer sells fruit, 30% of which is organically grown and 70% is conventionally grown. Sales of apples constitute 20% of the organically grown fruit and 45% of the conventionally grown fruit.

    A customer who has purchased fruit is chosen at random to take part in a survey.

    a   Find the probability that the customer bought apples.

    b   Given that the customer bought apples, find the probability that the apples were organically grown.

5 On any morning, the probability that I have to wait at the traffic lights on my way to school is 0.25.

a Find the probability that, on two consecutive mornings, I have to wait at the traffic lights

  i on exactly one morning

  ii on the second morning, given that I have to wait at the traffic lights on exactly one morning

b Find the probability that, on three consecutive mornings, I have to wait at least once.

6 Two golfers, Chris and Tayib, are attempting to qualify for a golf tournament. On past performance, the probability that Tayib will qualify is 0.8, the probability that Chris will qualify is p and the probability that *both* Tayib and Chris will qualify is 0.6. The event 'Chris qualifies' is independent of the event 'Tayib qualifies'.

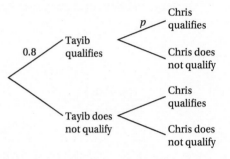

a Find p.

b Find the probability that just one qualifies.

c Given that just one qualifies, find the probability that it is Chris.

7 In a certain country, 52% of the population is male.

It is known that 16% of the males are left-handed and 12% of the females are left-handed. A person is chosen at random and found to be left-handed.

By drawing a tree diagram or otherwise, find the probability that the person is male, given that the person is left-handed.

8 Uday and Bonnie play each other at table tennis. Each game results in either a win for Uday or a win for Bonnie. If Uday wins a particular game, the probability of him winning the next game is 0.8, but, if he loses, the probability of him winning the next game is 0.3. The probability of Uday winning the first game is 0.7.

Find the probability that

a Uday wins the first game, given that he wins exactly one of the first two games

b Bonnie loses two games and wins one game in the first three games.

9 In an archery competition, Abhaya is allowed up to three attempts to hit the target. If he succeeds on any attempt, he does not make any more attempts. The probability that he will hit the target on the first attempt is 0.6. If he misses, the probability that he will hit the target on his second attempt is 0.7. If he misses on the second attempt, the probability that he will hit the target on his third attempt is 0.8.

a Find the probability that Abhaya will hit the target.

b Given that Abhaya hits the target, find the probability that he made at least two attempts.

10 It was found that 56% of passengers on a train bought a drink from the buffet car during their journey. Of those buying a drink from the buffet car, 45% were under 30 years old, 25% were between 30 and 65 years old and the rest were over 65 years old.

Of those not buying a drink from the buffet car, 23% were under 30 years old and 64% were over 65 years old.

Given that a randomly selected passenger is 42 years old, find the probability that this passenger bought a drink from the buffet car during the journey.

11 A large bookcase contains two types of book: hardback and paperback. The number of books of each type in each of four subject categories is shown in the table.

		Subject category				
		Crime	Romance	Science fiction	Thriller	Total
Type	Hardback	8	16	18	18	60
	Paperback	16	40	14	30	100
	Total	24	56	32	48	160

a A book is selected at random from the bookcase. Calculate the probability that the book is:

i a paperback;

ii not science fiction;

iii science fiction or hardback;

iv a thriller, given that it is a paperback.

b Three books are selected at random, without replacement, from the bookcase.

Calculate, to three decimal places, the probability that one is crime, one is romance and one is science fiction.

AQA MS1A June 2009

12 Gary and his neighbour Larry work at the same place.

On any day when Gary travels to work, he uses one of three options: his car only, a bus only or both his car and a bus. The probability that he uses his car, either on its own or with a bus, is 0.6. The probability that he uses both his car and a bus is 0.25.

a Calculate the probability that, on any particular day when Gary travels to work, he:

i does not use his car;

ii uses his car only;

iii uses a bus.

b On any day, the probability that Larry travels to work with Gary is 0.9 when Gary uses his car only, is 0.7 when Gary uses both his car and a bus, and is 0.3 when Gary uses a bus only.

Calculate the probability that, on any particular day when Gary travels to work, Larry travels with him.

AQA MS1A January 2009

13 A reliable estimate for the proportion of a population of fish with a certain disease is 60 per cent. A test for the presence of the disease in fish is possible. The test gives ones of three conclusions:

diseased, inconclusive, non-diseased.

For a **diseased** fish, the probabilities of these three conclusions are:

diseased	0.75
inconclusive	0.15
non-diseased	0.10

For **non-diseased** fish, the probabilities of these three conclusions are:

diseased	0.05
inconclusive	0.15
non-diseased	0.80

a A fish is selected at random. Using a tree diagram, or otherwise, calculate the probability that:

  i the fish has the disease and the test concludes that it is diseased;

  ii the test concludes that the fish has the disease;

  iii the test gives a correct conclusion.

b Three fish, all with the disease, are tested. Find the probability that the test concludes that two fish are diseased and one fish is non-diseased.

AQA MS1A June 2005

14 Which of the following is the more likely to occur?

  i a throw of six at least once in 4 throws of a single die

  ii a throw of double six at least once in 24 throws of a pair of dice?

Give numerical justification for your answer.

Assessment

1 Emma visits her local supermarket every Thursday to do her weekly shopping.

The event that she buys orange juice is denoted by J and the event that she buys bottled water is denoted by W. At each visit, Emma may buy neither, or one, or both of these items.

a Complete the table of probabilities completed below, for these events, where J' and W' denote the events 'not J' and 'not W' respectively.

	J	J'	Total
W			0.65
W'	0.15		
Total		0.30	1

[3]

b Hence, or otherwise, find the probability that, on any given Thursday, Emma buys either orange juice or bottled water, but not both. [2]

c Show that:

  i the events J and W are **not** mutually exclusive;

  ii the events J and W are **not** independent. [3]

AQA MS1A June 2011

2 Events A and B are such that $P(A) = 0.5$, $P(A$ and $B) = 0.2$ and $P(A$ or $B) = p$.
Find, in terms of p,

  a $P(B)$ [2]

  b $P(A$ given $B)$ [2]

  c If A and B are independent events find the value of p. [2]

3 Xavier, Yuri and Zara attend a sports centre for their judo club's practice
sessions. The probabilities of them arriving late are, independently, 0.3, 0.4
and 0.2 respectively.

  a Calculate the probability that for a particular practice session:

    i all three arrive late; [1]

    ii none of the three arrives late; [2]

    iii only Zara arrives late. [2]

  b Zara's friend, Wei, also attends the club's practice sessions. The
  probability that Wei arrives late is 0.9 when Zara arrives late, and is 0.25
  when Zara does not arrive late.

  Calculate the probability that for a particular practice session:

    i both Zara and Wei arrive late [2]

    ii either Zara or Wei, but not both, arrives late. [3]

AQA MS1B January 2006

4 Each school-day morning, three students, Rita, Said and Ting, travel
independently from their homes to the same school by one of three
methods: walk, cycle or bus. The table shows the probabilities of their
independent daily choices.

	Walk	Cycle	Bus
Rita	0.65	0.10	0.25
Said	0.40	0.45	0.15
Ting	0.25	0.55	0.20

  a Calculate the probability that, on any given school-day morning:

    i all 3 students walk to school; [2]

    ii only Rita travels by bus to school; [2]

    iii at least 2 of the 3 students cycle to school. [4]

  b Ursula, a friend of Rita, never travels to school by bus. The probability that:

  Ursula walks to school when Rita walks to school is 0.9;

  Ursula cycles to school when Rita cycles to school is 0.7.

  Calculate the probability that, on any given school-day morning, Rita
  and Ursula travel to school by:

    i the same method; [3]

    ii different methods. [1]

AQA MS1A January 2010

5 Rea, Suki and Tora take part in a shooting competition. The final round of the competition requires each of them to try to hit the centre of a target, placed at 100 metres, with a single shot. The independent probabilities that Rea, Suki and Tora hit the centre of this target with a single shot are 0.7, 0.6 and 0.8 respectively.

Find the probability that, in the final round of the competition, the centre of the target will be hit by:

a Tora only;   [2]

b exactly one of the three competitors;   [3]

c at least one of the three competitors.   [3]

AQA MS1A January 2007

6 Fred and his daughter, Delia, support their town's rugby team. The probability that Fred watches the game is 0.8. The probability that Delia watches the game is 0.9 when her father watches the game, and is 0.4 when her father does not watch the game.

a Calculate the probability that:

  i   both Fred and Delia watch a particular game;   [2]

  ii neither Fred nor Delia watch a particular game.   [2]

b Molly supports the same rugby team as Fred and Delia. The probability that Molly watches a game is 0.7 and is independent of whether or not Fred or Delia watches the game.

Calculate the probability that

  i   all three supporters watch a particular game;   [3]

  ii exactly two of the three supporters watch a particular game.   [4]

AQA MS1B June 2005

14 Discrete Random Variables

Introduction

In everyday life you constantly receive factual information from all manner of sources such as the internet, newspapers and television, but how do you make sense of it?

The information is a collection of observations known as data which is often difficult to interpret in its raw form.

This chapter starts by looking at ways of summarising discrete data by finding averages and also working out measures of spread to show the variability of the data.

These ideas are then linked with probability to explore models to describe real situations. The models are formed using discrete random variables and are called probability distributions.

Recap

You need to remember how to find the...
► mode, median, mean
► range
► quartiles and interquartile range.

You will also need the following, from Chapter 13
► probability methods.

Objectives

By the end of this chapter you should be able to ...

► Find measures of central tendency and spread for discrete data.
► Work with discrete random variables and their associated probability distributions.
► Calculate the mean, variance and standard deviation
  – of discrete random variables
  – of a simple function of a discrete random variable
  – of the sum or difference of two independent discrete random variables
  – of the sum of independent observations of a discrete random variable.

14.1 Discrete variables

A **variable** is a quantity that may take more than one value. When it is possible to make a list of its individual numerical values, the variable is said to be **discrete**.

Discrete variables can take only *exact* values such as

► the number of rally cars passing a checkpoint in a ten-minute interval
► the number of hits on an internet website in an hour
► the shoe sizes of children in a class
► the number of goals scored in a football match.

By contrast, a **continuous** variable, such as a length, mass or time, cannot be stated precisely but can be given only to a specified degree of accuracy.
For example, a height recorded as 144 cm, to the nearest cm, could have arisen from any value in the interval $143.5 \leq$ height < 144.5 cm.

In this chapter only discrete variables will be considered.

Measures of central tendency

Measures of central tendency are averages, which are typical or representative values of a set of data. The three main **averages** are the mode, mean and median.

Mode

> The mode is the value that occurs most often.

The mode is useful when you want to know the most popular value but not all sets of data have a mode and some may have more than one, so it is often not a helpful average.

Mean

The **mean** uses all the observations and so represents every item. It is the most useful average as it is the basis for much further work in statistics. However, a disadvantage is that it can be affected unduly by one or two extreme values (outliers).

The mean is denoted by $\bar{x}$, (read as "x bar") and is calculated by dividing the sum of all the observations by the number of observations.

For n observations,

$$\bar{x} = \frac{x_1 + x_2 + \ldots + x_n}{n} = \frac{\sum x_i}{n} \quad \text{for} \quad i = 1, 2, 3, \ldots, n$$

For data in a frequency distribution

$$\bar{x} = \frac{\sum x_i f_i}{\sum f_i} \quad \text{for} \quad i = 1, 2, 3, \ldots, n$$

> **Note**
>
> The symbol Σ is used to denote "the sum of". It is a Greek upper case letter and is read as "sigma".

Often the subscript i is omitted.

Example 1

The members of an orchestra were asked how many instruments each could play. Their replies are summarised in the frequency distribution.

Number of instruments, x	1	2	3	4	5
Frequency, f	11	10	5	3	1

Calculate the mean number of instruments played.

x	f	$x \times f$
1	11	11
2	10	20
3	5	15
4	3	12
5	1	5
	$\sum f = 30$	$\sum xf = 63$

$$\bar{x} = \frac{\sum xf}{\sum f}$$
$$= \frac{63}{30}$$
$$= 2.1$$

Note that the mean is not necessarily an integer, even if the data set consists only of integers.

Total number of people

Total number of instruments played

Median

The **median** is not influenced by extreme values, so it is a good average to use when there are one or two outliers in the data.

For a set of n numbers arranged in ascending order:

▶ when n is odd, the median is the middle value

▶ when n is even, the median is the mean of the two middle values.

This is summarised by saying that the median is the $\frac{1}{2}(n+1)^{\text{th}}$ value.

Example 2

Question

Find the median of these sets of numbers:

a 7, 7, 2, 3, 4, 2, 7, 9, 31   **b** 36, 41, 27, 32, 29, 39, 39, 43

Answer

Write the values in numerical order.

a 2   2   3   4 $\boxed{7}$ 7   7   9   31

Since there are 9 values, the median is the $\frac{1}{2}(9+1)^{\text{th}} = 5^{\text{th}}$ value, so median $= 7$

> The median is not influenced by the extreme value of 31.

b 27   29   32 $\boxed{36 \quad 39}$ 39   41   43

Since there are 8 values, the median is the $\frac{1}{2}(8+1)^{\text{th}} = 4.5^{\text{th}}$ value. This is the mean of the 4th and 5th values,

so median $= \dfrac{36+39}{2} = 37.5$

Measures of spread

Another way of summarising data is to give an idea of the variability, or spread. The three main **measures of spread** are the range, interquartile range and standard deviation.

Range

The **range** gives a quick snapshot of the overall spread of the data and is based entirely on the extreme values of the distribution.

Range = highest value − lowest value

Interquartile range

The **interquartile range** is based on the upper and lower quartiles. These are the values that, together with the median, split a distribution into four equal parts.

For data arranged in order,

▶ the lower quartile is the median of all the values *before* the median

▶ the upper quartile is the median of all the values *after* the median.

The difference between the quartiles gives the interquartile range. It tells you the range of the middle 50% of the data. It is unaffected by one or two extreme values, so it is a useful measure of spread when there are outliers.

Interquartile range = upper quartile − lower quartile

Discrete Random Variables

In the set of numbers below, the median is 35.

The lower quartile is 26 and the upper quartile is 42.

23 25 25 $\boxed{26}$ 29 30 32 $\boxed{35}$ 36 38 40 $\boxed{42}$ 43 45 86

         ↑   ↑   ↑

  lower quartile   median   upper quartile

So, interquartile range $= 42 - 26 = 16$.

Standard deviation and variance

Standard deviation gives a measure of the spread of the data *about the mean* and it is particularly important in later work. However, as with the mean, a disadvantage of using the standard deviation is that it can be unduly influenced by outliers.

When comparing distributions, the lower the standard deviation, the less variation there is and the more consistent the data are. Also, as a general rule, in most distributions the bulk of the observations lie within two standard deviations of the mean.

The standard deviation of a set of data is calculated using all the observations, as follows:

1 For each value x, calculate the deviation from the mean by finding $(x - \bar{x})$.

2 Square the deviation to give $(x - \bar{x})^2$.

3 Find the sum of these squares, written $\sum(x - \bar{x})^2$.

4 Divide by n, the number of observations, to give the variance $\dfrac{\sum(x - \bar{x})^2}{n}$.

5 Take the *positive* square root to give the standard deviation $\sqrt{\dfrac{\sum(x - \bar{x})^2}{n}}$.

Standard deviation is never negative.

This 'definition version' can be difficult to use, especially when $\bar{x}$ is not an integer, so it is more usual to use an alternative form. This is known as the 'calculation version'. The two versions are shown below.

The standard deviation of a set of n numbers with mean $\bar{x}$ is given by

Definition version

$$\text{s.d.} = \sqrt{\frac{\sum(x - \bar{x})^2}{n}}$$

Calculation version

$$\text{s.d.} = \sqrt{\frac{\sum x^2}{n} - \bar{x}^2} \quad \text{where } \bar{x} = \frac{\sum x}{n}$$

The variance is the square of the standard deviation, where

$$\text{variance} = \frac{\sum(x - \bar{x})^2}{n} \qquad \text{variance} = \frac{\sum x^2}{n} - \bar{x}^2 \quad \longleftarrow$$

The calculation version can be memorised as "the mean of the squares minus the square of the mean".

Remember

    variance $= (\text{standard deviation})^2$

    standard deviation $= \sqrt{\text{variance}}$

Example 3

The mean of the numbers 2, 3, 5, 6, 8 is 4.8. Calculate the standard deviation.

Method 1: Definition version		
x	$x - 4.8$	$(x - 4.8)^2$
2	−2.8	7.84
3	−1.8	3.24
5	0.2	0.04
6	1.2	1.44
8	3.2	10.24
		$\sum(x - \bar{x})^2 = 22.8$

Method 2: Calculation version	
x	x^2
2	4
3	9
5	25
6	36
8	64
	$\sum x^2 = 138$

$$\text{s.d.} = \sqrt{\frac{\sum(x - \bar{x})^2}{n}}$$

$$= \sqrt{\frac{22.8}{5}}$$

$$= 2.14 \ (3 \text{ sf})$$

$$\text{s.d.} = \sqrt{\frac{\sum x^2}{n} - \bar{x}^2}$$

$$= \sqrt{\frac{138}{5} - 4.8^2}$$

$$= 2.14 \ (3 \text{ sf})$$

Sometimes you have to work with **summary data**, rather than individual values.

Example 4

Find the standard deviation when

a $n = 50$ and $\sum(x - \bar{x})^2 = 34\,578$

b $n = 9$, $\sum x = 54$, $\sum x^2 = 438$

a $\text{s.d.} = \sqrt{\dfrac{\sum(x - \bar{x})^2}{n}} = \sqrt{\dfrac{34578}{50}} = 26.29\ldots = 26.3 \ (3 \text{ sf})$

b   You need to find the mean first.

$$\bar{x} = \frac{\sum x}{n} = \frac{54}{9} = 6$$

$$\text{s.d.} = \sqrt{\frac{\sum x^2}{n} - \bar{x}^2} = \sqrt{\frac{438}{9} - 6^2} = 3.559\ldots = 3.56 \ (3 \text{ sf})$$

When data are in a *frequency distribution*, the formula for the standard deviation is as follows:

Definition version

$$\text{s.d.} = \sqrt{\frac{\sum(x - \bar{x})^2 f}{\sum f}}$$

Calculation version

$$\text{s.d.} = \sqrt{\frac{\sum x^2 f}{\sum f} - \bar{x}^2} \qquad \text{where } \bar{x} = \frac{\sum xf}{\sum f}$$

Example 5

Question

The mean of this distribution is 2.9. Find the standard deviation.

x	1	2	3	4	5
f	3	4	8	2	3

Answer

x	x^2	f	$x^2 \times f$
1	1	3	3
2	4	4	16
3	9	8	72
4	16	2	32
5	25	3	75
		$\Sigma f = 20$	$\Sigma(x^2 f) = 198$

$$\text{s.d.} = \sqrt{\frac{\Sigma x^2 f}{\Sigma f} - \bar{x}^2}$$

$$= \sqrt{\frac{198}{20} - 2.9^2}$$

$$= \sqrt{1.49}$$

$$= 1.22 \text{ (3 sf)}$$

Do **not** add the x column
or the x^2 column.

Using a calculator in statistical mode

Although you can use your calculator in computation mode to find the totals and divide, you will find it very useful to be able to find the mean and standard deviation *directly* using the **statistical mode**.

An outline is shown below, but you may need to consult your calculator manual for the details.

▶ Set the statistical mode, sometimes written SD or STAT.
▶ Clear the statistical registers (memories).
▶ Input the data, either as individual values, or in a frequency table **in the order x then f**.

You should then have access to the following:
$\boxed{\bar{x}}$ (the mean), $\boxed{{}_x\sigma_n}$ (standard deviation)

If required, you can also check:
$\boxed{n}$ (the number of values), $\boxed{\Sigma x}$ (the sum of the values), $\boxed{\Sigma x^2}$ (the sum of the squares)
For data in a frequency table

$\quad \Sigma f$ is given by $\boxed{n}$, Σxf is given by $\boxed{\Sigma x}$ and $\Sigma x^2 f$ is given by $\boxed{\Sigma x^2}$

Note that the calculator key labelled $\boxed{{}_x\sigma_{n-1}}$ gives the value of s, the *unbiased estimate* of a *population standard deviation* based on a *sample*. This is not required until module S2.

Example 6

Question

Sweets are packed into bags with a nominal mass of 75 grams. Ten bags are picked at random from the production line and weighed. Their masses, in grams, are

$\quad\quad$ 76.0, 74.2, 75.1, 73.7, 72.0, 74.3, 75.4, 74.0, 73.1, 72.8

(continued)

(continued)

Question

a Use your calculator to find the mean and the standard deviation.

b It is later discovered that the scales were reading 3.2 grams below the correct mass.

   i What was the correct mean mass of the 10 bags?

   ii What was the correct standard deviation of the 10 bags?

Answer

a Using a calculator,

mean = 74.06 grams,

s.d. = 1.166... = 1.17 grams (3 sf)

b The correct readings are

        79.2, 77.4, 78.3, 76.9, 75.2, 77.5, 78.6, 77.2, 76.3, 76.0

Again, using a calculator,

   i Correct mean = 77.26 grams

   ii Correct s.d. = 1.166... = 1.17 grams (3 sf)

Notice that when each mass is increased by 3.2 grams,

▶ the mean mass is increased by 3.2 grams

▶ the standard deviation is unaltered.

When the two sets of data are shown on the same diagram, it is easy to see why the standard deviation remains the same, since the spread of the original data about the original mean is the same as the spread of the new data about the new mean.

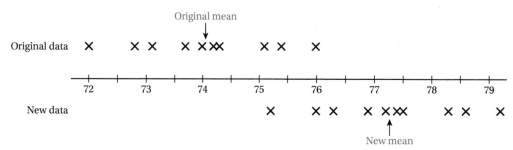

If each data value is increased by a constant a

▶   the mean is increased by a

▶   the standard deviation is <u>unaltered</u>.

Exercise 1

You may use a calculator in statistical mode unless instructed otherwise.

1 These are the test marks of 11 students.

        52, 61, 78, 49, 47, 79, 54, 58, 62, 73, 72

Find

a the median   **b** the lower quartile

c the upper quartile   **d** the interquartile range.

2 For each of the following sets of numbers, find

   a the range   **b** the median   **c** the interquartile range.

    i 192, 217, 189, 210, 214, 204

    ii 1267, 1896, 895, 3457, 2164, 2347, 2347, 2045

3 In an experiment to estimate the value of π, Jon measured the circumference and diameter of several tins. He then divided the circumference by the diameter for each tin. His results are recorded below.

$$3.05, 3.45, 3.19, 2.98, 2.85, 3.04, 3.28, 3.45, 4.87, 3.05$$

   a State the mode.

   b Find the median.

   c Calculate the mean.

   d What percentage of Jon's results were higher than the true value of π?

4 **a** For each of the following sets of numbers, calculate the mean and the standard deviation.

    i Set A: 2, 4, 5, 6, 8   **ii** Set B: 6, 8, 9, 11   **iii** Set C: 11, 14, 17, 23, 29

   b It was later discovered that all the numbers in Set C had been overstated by 2. State the correct mean and standard deviation for this set of numbers.

5 Find the mean and standard deviation in this frequency distribution.

Do not use the statistical mode on your calculator and show all your working.

x	10	11	12	13
f	3	7	11	2

6 The score for a round of golf for each of 50 club members was recorded and the results are summarised in the frequency table.

Score, x	66	67	68	69	70	71	72	73
Frequency, f	2	5	10	12	9	6	4	2

   a State the mode.

   b Calculate the mean score.

   c Calculate the standard deviation.

7 Thida plays a computer game where she fires at a target. Her score is 1 if she hits the target and 0 if she misses it. She has 30 attempts and hits the target 18 times.

   a Find her mean score for the 30 attempts.

   b Find the variance of her scores for the 30 attempts.

8 Thirty-one people completed a jigsaw in the following times (x minutes).

    11 53 72 48 48 49 39 87 73 23 120 24 61 36 66 67

    86 79 65 47 36 133 78 81 70 75 53 42 42 72 144

   a Calculate   **i** the mean   **ii** the standard deviation.

   b The following rule can be used to identify unusually high or low values, called outliers.

    'An outlier is more than 2 standard deviations away from the mean'.

    Use this rule to identify any outliers.

9 Cartons of orange juice are advertised as containing 1 litre.

The actual volume of orange juice in a carton is denoted by x litres. A sample of 100 cartons gave the following results:

$$\sum x = 101.4 \text{ and } \sum x^2 = 102.83$$

Find the mean and standard deviation of the volume of orange juice in the cartons in the sample.

10 For a set of 10 numbers, $\sum x = 290$ and $\sum x^2 = 8469$.

Find the mean and the standard deviation.

11 For a particular set of data, $n = 100$, $\sum x = 584$ and $\sum x^2 = 23\,781$.

Find the mean and the variance.

12 For a set of 9 numbers, $\sum(x - \overline{x})^2 = 234$, where $\overline{x}$ is the mean.

Find the standard deviation of the numbers.

13 For a set of 12 numbers with mean $\overline{x}$, it is given that $\sum(x - \overline{x})^2 = 60$.

   a Find the standard deviation.

   b If $\sum x^2 = 285$, find $\overline{x}$.

14 A machine cuts lengths of wood. A sample of 20 rods gave the following results for the length, x cm.

$$\sum xf = 997 \quad \sum x^2 f = 49\,711$$

   a Find the mean length of the 20 rods.

   b Find the variance of the lengths of the 20 rods.

15 For a particular set of observations,

$$\sum f = 20, \sum x^2 f = 16\,143, \sum xf = 563.$$

Calculate the standard deviation.

16 The marks of 25 students in a test had a mean of 74 and a standard deviation of 8.

   a Find the total of the marks, $\sum x$.

   b Show that $\sum x^2 = 138\,500$.

It was later discovered that a mark of 86 had been entered incorrectly as 68.

   c Calculate the mean and standard deviation of the corrected set of marks.

17 Applicants for a job were asked to carry out a task to assess their practical skills.

The times, in seconds, taken by 19 applicants were as follows:

  61, 229, 164, 76, 74, 49, 67, 86, 70, 82, 48, 74, 61, 59, 72, 81, 102, 61, 73

   a Find the mode, the median and the mean.

   b Find the range, interquartile range and standard deviation.

   c State, with a reason, which of the measures you have calculated you consider most appropriate

      i as a measure of central tendency

      ii as a measure of variability.

18 The times, in seconds, taken by 20 people to solve a simple numerical puzzle were

$$17 \quad 19 \quad 22 \quad 26 \quad 28 \quad 31 \quad 34 \quad 36 \quad 38 \quad 39$$

$$41 \quad 42 \quad 43 \quad 47 \quad 50 \quad 51 \quad 53 \quad 55 \quad 57 \quad 58$$

a Calculate the mean and standard deviation of these times.

b In fact, 23 people solved the puzzle. However, 3 of them failed to solve it within the allotted time of 60 seconds. Calculate the median and the interquartile range of the times taken by all 23 people.

c For the times taken by all 23 people, explain why:

    i the mode is **not** an appropriate measure;

    ii the range is not an appropriate numerical measure.

AQA MS1A January 2007

14.2 Discrete random variables

A **discrete random variable** is a discrete variable whose value is determined by chance. It can take individual values, each with a given probability. The values of the variable are usually the outcome of an experiment.

These are some examples of discrete random variables:

	Possible values
The score when you throw an ordinary fair cubical die.	1, 2, 3, 4, 5, 6
The number of heads when you toss a fair coin 3 times.	0, 1, 2, 3
Your profit in dollars when you play a game with an entry fee of $1 and prizes of $5 and $10.	−1, 4, 9
The number of times you toss a coin until a tail occurs.	1, 2, 3, 4, … to infinity

Notation

Consider throwing a fair cubical die. For convenience, an upper case letter, X say, is used as shorthand notation for 'the score on the die'. The values that X can take (the possible scores) are 1, 2, 3, 4, 5, 6.

The probability that X takes the value 4, for example, is written $P(X = 4)$. Since the die is fair, each number is equally likely to occur, so $P(X = 4) = \dfrac{1}{6}$.

In general:

▶ Random variables are denoted by upper case letters, such as X, Y, R, …

▶ Particular values of the variable are denoted by lower case letters, such as x, y, r, …

▶ The probability that the variable X takes a particular value x is written $P(X = x)$.

▶ Alternatively, the x values may be denoted by $x_1, x_2, x_3, \dots, x_n$, and their probabilities summarised by writing p_i where $i = 1, 2, 3, \dots, n$, such that $p_1 = P(X = x_1)$, $p_2 = P(X = x_2)$, and so on.

Probability distribution

A list of all possible values of the discrete random variable X, together with their associated probabilities, is called a **probability distribution**. It is often helpful to show the probability distribution in a table.

The probability distribution of X, the score on a fair cubical die, is shown below and illustrated by a **vertical line graph**.

x	1	2	3	4	5	6
$P(X=x)$	$\frac{1}{6}$	$\frac{1}{6}$	$\frac{1}{6}$	$\frac{1}{6}$	$\frac{1}{6}$	$\frac{1}{6}$

This probability distribution provides a model for the scores obtained when the die is thrown a number of times.

Suppose the die is thrown 120 times. The expected frequency for each score is $120 \times \frac{1}{6} = 20$.

When the experiment was performed 120 times the number of times each score occurred was noted. These observed frequencies, together with the expected frequencies according to the probability distribution, are shown below.

Score x	1	2	3	4	5	6
Observed frequency	15	22	23	19	23	18
Expected frequency	20	20	20	20	20	20

Notice that the observed frequencies are close to the expected frequencies, which is what you would expect if the die is fair.

Sum of probabilities

Consider the discrete random variable X with the following probability distribution.

x	x_1	x_2	x_3	...	x_n
$P(X=x)$	p_1	p_2	p_3	...	p_n

For a random variable, the sum of all the possible probabilities is:

$$p_1 + p_2 + p_3 + \ldots + p_n = 1$$

so $$\sum p_i = 1 \quad \text{for } i = 1, 2, 3, \ldots, n$$

This can also be written

$$\sum_{\text{all } x} P(X = x) = 1$$

The subscripts are often omitted and this is written $\sum p = 1$.

Example 7

Zhu plays a game with a biased five-sided spinner marked with the numbers 1, 2, 3, 4 and 5.

When she spins the spinner, her score, X, is the number on which the spinner lands. The probability distribution of X is shown in the table.

x	1	2	3	4	5
$P(X=x)$	0.15	0.24	a	0.25	0.19

a Find the value of a.

b Find the probability that the score is at least 4.

c Find the probability that the score is less than 5.

d Find $P(2 < X \leq 4)$.

e Write down the most likely score, that is the mode.

a Since $\sum P(X=x) = 1$,

$0.15 + 0.24 + a + 0.25 + 0.19 = 1$

$a + 0.83 = 1$

$a = 0.17$

b $P(X \text{ is at least } 4) = P(X \geq 4)$

$= P(X=4) + P(X=5)$

$= 0.25 + 0.19$

$= 0.44$

c $P(X \text{ is less than } 5) = P(X < 5)$

$= 1 - P(X=5)$

$= 1 - 0.19$

$= 0.81$

Alternatively:

$P(X < 5) = P(X=1) + P(X=2) + P(X=3) + P(X=4)$

$= 0.15 + 0.24 + 0.17 + 0.25$

$= 0.81$

d $P(2 < X \leq 4) = P(X=3) + P(X=4)$

$= 0.17 + 0.25$

$= 0.42$

$P(2 < X \leq 4)$ means that X is greater than 2 but at most 4, so X can take the values 3 or 4.

e The most likely score is the value of X with the greatest probability, so the mode is 4.

Example 8

Jeremias takes out the two used batteries in his torch to replace them with new ones. Unfortunately, he mixes them up with three new batteries. All five batteries are identical in appearance.

Jeremias selects two of the batteries at random. Draw up a probability distribution table for X, the number of **new** batteries that Jeremias selects.

X is the number of <u>new</u> batteries that Jeremias selects. He could select no new batteries, 1 new battery or 2 new batteries, so X can take the values 0, 1, and 2.

To find the corresponding probabilities, it is helpful to draw a tree diagram.

Let N be the event 'a new battery is selected' and U the event 'a used battery is selected'.

The outcomes and probabilities are shown in the tree diagram:

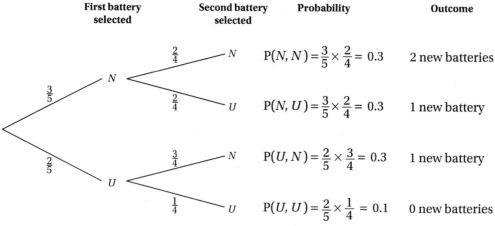

First battery selected	Second battery selected	Probability	Outcome
N (prob $\frac{3}{5}$)	N (prob $\frac{2}{4}$)	$P(N, N) = \frac{3}{5} \times \frac{2}{4} = 0.3$	2 new batteries
	U (prob $\frac{2}{4}$)	$P(N, U) = \frac{3}{5} \times \frac{2}{4} = 0.3$	1 new battery
U (prob $\frac{2}{5}$)	N (prob $\frac{3}{4}$)	$P(U, N) = \frac{2}{5} \times \frac{3}{4} = 0.3$	1 new battery
	U (prob $\frac{1}{4}$)	$P(U, U) = \frac{2}{5} \times \frac{1}{4} = 0.1$	0 new batteries

From the tree diagram,

$P(X = 0) = P(\text{no new batteries}) = 0.1$

$P(X = 1) = P(1 \text{ new battery}) = 0.3 + 0.3 = 0.6$

$P(X = 2) = P(2 \text{ new batteries}) = 0.3$

Probability distribution table:

x	0	1	2
$P(X = x)$	0.1	0.6	0.3

Probability functions

Sometimes the probability distribution of X can be defined in terms of x by a **probability function**.

Example 9

The discrete random variable R has probability function

$P(R = r) = kr$, for $r = 5, 10, 15, 20, 25$.

Find the value of k.

The probability distribution has been given as a formula.

Substitute the values of r to find the probabilities in terms of k and form a probability distribution table.

$P(R = r) = kr$

$P(R = 5) = k \times 5 = 5k$, $P(R = 10) = k \times 10 = 10k$, and so on.

Probability distribution table for R:

r	5	10	15	20	25
$P(R = r)$	$5k$	$10k$	$15k$	$20k$	$25k$

Now use the fact that the sum of the probabilities is 1.

(continued)

(continued)

Since $\sum P(R=r)=1$,

$$1 = 5k + 10k + 15k + 20k + 25k$$

$$= 75k$$

$$k = \frac{1}{75}$$

Example 10

Paavo throws two fair tetrahedral dice, each with faces labelled 1, 2, 3, 4. The random variable X, the sum of the numbers on which the dice land, has probability distribution given by

$$P(X=x) = \begin{cases} \dfrac{x-1}{16} & x=2,3,4,5 \\[2mm] \dfrac{9-x}{16} & x=6,7,8 \end{cases}$$

a   Write out the probability distribution table.

b   Find the probability that the score is at least 4.

c   Given that the score is at least 4, find the probability that it is less than 7.

a   Substitute the x-values into the appropriate formula to get the probabilities.

Probability distribution of X

x	2	3	4	5	6	7	8
$P(X=x)$	$\dfrac{1}{16}$	$\dfrac{2}{16}$	$\dfrac{3}{16}$	$\dfrac{4}{16}$	$\dfrac{3}{16}$	$\dfrac{2}{16}$	$\dfrac{1}{16}$

Check the sum of the probabilities:

$$\sum P(X=x) = \frac{1}{16}(1+2+3+4+3+2+1) = 1$$

b $P(X \text{ is at least } 4) = P(X \geq 4)$

$$= 1 - P(X < 4)$$

$$= 1 - \left(\frac{1}{16} + \frac{2}{16}\right)$$

$$= \frac{13}{16}$$

You could add the probabilities for $X = 4, 5, 6, 7, 8$.

c $P(\text{score is less than 7, given it is at least 4})$

$$= \frac{P(X \text{ is 4, 5 or 6})}{P(X \geq 4)} = \frac{\frac{10}{16}}{\frac{13}{16}} = \frac{10}{13}$$

Recall: Conditional probability section 13.3.

Exercise 2

1 The discrete random variable X has the probability distribution shown in the table.

x	1	2	3	4	5
$P(X=x)$	0.2	0.25	0.4	c	0.05

Find

    a   the value of c **b** $P(1 \le X \le 3)$

    c   the probability that X is at least 3   **d** $P(2 < X < 5)$

    e   the probability that X is at most 2   **f** $P(X$ is greater than the mode of X).

2 The discrete random variable Y has the following probability distribution.

y	10	15	20	25	30
$P(Y=y)$	a	$5a$	$\dfrac{7}{16}$	$\dfrac{1}{32}$	$\dfrac{1}{32}$

    a Find the value of a.

    b Find $P(Y > 18)$.

    c Find $P(Y = 25 \mid Y > 15)$.

3 The discrete random variable R has the following probability distribution function, where k is a constant:

$$P(R=r) = k(r-10) \quad \text{for} \quad r = 12, 13, 14$$

    a Draw up a probability distribution table for R in terms of k.

    b Find the value of k.

    c Find the probability that R is an even number.

4 The discrete random variable X can take the values 0, 1, 2, 3 and 4 only and

$$P(X=x) = k(x^2 + 2)$$

    a Find the value of the constant k.

    b Find $P(1 < X < 4)$.

5 The discrete random variable X has probability function

$$P(X=x) = \frac{x+1}{k} \quad \text{for } x = 0, 1, 3, 4.$$

    a Find the value of the constant k.

    b Find $P(X \le 1 \mid X > 0)$.

6 A discrete random variable X has probability function

$$P(X=x) = \begin{cases} 0.1 & x = -1, 0, 5 \\ a & x = 1, 3 \\ 0.3 & x = 4 \end{cases}$$

    a Write out the probability distribution table in terms of a.

    b Find a.

    c Find $P(X \ge 3)$.

7 A drawer contains 8 brown socks and 4 blue socks. Liam takes two socks at random from the drawer, one after the other.

a Show that the probability that Liam takes one brown sock and one blue sock is $\dfrac{16}{33}$.

b The discrete random variable B is the number of *brown* socks taken. Draw up a probability distribution table for B.

8 Elena and Toimi are playing a game in which they try to throw balls into a bucket. The probability that the ball lands in the bucket is 0.4 for each attempt.

a Elena has two attempts.

   i Show that the probability that exactly one ball lands in the bucket is 0.48.

   ii Draw up a probability distribution table for X, the number of balls that land in the bucket.

b Toimi has three attempts.

   i By drawing a tree diagram, or otherwise, show that the probability that exactly one ball lands in the bucket is 0.432.

   ii Draw up a probability distribution table for Y, the number of balls that land in the bucket.

   iii Toimi wins a prize if at least two balls land in the bucket. What is the probability that he wins a prize?

14.3 E(X), the expectation of X

The **expectation**, or **expected value**, of a random variable X is the result that you would *expect* to get if you took a very large number of values of X and found their mean.

The expectation of X is written E(X) and is denoted by the symbol μ. It is also called the **expected mean**, or just the **mean** of X.

> **Note**
>
> E(X) is read as "E of X" and μ is the Greek letter mu, read as "mew".

Practical approach

Consider again the observed frequency distribution showing the scores on the fair cubical die when it was thrown 120 times.

Score x	1	2	3	4	5	6	
Frequency f	15	22	23	19	23	18	Total $N = 120$

The mean score is $\bar{x}$, where

$$\bar{x} = \frac{\sum x_i f_i}{N} = \frac{1\times15+2\times22+3\times23+4\times19+5\times23+6\times18}{120} = 3.5583...$$

You could write the formula in a different way as follows:

$$\bar{x} = \sum\left(x_i \times \frac{f_i}{N}\right)$$

$$= x_1 \times \frac{f_1}{N} + x_2 \times \frac{f_2}{N} + ... + x_n \times \frac{f_n}{N}$$

$$= 1\times\frac{15}{120} + 2\times\frac{22}{120} + ... + 6\times\frac{18}{120}$$

$$= 3.5583...$$

The fractions $\dfrac{15}{120}, \dfrac{22}{120}, \ldots \dfrac{18}{120}$ are the relative frequencies of the scores of 1, 2, ... , 6. Notice that they are quite close to $\dfrac{20}{120} = \dfrac{1}{6}$.

If you throw the die a *large number of times*, each of the fractions $\dfrac{f_i}{N}$ should be *very* close to $\dfrac{1}{6}$, the limiting value of the relative frequency of a particular score on the die.

In general, as N becomes very large,

▶ the relative frequencies $\dfrac{f_i}{N}$ tend to the probabilities p_i

▶ $\bar{x} = \dfrac{\Sigma x_i f_i}{N} = \Sigma\left(x_i \times \dfrac{f_i}{N}\right)$ tends to the expected mean $\mu = \mathrm{E}(X) = \Sigma x_i p_i$

Theoretical approach

The probability distribution for X, the score on the fair cubical die, is shown below.

x	1	2	3	4	5	6
$P(X=x)$	$\dfrac{1}{6}$	$\dfrac{1}{6}$	$\dfrac{1}{6}$	$\dfrac{1}{6}$	$\dfrac{1}{6}$	$\dfrac{1}{6}$

The expectation, or expected mean, is obtained by multiplying each score by its probability,

so $\quad \mu = \mathrm{E}(X) = 1 \times \dfrac{1}{6} + 2 \times \dfrac{1}{6} + 3 \times \dfrac{1}{6} + 4 \times \dfrac{1}{6} + 5 \times \dfrac{1}{6} + 6 \times \dfrac{1}{6} = 3.5$

The expectation can be thought of as the average value when the number of experiments increases indefinitely.

Calculating E(X)

If the discrete random variable X has the following probability distribution:

x	x_1	x_2	x_3	...	x_n
$P(X=x)$	p_1	p_2	p_3	...	p_n

to calculate $\mu = \mathrm{E}(X)$:

▶ multiply each value x_i by its corresponding probability p_i
▶ add these products together.

$$\mu = \mathrm{E}(X) = x_1 p_1 + x_2 p_2 + x_3 p_3 + \ldots + x_n p_n$$
$$= \Sigma x_i p_i \quad \text{for} \quad i = 1, 2, 3, \ldots, n$$

This may also be written

$$\mu = \mathrm{E}(X) = \Sigma x\, P(X=x)$$

In general,

▶ a *practical* approach results in a *frequency distribution* and an *experimental mean* $\bar{x}$

▶ a *theoretical* approach uses a *probability distribution* and results in an *expected mean* μ.

Example 11

A random variable X has mean μ and probability distribution as shown.

x	-2	-1	0	1	2
$P(X=x)$	0.3	0.1	0.15	0.4	0.05

a Find the value of μ.

b Find $P(X < \mu)$.

a $\mu = E(X) = \sum x P(X=x)$

$\quad = (-2) \times 0.3 + (-1) \times 0.1 + 0 \times 0.15 + 1 \times 0.4 + 2 \times 0.05$

$\quad = -0.2$

b $P(X < \mu) = P(X < -0.2)$

$\quad = P(X=-1) + P(X=-2)$

$\quad = 0.3 + 0.1$

$\quad = 0.4$

Example 12

The discrete random variable X has the probability distribution shown in the table and illustrated in the vertical line graph. Find the mean of X.

x	1	2	3	4	5
$P(X=x)$	0.1	0.2	0.4	0.2	0.1

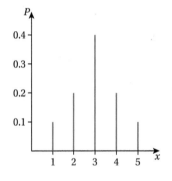

$\mu = E(X) = \sum x P(X=x)$

$\quad = 1 \times 0.1 + 2 \times 0.2 + 3 \times 0.4 + 4 \times 0.2 + 5 \times 0.1$

$\quad = 3$

Note about symmetry

In Example 12 above, the distribution is symmetrical about $x = 3$. If you spot this, you can write down straight away that $\mu = 3$.

Another example is the distribution of the score, X, on a fair die, described on page 231.

By symmetry, $\mu = 3.5$, confirming the calculation for $E(X)$ on page 238.

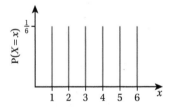

Example 13

The discrete random variable X has the following probability distribution.

x	2	3	4
$P(X=x)$	a	0.4	b

a Write down an equation relating a and b.

b Given that the mean of X is 3.1, find the values of a and b.

a Use the fact that the sum of the probabilities is 1.

$$\sum P(X=x) = 1$$

So $a + 0.4 + b = 1$

$$a + b = 0.6 \qquad\qquad (1)$$

b Use the fact that the mean is 3.1 to form a second equation in a and b.

$$\mu = \sum xP(X=x)$$
$$3.1 = 2 \times a + 3 \times 0.4 + 4 \times b$$
$$3.1 = 2a + 1.2 + 4b$$
$$2a + 4b = 1.9 \qquad\qquad (2)$$

Now solve the simultaneous equations.

$$a + b = 0.6 \qquad\qquad (1)$$
$$2a + 4b = 1.9 \qquad\qquad (2)$$
$$(1) \times 2 \quad 2a + 2b = 1.2 \qquad\qquad (3)$$
$$(2) - (3) \quad\quad 2b = 0.7$$
$$b = 0.35$$

Substitute in (1) $a + 0.35 = 0.6$

$$a = 0.25$$

So, $a = 0.25$, $b = 0.35$.

E(g(X)), the expectation of g(X)

The definition of expectation (expected value or mean) can be extended to *any function* of X, such as $10X$, X^2, $\dfrac{1}{X}$, $X - 4$.

> If $g(X)$ is any function of the discrete random variable X, then
>
> $E(g(X)) = \sum g(x_i)p_i$

For example

$$E(X-3) = \sum(x-3)P(X=x)$$

$$E(X^2) = \sum x^2 P(X=x) \qquad \longleftarrow \text{Notice that the } x \text{ values are squared but the probabilities are not.}$$

$$E(X^{-2}) = E\left(\frac{1}{X^2}\right) = \sum \frac{1}{x^2}P(X=x)$$

Example 14

The table below shows the probability distribution of the discrete random variable X.

x	1	2	3	4
$P(X=x)$	0.05	0.15	0.6	0.2

Calculate   **a** $E(X)$ **b** $E\left(\dfrac{1}{X}\right)$ **c** $E(X-3)$

a $\quad E(X) = \sum x P(X=x)$

$\qquad = 1 \times 0.05 + 2 \times 0.15 + 3 \times 0.6 + 4 \times 0.2$

$\qquad = 2.95$

b $\quad E\left(\dfrac{1}{X}\right) = \sum \dfrac{1}{x} P(X=x)$

$\qquad = \dfrac{1}{1} \times 0.05 + \dfrac{1}{2} \times 0.15 + \dfrac{1}{3} \times 0.6 + \dfrac{1}{4} \times 0.2$

$\qquad = 0.375$

$E\left(\dfrac{1}{X}\right)$ is **not** equal to $\dfrac{1}{E(X)}$

c $\quad E(X-3) = \sum (x-3) P(X=x)$

$\qquad = (-2) \times 0.05 + (-1) \times 0.15 + 0 \times 0.6 + 1 \times 0.2$

$\qquad = -0.05$

Example 15

A cube has side of length X, where X has probability distribution as shown.

x	1	2	3
$P(X=x)$	0.3	0.5	0.2

a Find the mean of S, the surface area of the cube.

b Find the mean of V, the volume of the cube.

a If S is the surface area of the cube, then $S = 6X^2$.

x	1	2	3
$6x^2$	6	24	54
$P(X=x)$	0.3	0.5	0.2

$E(S) = E(6X^2)$

$\qquad = \sum 6x^2 P(X=x)$

$\qquad = 6 \times 0.3 + 24 \times 0.5 + 54 \times 0.2$

$\qquad = 24.6$

b If V is the volume of the cube, then $V = X^3$.

x	1	2	3
x^3	1	8	27
$P(X=x)$	0.3	0.5	0.2

$E(V) = \sum x^3 P(X=x)$

$\qquad = 1 \times 0.3 + 8 \times 0.5 + 27 \times 0.2$

$\qquad = 9.7$

Mean of a simple function of X

Example 16

The random variable X has the following probability distribution:

x	1	2	3
$P(X=x)$	0.1	0.6	0.3

Find   a $E(3)$ b $E(X)$ c $E(5X)$ d $E(5X+3)$

a $E(3) = \sum 3P(X=x)$

$= 3\sum P(X=x)$

$= 3 \times 1$

$= 3$

b $E(X) = \sum x\, P(X=x)$

$= 1 \times 0.1 + 2 \times 0.6 + 3 \times 0.3$

$= 2.2$

c $E(5X) = \sum 5x\, P(X=x)$

$= 5\sum x\, P(X=x)$

$= 5E(X)$

$= 11$

d $E(5X+3) = \sum(5x+3)P(X=x)$

$= \sum 5xP(X=x) + \sum 3P(X=x)$

$= 5E(X) + 3$

$= 14$

In general, for random variable X and constants a and b,

$E(a) = a$ The expected value of a constant is the constant itself.

$E(aX+b) = aE(X)+b$

$E(aX-b) = aE(X)-b$

For example

$E(2X+3) = 2E(X)+3$

$E(4-3X) = 4-3E(X)$

The rule also applies to functions such as

$E(5X^2) = 5E(X^2)$

$E((X+4)^2) = E(X^2+8X+16) = E(X^2) + 8E(X) + 16$

Take special care with the following

$E(10X^{-1}) = E\left(\dfrac{10}{X}\right) = 10\,E\left(\dfrac{1}{X}\right)$ $E\left(\dfrac{10}{X}\right)$ is **not** equal to $\dfrac{10}{E(X)}$

$E(100X^{-2}) = E\left(\dfrac{100}{X^2}\right) = 100\,E\left(\dfrac{1}{X^2}\right)$ $E\left(\dfrac{100}{X^2}\right)$ is **not** equal to $\left[E\left(\dfrac{10}{X}\right)\right]^2$

Example 17

Question

A bag contains some green and yellow counters. In a game, a player takes out two counters. The probability distribution of Y, the number of yellow counters taken out, is shown in the table.

y	0	1	2
$P(Y=y)$	$\dfrac{2}{15}$	$\dfrac{8}{15}$	$\dfrac{1}{3}$

The player's score is calculated by multiplying the number of yellow counters taken out of the bag by 10, then adding 4.

a Find $E(Y)$.

b Find the player's expected score when two counters are taken from the bag.

Answer

a $E(Y) = \sum y\, P(Y=y)$

$\qquad = 0 \times \dfrac{2}{15} + 1 \times \dfrac{8}{15} + 2 \times \dfrac{1}{3}$

$\qquad = 1.2$

b If S is the score, then $S = 10Y + 4$

$\quad E(S) = E(10Y + 4)$

$\qquad\quad = 10E(Y) + 4$

$\qquad\quad = 10 \times 1.2 + 4$

$\qquad\quad = 16$

The expected score is 16.

Example 18

Question

A random variable X has mean μ. Show that

$\quad E((X-\mu)^2) = E(X^2) - \mu^2$.

Answer

Expand the function

$E((X-\mu)^2) = E(X^2 - 2X\mu + \mu^2)$

$\qquad\qquad\;\; = E(X^2) - E(2\mu X) + \mu^2$

$\qquad\qquad\;\; = E(X^2) - 2\mu\, E(X) + \mu^2$

$\qquad\qquad\;\; = E(X^2) - 2\mu \times \mu + \mu^2 \quad\longleftarrow$ since $E(X) = \mu$.

$\qquad\qquad\;\; = E(X^2) - \mu^2$

Exercise 3

1 The probability distribution of the discrete random variable X is shown in the table below.

x	1	2	3	4	5
$P(X=x)$	0.1	0.3	0.3	0.2	0.1

Find **a** $E(X)$ **b** $P(X > E(X))$

2 The discrete random variable, X, has the probability function given by the following

$\quad P(X=x) = \dfrac{1}{20}x,$ for $x = 5, 7$ and 8.

Find

a $E(X)$ **b** $E(5X+1)$

3 The probability distribution of the discrete random variable Y is shown in the table below.

y	−2	0	1	6
$P(Y=y)$	0.1	0.3	0.4	0.2

Find   **a** $E(Y)$ **b** $E(4Y)$ **c** $E(2Y-1)$

4 A bag contains five green counters and six red counters. Two counters are taken at random from the bag, one at a time, and not replaced.
The number of red counters taken is denoted by X.

a Draw up a probability distribution table for X.

b Calculate the mean number of red counters taken.

5 The probability distribution of the discrete random variable X is given in the table.

x	10	20	30
$P(X=x)$	a	0.5	b

a Write down an equation satisfied by a and b.

b Given that $E(X) = 17$, find a second equation in a and b.

c Find the values of a and b.

d Find $E(50 - 2X)$.

6 A discrete random variable X can take the values 10 and 20 only.
The mean of X is 16.

Write out the probability distribution of X.

7 The discrete random variable X has the probability distribution given by the following, where c is a constant.

$$P(X=x)=\begin{cases} \left(\dfrac{1}{2}\right)^x & x=1,2,3,4,5 \\ c & x=6 \\ 0 & \text{otherwise} \end{cases}$$

a Find the value of c.

b Find the mode of X.

c Find the mean of X.

8 The random variable W has the probability distribution shown in the table.

w	1	4	9
$P(W=w)$	0.3	0.4	0.3

Find   **a** $E(W)$ **b** $E\left(\dfrac{1}{W}\right)$ **c** $E(W^2)$ **d** $E\left(\sqrt{W}\right)$

9 The discrete random variable X has the probability distribution

x	2	3	4
$P(X=x)$	$\frac{1}{4}$	$\frac{1}{2}$	$\frac{1}{4}$

a State the value of $E(X)$. **b** Find $E(X^2)$.

c Find $E(X^2+6X)$. **d** Find $E((X-3)^2)$.

10 The random variable X has mean 12. The random variable $aX+2$ has mean 5, where a is a constant.

Find the value of a.

11 The discrete random variable R has the following probability distribution:

r	1	2	4
$P(R=r)$	$\frac{1}{4}$	$\frac{1}{2}$	$\frac{1}{4}$

a Calculate exact values for $E(R)$ and $Var(R)$.

b **i** By tabulating the probability distribution for $X=\frac{1}{R^2}$, show that $E(X)=\frac{25}{64}$

ii Find the value of the mean of the **area** of the rectangle which has

sides of length $\frac{8}{R}$ and $\left(R+\frac{8}{R}\right)$.

AQA MS2B June 2005

14.4 The variance and standard deviation of X

The standard deviation of X is a measure of spread of X about the mean μ. It is denoted by σ and calculated by first finding σ^2, the variance of X.

Practical approach

Note

σ is a lower case Greek letter, read as "sigma".

For discrete data the variance of a frequency distribution with mean $\bar{x}$ is given by

$$\text{variance} = \frac{\Sigma(x_i-\bar{x})^2 f}{\Sigma f_i}$$

$$= \Sigma(x_i-\bar{x})^2 \times \frac{f_i}{N} \quad \text{writing } N \text{ for } \Sigma f_i$$

$$= (x_i-\bar{x})^2 \times \frac{f_1}{N}+(x_2-\bar{x})^2 \times \frac{f_2}{N}+...+(x_n-\bar{x})^2 \times \frac{f_n}{N}$$

Now, as n becomes very large,

▶ the relative frequencies $\frac{f_i}{N}$ tend to the probabilities p_i

▶ $\bar{x}$ tends to μ, the expected mean of X

▶ the variance tends to σ^2, the variance of X.

Theoretical approach

The **variance of X**, written **Var(X)**, is the expected value of $(X-\mu)^2$,

so $Var(X) = E((X-\mu)^2)$

However, this 'definition' version can be difficult to work with.

You saw in Example 18 that

$$E((X-\mu)^2) = E(X^2) - \mu^2$$

This leads to the 'calculation' version which is more generally used

$$Var(X) = E(X^2) - \mu^2$$

Summarising:

Definition version	Calculation version
$Var(X) = \sigma^2 = E((X-\mu)^2)$	$Var(X) = \sigma^2 = E(X^2) - \mu^2$ where $\mu = E(X)$
$= \sum(x_i - \mu)^2 p_i$	$= \sum x_i^2 p_i - \mu^2$
$= \sum(x-\mu)^2 P(X=x)$	$= \sum x^2 P(X=x) - \mu^2$

The calculation version is sometimes written

$$Var(X) = E(X^2) - (E(X))^2 \longleftarrow \text{'The mean of the squares minus the square of the mean'.}$$

The **standard deviation of X** is denoted by σ and is given by the square root of the variance:

$$\sigma = \sqrt{Var(X)}$$

Example 19

Question

The discrete random variable X has the probability distribution shown in the table.

x	1	2	3	4
$P(X=x)$	0.1	0.3	0.45	0.15

a   Calculate

  i $E(X)$ **ii** $E(X^2)$ **iii** $Var(X)$

b   The mean of X is μ and the standard deviation is σ.
  Find $P(\mu - \sigma < X < \mu + \sigma)$.

Answer

a   **i** $E(X) = \sum x P(X=x)$

       $= 1 \times 0.1 + 2 \times 0.3 + 3 \times 0.45 + 4 \times 0.15$

       $= 2.65$

  ii $E(X^2) = \sum x^2 P(X=x)$ Do not square the probabilities.

       $= 1^2 \times 0.1 + 2^2 \times 0.3 + 3^2 \times 0.45 + 4^2 \times 0.15$

       $= 7.75$

  iii $Var(X) = E(X^2) - \mu^2$

       $= 7.75 - (2.65)^2$

       $= 0.7275$

b $\mu = E(X) = 2.65$ and $\sigma = \sqrt{0.7275}$

  $\mu - \sigma = 2.65 - \sqrt{0.7275} = 1.80\,(3\,sf)$ and $\mu + \sigma = 2.65 + \sqrt{0.7275} = 3.50\,(3sf)$

(continued)

(continued)

$$P(\mu - \sigma < X < \mu + \sigma) = P(1.80 < X < 3.50)$$
$$= P(X = 2) + P(X = 3)$$
$$= 0.3 + 0.45$$
$$= 0.75$$

This is the probability that X lies within one standard deviation of the mean.

Example 20

Given that $E(Y) = 10$ and $E((Y - 10)^2) = 4$, find the standard deviation of the discrete random variable Y.

Now $\mathrm{Var}(Y) = E((Y - \mu)^2)$

Since $\mu = E(Y) = 10$,

       $\mathrm{Var}(Y) = E((Y - 10)^2)$

So, $\mathrm{Var}(Y) = 4$

       standard deviation $= \sqrt{4} = 2$

Variance of a simple function of X

In general, for the random variable X and constants a and b

$\mathrm{Var}(a) = 0$ ⟵—— A constant does not vary, so its variance is zero.

$\mathrm{Var}(aX) = a^2 \mathrm{Var}(X)$ ⟵—— Note a^2 here.

$\mathrm{Var}(aX + b) = a^2 \mathrm{Var}(X)$

$\mathrm{Var}(aX - b) = a^2 \mathrm{Var}(X)$

Note

For the standard deviation, find the square root of the variance.

For example

$\mathrm{Var}(5X) = 5^2 \mathrm{Var}(X) = 25 \mathrm{Var}(X)$

$\mathrm{Var}(3X - 7) = 3^2 \mathrm{Var}(X) = 9 \mathrm{Var}(X)$

The rule applies to functions such as

$$\mathrm{Var}\left(\frac{4}{X}\right) = \mathrm{Var}\left(4 \times \frac{1}{X}\right) = 4^2 \mathrm{Var}\left(\frac{1}{X}\right) = 16 \mathrm{Var}\left(\frac{1}{X}\right)$$

Note that there is no quick way of finding $\mathrm{Var}\left(\dfrac{1}{X}\right)$. You would need to calculate it using

$$\mathrm{Var}\left(\frac{1}{X}\right) = E\left(\frac{1}{X^2}\right) - \left(E\left(\frac{1}{X}\right)\right)^2$$

Take special care:

$\mathrm{Var}\left(\dfrac{4}{X}\right)$ does **not** equal $\dfrac{16}{\mathrm{Var}(X)}$

Example 21

The discrete random variable X has probability distribution shown in the table.

x	1	2	3	4
$P(X = x)$	0.2	0.3	0.3	0.2

Find

a $E(X)$

b $\mathrm{Var}(X)$

c $\mathrm{Var}\left(\dfrac{1}{4}X - 9\right)$

a $E(X) = 2.5$ (by symmetry)

b $E(X^2) = \sum x^2 P(X = x)$

$$= 1^2 \times 0.2 + 2^2 \times 0.3 + 3^2 \times 0.3 + 4^2 \times 0.2$$

$$= 7.3$$

$Var(X) = E(X^2) - [E(X)]^2$

$$= 7.3 - (2.5)^2$$

$$= 1.05$$

c $Var\left(\dfrac{1}{4}X - 9\right) = \left(\dfrac{1}{4}\right)^2 Var(X)$

$$= \dfrac{1}{16} \times 1.05$$

$$= 0.003125$$

Example 22

Exam marks, X have mean 66 and standard deviation 6.

The marks are scaled using the formula

$\quad Y = aX + b$

so that the scaled marks, Y, have mean 60 and standard deviation 5.

a Find the values of a and b.

b Find the scaled mark corresponding to a mark in the examination of 72.

a If $\quad Y = aX + b$

then $\quad E(Y) = aE(X) + b$

$$= a \times 66 + b$$

But $E(Y) = 60$,

so $\quad 60 = 66a + b$ (1)

Now $Var(Y) = a^2 Var(X)$

$$= a^2 \times 6^2$$

$$= 36a^2$$

But you know that $Var(Y) = 5^2 = 25$,

so $\quad 36a^2 = 25$

$$a^2 = \dfrac{25}{36}$$

$$a = \sqrt{\dfrac{25}{36}} = \dfrac{5}{6}$$

Substitute into equation (1)

$$60 = 66 \times \dfrac{5}{6} + b$$

$$60 = 55 + b$$

$$b = 5$$

So $a = \dfrac{5}{6}$ and $b = 5$.

b The formula for the scaling is $Y = \dfrac{5}{6}X + 5$.

When $x = 72$,

$$y = \dfrac{5}{6} \times 72 + 5 = 65$$

The scaled mark is 65.

Example 23

The number of fish, X, caught by Pearl when she goes fishing can be modelled by a discrete probability distribution:

x	1	2	3	4	5	6	≥ 7
$P(X=x)$	0.01	0.05	0.14	0.30	k	0.12	0

a   Find the value of k.

b   Find:

   i $E(X)$;

   ii $Var(X)$.

c   When Pearl sells her fish, she earns a profit, in pounds, given by $Y = 5X + 2$.

  Find:

   i $E(Y)$;

   ii   the standard deviation of Y.

AQA MS2B January 2007

a   Since $\sum P(X=x) = 1$,

$$0.01 + 0.05 + 0.14 + 0.30 + k + 0.12 = 1$$

$$0.62 + k = 1$$

$$k = 0.38$$

b   i $E(X) = \sum x\, P(X=x)$

$$= 1 \times 0.01 + 2 \times 0.05 + 3 \times 0.14 + 4 \times 0.30 + 5 \times 0.38 + 6 \times 0.12$$

$$= 4.35$$

  ii $E(X^2) = \sum x^2\, P(X=x)$

$$= 1^2 \times 0.01 + 2^2 \times 0.05 + 3^2 \times 0.14 + 4^2 \times 0.30 + 5^2 \times 0.38 + 6^2 \times 0.12$$

$$= 20.09$$

$$Var(X) = E(X^2) - [E(X)]^2$$

$$= 20.09 - 4.35^2$$

$$= 1.1675$$

c $Y = 5X + 2$

  i $E(Y) = 5E(X) + 2$

$$= 5 \times 4.35 + 2$$

$$= 23.75$$

  ii $Var(Y) = 5^2\, Var(X)$

$$= 25 \times 1.1675$$

$$= 29.1875$$

$$\text{s.d. of } Y = \sqrt{29.1875}$$

$$= 5.402...$$

$$= 5.40 \ (3\ \text{sf})$$

Exercise 4

1 The discrete random variable X has the following probability distribution.

x	3	5	7	9
$P(X=x)$	0.2	0.3	0.4	0.1

Calculate

a $E(X)$ **b** $Var(X)$

2 Find $Var(X)$ for each of the following probability distributions.

a

x	−3	−2	0	2	3
$P(X=x)$	0.3	0.3	0.2	0.1	0.1

b

x	1	3	5	7	9
$P(X=x)$	$\frac{1}{6}$	$\frac{1}{4}$	$\frac{1}{6}$	$\frac{1}{4}$	$\frac{1}{6}$

3 The discrete random variable R has the following probability distribution.

r	−2	−1	0	1	2
$P(R=r)$	0.05	c	0.43	3c	0.12

a Find the value of c.

b Find $E(R)$.

c Find the standard deviation of R.

4 Two boxes each contain three cards. The first box contains cards labelled 1, 3 and 5. The second box contains cards labelled 2, 6 and 8. In a game, a player picks a card at random from each box.

The score, X, is the sum of the numbers on the two cards.

a Using a possibility space, or otherwise, list the six possible values of X and calculate the corresponding probabilities.

b Calculate the expected score.

c Calculate the standard deviation of X.

5 A discrete random variable X has the following probability distribution. The mean of X is 6.

x	1	3	6	n	12
$P(X=x)$	0.1	0.3	k	0.25	0.15

a Find the value of k.

b Find the value of n.

c Find the variance of X.

6 The discrete random variable X has probability distribution given by the following, where k is a constant.

$$P(X=x)= \begin{cases} \dfrac{kx}{(x^2+1)} & x=2,3 \\[2ex] \dfrac{2kx}{(x^2-1)} & x=4,5 \end{cases}$$

a Show that $k = \dfrac{20}{33}$.

b Find the probability that X is less than 3 or greater than 4.

c Find $E(X)$.

d Find $Var(X)$.

7 The random variable R has probability distribution

r	1	2	3	4
$P(R=r)$	0.1	0.4	0.4	0.1

Find $Var(3R+2)$.

8 A computer is programmed to produce a sequence of integers, R, in the range 0 to 5 inclusive.

R has probability function given by

$$P(R=r) = \begin{cases} \dfrac{1}{2} & r=0 \\[2mm] \dfrac{r}{30} & r=1,2,3,4,5 \end{cases}$$

a Calculate $E(R)$.

b Calculate

   i $Var(R)$ **ii** $Var(4R)$ **iii** $Var(3-2R)$

9 The random variable X has mean 20 and variance 4.

Find the mean and variance of the following random variables

a $Y=6X+1$ **b** $D=3X-2$ **c** $W=\dfrac{1}{2}X-4$ **d** $V=-4X$

10 The random variable Y has standard deviation 3. The random variable $aY+9$ has standard deviation 7.5, where a is a constant. Find the value of a.

11 Test marks, X, have mean 60 and standard deviation 5.

The marks are scaled using the formula $Y=aX+b$.

The scaled marks have mean 80 and standard deviation 7.5.

a Find the values of a and b.

b Find the scaled mark corresponding to a mark of 50.

12 The discrete random variable X has probability distribution

x	1	2	3
$P(X=x)$	$\dfrac{1}{10}$	$\dfrac{3}{5}$	$\dfrac{3}{10}$

a Find $E\left(\dfrac{1}{X}\right)$.   **b** Show that $E\left(\dfrac{1}{X^2}\right)=\dfrac{17}{60}$.   **c** Find $Var\left(\dfrac{1}{X}\right)$.

14.5 Sum or difference of two independent random variables

Suppose you have two *independent* random variables, X and Y and form a new variable that is a **linear combination** of X and Y. It will be in the form $aX \pm bY$, where a and b are constants.

Examples are $X + Y$, $X - Y$, $3X + 2Y$, $5X - 4Y$, $\frac{1}{2}X + 4Y$.

The mean and variance of $aX \pm bY$ can be written in terms of the mean and variance of X and Y as follows:

> For random variables X and Y and constants a and b,
>
> $$E(aX + bY) = aE(X) + bE(Y)$$
>
> $$E(aX - bY) = aE(X) - bE(Y)$$
>
> When X and Y are independent,
>
> $$\text{Var}(aX + bY) = a^2\,\text{Var}(X) + b^2\,\text{Var}(Y)$$
>
> $$\text{Var}(aX - bY) = a^2\,\text{Var}(X) + b^2\,\text{Var}(Y)$$ ← The + sign is very important here. Variances are *always* added.

So, for example,

$$E(2X + 3Y) = 2E(X) + 3E(Y)$$

$$E(4X - 5Y) = 4E(X) - 5E(Y)$$

$$\text{Var}(X + Y) = \text{Var}(X) + \text{Var}(Y)$$

$$\text{Var}(X - Y) = \text{Var}(X) + \text{Var}(Y)$$

$$\text{Var}(4X + 7Y) = 4^2\,\text{Var}(X) + 7^2\,\text{Var}(Y) = 16\text{Var}(X) + 49\text{Var}(Y)$$

$$\text{Var}\left(3Y - \frac{1}{2}X\right) = 3^2\,\text{Var}(Y) + \left(\frac{1}{2}\right)^2\,\text{Var}(X) = 9\,\text{Var}(Y) + \frac{1}{4}\text{Var}(X)$$

> **Note**
>
> $\text{Var}(X + Y) = \text{Var}(X - Y)$

Example 24

Question

A restaurant charges \$17 for a 2-course meal and \$20 for a 3-course meal. On any day, the number, X, of 2-course meals served has mean 23 and standard deviation 5 and the number, Y, of 3-course meals has mean 18 and standard deviation 3.5.

a   The daily takings, in dollars, for the meals served in the restaurant is W. Write an expression for W in terms of X and Y.

b   Find the mean daily takings from meals served.

c   Find the standard deviation of the daily takings. Give your answer to the nearest dollar.

Answer

a   The amount taken for 2-course meals is $17X$ and the amount taken for 3-course meals is $20Y$, so $W = 17X + 20Y$

b $E(W) = E(17X + 20Y)$

   $\qquad = 17E(X) + 20E(Y)$

   $\qquad = 17 \times 23 + 20 \times 18$

   $\qquad = 751$

   The mean daily takings from meals is \$751.

c $\text{Var}(W) = \text{Var}(17X + 20Y)$

   $\qquad = 17^2\text{Var}(X) + 20^2\,\text{Var}(Y)$

   $\qquad = 17^2 \times 5^2 + 20^2 \times 3.5^2$

   $\qquad = 12125$

   s.d. of $W = \sqrt{12125} = 110.11\ldots$

   The standard deviation of the daily takings is \$110 (nearest dollar).

Sum of independent observations of a discrete random variable

Suppose you have a random variable X with mean μ and variance σ^2.

Now take n independent observations of X and call them $X_1, X_2, ..., X_n$.

Since the observations are all from the same distribution of X,

$$E(X_1) = \mu \qquad\qquad Var(X_1) = \sigma^2$$
$$E(X_2) = \mu \qquad\qquad Var(X_2) = \sigma^2$$
$$... \qquad\qquad ...$$
$$E(X_n) = \mu \qquad\qquad Var(X_n) = \sigma^2$$

Form a new variable, $X_1 + X_2 + ... + X_n$. This is *sum* of the n observations.

Now
$$E(X_1 + X_2 + ... + X_n)$$
$$= E(X_1) + E(X_2) + ... + E(X_n)$$
$$= \mu + \mu + ... \mu$$
$$= n\mu$$

$$Var(X_1 + X_2 + ... + X_{12})$$
$$= Var(X_1) + Var(X_2) + ... + Var(X_n)$$
$$= \sigma^2 + \sigma^2 + ... + \sigma^2$$
$$= n\sigma^2$$

In general

$$E(X_1 + X_2 + ... + X_n) = nE(X)$$
$$Var(X_1 + X_2 + ... + X_n) = nVar(X)$$

> **Note**
>
> These can be written
> $E(\Sigma X_i) = \Sigma\, E(X_i)$
> $Var(\Sigma X_i) = \Sigma\, Var(X_i)$

Example 25

Question

Paper clips are packed by a machine into boxes. The mean number of paper clips in a box is 54 and the standard deviation is 3.

Ten boxes are selected at random from the production line. Find the mean and standard deviation of the total number of paper clips in the 10 boxes.

Answer

Let X be the number of paper clips in a box.

$E(X) = 54$, $Var(X) = 3^2 = 9$

Let T be the total number of paper clips in 10 boxes.

$T = X_1 + X_2 + X_3 + ... + X_{10}$

$E(T) = 10E(X) = 10 \times 54 = 540$

$Var(T) = 10Var(X) = 10 \times 9 = 90$

s.d of $T = \sqrt{90} = 9.486...$

So, for the total number of paper clips in 10 boxes,
mean $= 540$, standard deviation $= 9.49$ (3 sf).

Example 26

On a multiple choice examination paper, each question has five multiple choice answers given, only one of which is correct. For each question, candidates gain 4 marks for a correct answer and lose 1 mark for an incorrect answer.

a James guesses the answer to each question.

   i Complete the following table for the probability distribution of X, the number of marks obtained by James for each question.

x	4	−1
$P(X=x)$		

   ii Hence find E(X).

b Karen is able to eliminate two of the incorrect answers from the five alternative answers given for each question before guessing the answer from those remaining.

Given that the examination paper consists of 24 questions, calculate Karen's expected total mark.

AQA MS2B June 2007

a **i** $P(X=4) = P(\text{correct answer}) = \dfrac{1}{5}$

$P(X=-1) = P(\text{incorrect answer}) = \dfrac{4}{5}$

x	4	−1
$P(X=x)$	$\dfrac{1}{5}$	$\dfrac{4}{5}$

ii $E(X) = \sum x\, P(X=x)$

$\quad = 4 \times \dfrac{1}{5} + (-1) \times \dfrac{4}{5}$

$\quad = 0$

b Let Y be the number of marks obtained by Karen for each question.

Since she is choosing from 3 alternatives,

$P(Y=4) = \dfrac{1}{3}$ and $P(Y=-1) = \dfrac{2}{3}$

First find Karen's expected mark for a question.

$E(Y) = \sum y\, P(Y=y) = 4 \times \dfrac{1}{3} + (-1) \times \dfrac{2}{3} = \dfrac{2}{3}$

Let T be Karen's total mark for the questions on the examination paper.

So $\quad T = Y_1 + Y_2 + \dots + Y_{24}$

$E(T) = 24\,E(Y)$

$\quad = 24 \times \dfrac{2}{3}$

$\quad = 16$

Karen's expected mark is 16.

> **Note**
>
> T is the sum of 24 independent observations

Comparing the distributions of 2X and $X_1 + X_2$

Confusion sometimes arises between the random variables $2X$ and $X_1 + X_2$, where $2X$ is a **multiple** of a single observation of X, and $X_1 + X_2$, is the **sum** of two independent observations of X. You will see from the following example that their distributions are different.

A fair tetrahedral die is thrown. Let X be the number on which it lands.

x	1	2	3	4
$P(X=x)$	$\frac{1}{4}$	$\frac{1}{4}$	$\frac{1}{4}$	$\frac{1}{4}$

$E(X) = 2.5$ (by symmetry)

$Var(X) = E(X^2) - [E(X)]^2$

$\qquad = 1^2 \times \frac{1}{4} + 2^2 \times \frac{1}{4} + 3^2 \times \frac{1}{4} + 4^2 \times \frac{1}{4} - 2.5^2$

$\qquad = 1.25$

Throw the die <u>once</u> and consider $D = 2X$, where D is 'double the number on which the die lands'.

d	2	4	6	8
$P(D=d)$	$\frac{1}{4}$	$\frac{1}{4}$	$\frac{1}{4}$	$\frac{1}{4}$

> **Note**
>
> $2X$ is a *multiple*.

$E(D) = 5$ (by symmetry)

$Var(D) = 2^2 \times \frac{1}{4} + 4^2 \times \frac{1}{4} + 6^2 \times \frac{1}{4} + 8^2 \times \frac{1}{4} - 5^2 = 5$

So, as expected,

$E(D) = E(2X) = 2E(X)$

$Var(D) = Var(2X) = 2^2 Var(X) = 4 Var(X)$

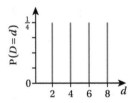

Now throw the die <u>twice</u> and consider $S = X_1 + X_2$, where S is 'the sum of the two numbers on which the die lands'.

s	2	3	4	5	6	7	8
$P(S=s)$	$\frac{1}{16}$	$\frac{2}{16}$	$\frac{3}{16}$	$\frac{4}{16}$	$\frac{3}{16}$	$\frac{2}{16}$	$\frac{1}{16}$

> **Note**
>
> $X_1 + X_2$ is a *sum*.

$E(S) = 5$ (by symmetry)

$Var(S) = \left(2^2 \times \frac{1}{16} + 3^2 \times \frac{2}{16} + \dots + 8^2 \times \frac{1}{16}\right) - 5^2 = 2.5$

So, as expected,

$E(S) = E(X_1 + X_2) = 2E(X)$

$Var(S) = Var(X_1 + X_2) = 2Var(X)$

Notice that the means of the two distributions are the same, but the variances are not. The random variable 'double the number' has a larger variance than the random variable 'the sum of the two numbers'.

In general, take care to distinguish between situations when a single observation has been multiplied by a constant (this is a **multiple**, as in $2X$ or $3X$ or $10X$) and when several independent observations of the same random variable are added (this is a **sum**, as in $X_1 + X_2$ or $X_1 + X_2 + X_3$ or $X_1 + X_2 + \ldots + X_{10}$).

The general results are summarised below.

	Multiple	**Sum**
Mean	$E(nX) = n\,E(X)$	$E(X_1 + X_2 + \ldots + X_n) = n\,E(X)$
Variance	$\text{Var}(nX) = n^2\,\text{Var}(X)$	$\text{Var}(X_1 + X_2 + \ldots + X_n) = n\,\text{Var}(X)$

Exercise 5

1 X and Y are independent random variables such that

   $E(X) = 8$ and $E(Y) = 10$

   $\text{Var}(X) = 2$ and $\text{Var}(Y) = 3$

   Find

   a $E\left(\dfrac{1}{2}X + 6Y\right)$

   b $\text{Var}(2X + 7Y)$

   c the standard deviation of W, where $W = 4X - 3Y$.

2 Independent random variables X and Y have probability distributions as shown in the tables below.

x	1	2	3	4
$P(X=x)$	0.1	0.2	0.4	0.3

y	0	1	2
$P(Y=y)$	0.5	0.4	0.1

   a Find   **i** $E(X)$ and $\text{Var}(X)$ **ii** $E(Y)$ and $\text{Var}(Y)$

   b A new variable, T is formed, where $T = 2X + 5Y$.

   Find the mean and standard deviation of T.

   c Five independent observations are taken from Y. The sum of these observations is denoted by S.

   i Find $E(S)$.   **ii** Find $\text{Var}(S)$.

3 Nikhil is playing a game with unusual shaped fair dice. He has a dodecahedron which has 12 faces numbered from 1 to 12 and an octahedron with 8 faces, numbered 1 to 8. His turn consists of rolling the two dice together.

   The score on the dodecahedron is D and the score on the octahedron is C.

   a State the values of $E(D)$ and $E(C)$.

   Nikhil is awarded T points for his turn, where $T = 6D + 4C$.

   b Find the expected number of points he obtains for his turn.

4 On a particular motorway, the number of vehicles passing an electronic counter each minute is recorded. The number travelling north, N, has a mean of 5 and a standard deviation of 2. The number of vehicles travelling south, S, has a mean of 6 and a standard deviation of 1.5. Find the mean and standard deviation of T, the total number of vehicles passing the counter in a minute.

5 The random variable X has mean 20 and variance 4.

The variable Y is the sum of 5 independent observations of X.

Find the mean and standard deviation of Y.

6 The random variable X has mean 20 and standard deviation 3.

  a Find the mean and standard deviation of $4X$.

  b X_1, X_2, X_3 and X_4 are independent observations of X.
     Find the mean and standard deviation of $X_1 + X_2 + X_3 + X_4$.

7 The random variable X has mean 12 and standard deviation 4.

  a Find the mean and standard deviation of Y where $Y = 6X$.

  b Find the mean and standard deviation of S where S is the sum of 6 independent observations of X.

14.6 Further applications

Example 27

Johann supplies free-range eggs to a health food shop. The eggs are supplied in boxes of six. Simone, the shop owner, checks a sample of these boxes for cracked eggs. She finds that the number, X, of cracked eggs in a box may be modelled by the following probability distribution.

x	$P(X=x)$
0	0.925
1	0.061
2	0.010
3	0.000
4	0.000
5	0.000
6	0.004

a i Find the mean number of cracked eggs in a box.

   ii Verify that the standard deviation of the number of cracked eggs in a box is 0.484, correct to three significant figures.

b Christos also samples free-range eggs in boxes of six. The mean number of cracked eggs in his boxes is 0.25 with a standard deviation of 0.21.

   Compare the distributions of the numbers of cracked eggs in boxes supplied by Johann and Christos.   AQA SS02 January 2009

a **i** $E(X) = \sum x\,P(X = x)$

$= 0 \times 0.925 + 1 \times 0.061 + 2 \times 0.010 + 3 \times 0.000 + 4 \times 0.000 + 5 \times 0.000 + 6 \times 0.004$

$= 0.105$

Mean number of cracked eggs in a box $= 0.105$.

ii $E(X^2) = \sum x^2 P(X = x)$

$= 1^2 \times 0.061 + 2^2 \times 0.010 + 6^2 \times 0.004$

$= 0.245$

$Var(X) = E(X^2) - [E(X)]^2$

$= 0.245 - 0.105^2$

$= 0.2339...$

Now find the square root of the variance.

s. d. of $X = \sqrt{0.2339...} = 0.4837... = 0.484$ (3 sf)

Standard deviation of the number of cracked eggs in a box $= 0.484$.

b Johann: Mean $= 0.105$ Standard deviation $= 0.484$

Christos: Mean $= 0.25$ Standard deviation $= 0.21$

Christos's boxes have, on average, more cracked eggs than Johann's, but the number of cracked eggs in Christos's boxes is less variable.

Example 28

A house has a total of five bedrooms, at least one of which is always rented.

The probability distribution for R, the number of bedrooms that are rented at any given time, is given by

$$P(R = r) = \begin{cases} 0.5 & r = 1 \\ 0.4(0.6)^{r-1} & r = 2, 3, 4 \\ 0.0296 & r = 5 \end{cases}$$

a Complete the table

r	1	2	3	4	5
$P(R = r)$	0.5				0.0296

b Find the probability that fewer than 3 bedrooms are **not** rented at any given time.

c **i** Find the value of $E(R)$.

ii Show that $E(R^2) = 4.8784$ and hence find the value of $Var(R)$.

d Bedrooms are rented on a monthly basis.

The monthly income, $£M$, from renting bedrooms in the house may be modelled by

$$M = 1250R - 282$$

Find the mean and standard deviation of M.

AQA MS2B June 2012

(continued)

(continued)

a R is the number of bedrooms that are rented.

When $r = 2$, $P(R = 2) = 0.4(0.6)^1 = 0.24$

When $r = 3$, $P(R = 3) = 0.4(0.6)^2 = 0.144$

When $r = 3$, $P(R = 4) = 0.4(0.6)^3 = 0.0864$

r	1	2	3	4	5
$P(R=r)$	0.5	0.24	0.144	0.0864	0.0296

b P(fewer than 3 bedrooms are not rented)

$= P(0, 1 \text{ or } 2 \text{ are not rented})$

$= P(3, 4 \text{ or } 5 \text{ are rented})$

$= 0.144 + 0.0864 + 0.0296$

$= 0.26$

c i mean $= E(R) = \sum r\, P(R = r)$

$= 1 \times 0.5 + 2 \times 0.24 + 3 \times 0.144 + 4 \times 0.0864 + 5 \times 0.0296$

$= 1.9056$

ii First find $E(R^2)$.

$E(R^2) = \sum r^2\, P(R = r)$

$= 1^2 \times 0.5 + 2^2 \times 0.24 + 3^2 \times 0.144 + 4^2 \times 0.0864 + 5^2 \times 0.0296$

$= 4.8784$

$Var(R) = E(R^2) - [E(R)]^2$

$= 4.8784 - 1.9056^2$

$= 1.2470... = 1.25 \text{ (3 sf)}$

d $E(M) = E(1250R - 282)$

$= 1250\, E(R) - 282$

$= 1250 \times 1.9056 - 282$

$= 2100$

So, mean of M is 2100.

$Var(M) = Var(1250R - 282)$

$= 1250^2\, Var(R)$

$= 1250^2 \times 1.2470...$

$= 1948576$

s.d. of $M = \sqrt{1948576} = 1395.9... = 1400 \text{ (3 sf)}$

Example 29

The discrete random variable X has

$E(X) = 4$ and $E(X^2) = 17.2$.

a Find the value of $Var(X)$.

b A circle has radius $(X + 8)$.

Find, in terms of π, values for the mean and variance of the circumference, C, of the circle.

(continued)

c The area, S, of the circle, with radius $(X+8)$ is to be expressed in the form

$$\pi(X^2+aX+b),$$

where a and b are positive constants.

i Find the values of a and b.

ii Hence find, in terms of π, the value of $E(S)$.

AQA MAS1 January 2005

a $\mathrm{Var}(X)=\mathrm{E}(X^2)-[\mathrm{E}(X)]^2$

$\qquad = 17.2 - 4^2$

$\qquad = 1.2$

b $C = 2\pi(X+8)$, so $C = 2\pi X + 16\pi$

So $\mathrm{E}(C) = \mathrm{E}(2\pi X + 16\pi)$

$\qquad = 2\pi\mathrm{E}(X) + 16\pi$

$\qquad = 2\pi \times 4 + 16\pi$

$\qquad = 24\pi$

$\mathrm{Var}(C) = \mathrm{Var}(2\pi X + 16\pi)$

$\qquad = (2\pi)^2\mathrm{Var}(X)$

$\qquad = 4\pi^2 \times 1.2$

$\qquad = 4.8\pi^2$

c $S = \pi(X+8)^2$

$\qquad = \pi(X^2 + 16X + 64),$

so $a = 16$ and $b = 64$.

$\mathrm{E}(S) = \mathrm{E}[\pi(X^2 + 16X + 64)]$

$\qquad = \pi(\mathrm{E}(X^2) + 16\mathrm{E}(X) + 64)$

$\qquad = \pi(17.2 + 16 \times 4 + 64)$

$\qquad = 145.2\pi$

Summary

Discrete data

Mode

The mode is the value that occurs most often.

Mean

For raw data $\bar{x} = \dfrac{\sum x}{n}$

For data in a frequency table $\bar{x} = \dfrac{\sum xf}{\sum f}$

Note

These formulae are **not** given in the examination.

Median

For a set of n numbers arranged in ascending order:

► when n is odd, the median is the middle value
► when n is even, the median is the mean of the two middle values.

This is summarised by saying that the median is the $\frac{1}{2}(n+1)^{\text{th}}$ value.

Quartiles

► The lower quartile is the median of all the values before the median.
► The upper quartile is the median of all the values after the median.

List the values in order and count through the data values to find the median and quartiles.

Ranges

Range = highest value − lowest value

Interquartile range (IQR) = upper quartile − lower quartile

Standard deviation (s.d.)

For raw data

Definition version

$$\text{s.d.} = \sqrt{\frac{\sum(x-\bar{x})^2}{n}}$$

Calculation version

$$\text{s.d.} = \sqrt{\frac{\sum x^2}{n} - \bar{x}^2}$$

For data in a frequency distribution

$$\text{s.d.} = \sqrt{\frac{\sum(x-\bar{x})^2 f}{\sum f}}$$

$$\text{s.d.} = \sqrt{\frac{\sum x^2 f}{\sum f} - \bar{x}^2}$$

> **Note**
> These formulae are **not** given in the examination.

Variance

Variance = (standard deviation)^2

Discrete random variables

A list of all possible values of the discrete random variable X, together with their associated probabilities, is called a probability distribution.

The sum of the probabilities of all possible values of X is 1.

so $\quad \sum P(X=x) = 1 \quad$ or $\quad \sum p_i = 1 \quad i = 1, 2, ..., n$

$E(X)$, the expectation of X

$$\mu = E(X) = \sum x_i p_i = \sum x P(X=x)$$

$E(g(X))$, the expectation of $g(X)$

$$E(g(X)) = \sum g(x_i) p_i = \sum g(x) P(X=x)$$

Var(X), the variance of X

Definition version

$$\text{Var}(X) = \sigma^2 = \text{E}[(X - \mu)^2]$$
$$= \Sigma(x_i - \mu)^2 p_i$$
$$= \Sigma(x - \mu)^2 \text{P}(X = x)$$

Calculation version

$$\text{Var}(X) = \sigma^2 = \text{E}(X^2) - \mu^2$$
$$= \Sigma x_i^2 p_i - \mu^2$$
$$= \Sigma x^2 \text{P}(X = x) - \mu^2$$

Standard deviation of X

$$\sigma = \text{standard deviation} = \sqrt{\text{Var}(X)}$$

Simple functions of a random variable X

For any random variable X and constants a and b,

$$\text{E}(aX) = a\text{E}(X)$$
$$\text{E}(aX + b) = a\text{E}(X) + b$$
$$\text{E}(aX - b) = a\text{E}(X) - b$$

$$\text{Var}(aX) = a^2\text{Var}(X)$$
$$\text{Var}(aX + b) = a^2\text{Var}(X)$$
$$\text{Var}(aX - b) = a^2\text{Var}(X)$$

Sum and difference of two independent random variables

For two variables X and Y,

$$\text{E}(aX + bY) = a\text{E}(X) + b\text{E}(Y)$$
$$\text{E}(aX - bY) = a\text{E}(X) - b\text{E}(Y)$$

If X and Y are independent,

$$\text{Var}(aX + bY) = a^2\text{Var}(X) + b^2\text{Var}(Y)$$
$$\text{Var}(aX - bY) = a^2\text{Var}(X) + b^2\text{Var}(Y) \qquad \leftarrow \text{Remember the + sign here}$$

Sum of n independent observations of a random variable

For n independent observations of the random variable X,

$$\text{E}(X_1 + X_2 + \cdots + X_n) = n\text{E}(X)$$
$$\text{Var}(X_1 + X_2 + \cdots + X_n) = n\text{Var}(X)$$

Review

1. Lizzie, the receptionist at a dental practice, was asked to keep a weekly record of the number of patients who failed to turn up for an appointment. Her records for the first 15 weeks were as follows.

   $$20 \quad 26 \quad 32 \quad a \quad 37 \quad 14 \quad 27 \quad 34 \quad 15 \quad 18 \quad b \quad 25 \quad 37 \quad 29 \quad 25$$

   Unfortunately, Lizzie forgot to record the actual values for two of the 15 weeks, so she recorded them as a and b. However, she did remember that $a < 10$ and that $b > 40$.

   a   Calculate the median and the interquartile range of these 15 values.

   b   Give reasons why, for these data:

      i   the mode is **not** an appropriate measure of average;

      ii   the standard deviation **cannot** be used as a measure of spread.

   c   Subsequent investigations revealed that the missing values were 8 and 43.

      Calculate the mean and standard deviation of the 15 values.

   AQA MS1B January 2010

2 A fair tetrahedral die has four faces numbered 1, 2, 3 and 6. The die and a fair coin are tossed together. If the coin shows heads the score S is equal to double the number on the hidden face of the die. If the coin shows tails then the score S is equal to the number on the hidden face of the die.

a Copy and complete the possibility space showing the possible outcomes.

		Die			
		1	2	3	6
Coin	H			6	
	T		2		

b Show that $P(S = 6) = \dfrac{1}{4}$.

c Draw up the probability distribution table for S.

d Show that the expected value of S is 4.5.

e Calculate the variance of S.

3 The discrete random variable D is the number of parcels delivered in a day to a particular house. The probability distribution for D is shown below.

d	0	1	2	3	4
$P(D = d)$	0.1	0.4	0.25	0.15	a

a Find the value of a.

b Write down the most likely number of parcels delivered.

c Find the mean number of parcels delivered.

d Find the probability that the number of parcels delivered is fewer than the mean.

e Find the variance of D.

4 The number of spelling mistakes, X, that Jo makes when writing an essay can be modelled by the following probability distribution.

x	0	1	2	3	4
$P(X = x)$	0.23	0.31	0.27	0.14	0.05

The mean number of spelling mistakes is μ and the standard deviation is σ.

a Find the values of μ and σ.

b Find $P(X > \mu + \sigma)$.

5 The random variable X takes values 2, 4, 6 and 8.

The probability distribution of X is illustrated in the vertical line graph.

Find Var(X).

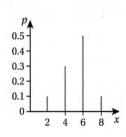

6 A discrete random variable X has the following probability distribution.

x	0	1	2	3
$P(X=x)$	c	$2c$	$3c$	$4c$

  a Find the value of the constant c.

  b Find $E(X)$ and $Var(X)$.

  c Find $P(X > E(X))$.

7 The probability distribution of the random variable X is shown in the following table.

The mean of X is 3.3.

x	1	2	3	4	5
$P(X=x)$	0.1	a	a	b	a

  a Write down two equations involving a and b and hence find the values of a and b.

  b Calculate the variance of X.

  c Find $E(4X+2)$.

  d Find $Var(4-5X)$.

8 Aiden takes his car to a garage for its MOT test. The probability that his car will need to have X tyres replaced is shown in the table.

x	0	1	2	3	4
$P(X=x)$	0.1	0.35	0.25	0.2	0.1

  a Show that the mean of X is 1.85 and calculate the variance of X.

  b The charge for the MOT test is £c and the cost of **each** new tyre is £n. The total amount that Aiden must pay the garage is £T.

  i Express T in terms of c, n and X.

  ii Hence, using your results from part **a**, find expressions for $E(T)$ and $Var(T)$.

AQA MS2B January 2013

9 a The random variable X has mean 12.

  The random variable Y has mean 5, where $Y = aX + 2$.

  Find the value of a.

  b The random variable X has standard deviation 4.

  The random variable W has standard deviation 40, where $W = 1 - cX$.

  Find the value of c.

10 X is a discrete random variable, where $E(X) = 20$ and $Var(X) = 4$.

The variable Y is the sum of n independent observations of X, where $E(Y) = 160$. Find

  a the value of n

  b $Var(Y)$.

Assessment

1 Henrietta lives on a small farm where she keeps some hens. For a period of 35 weeks during the hens' first laying season, she records, each week, the total number of eggs laid by the hens.

Her records are shown in the table.

Total number of eggs laid in a week (x)	Number of weeks (f)
66	1
67	2
68	3
69	5
70	7
71	8
72	4
73	2
74	2
75	1
TOTAL	35

  a For these data, calculate the values for the mean and the standard deviation. [4]

  b Each week, for the 35 weeks, Henrietta sells 60 eggs to a local shop, keeping the remainder for her own use.

    State values for the mean and the standard deviation of the number of eggs that she keeps. [2]

AQA MS1A June 2014

2 The Globe Express agency organizes trips to the theatre. The cost, £X, of these trips can be modelled by the following probability distribution:

x	40	45	55	74
$P(X=x)$	0.30	0.24	0.36	0.10

  a Calculate the mean and standard deviation of X. [4]

  b For special celebrity charity performances, Globe Express increases the cost of the trips to £Y, where

$$Y = 10X + 250$$

    Determine the mean and standard deviation of Y. [2]

AQA MS2B January 2006

3 A box contains a large number of pea pods. The number of peas in a pod may be modelled by the random variable X. The probability distribution of X is tabulated below.

x	2 or fewer	3	4	5	6	7	8 or more
$P(X=x)$	0	0.1	0.2	a	0.3	b	0

a 2 pods are picked randomly from the box. Find the probability that the number of peas in **each** pod is at most 4. [2]

b It is given that $E(X) = 5.1$.

    i Determine the values of a and b. [4]

    ii Hence show that $\text{Var}(X) = 1.29$. [2]

    iii Some children play a game with the pods, randomly picking a pod and scoring points depending on the number of peas in the pod. For each pod picked, the number of points scored, N, is found by doubling the number of peas in the pod and then subtracting 5.

       Find the mean and the standard deviation of N. [3]

AQA MS2B June 2014

4 In a computer game, players try to collect five treasures. The number of treasures that Isaac collects in one play of the game is represented by the discrete random variable X.

The probability distribution of X is defined by

$$P(X=x) = \begin{cases} \dfrac{1}{x+2} & x=1, 2, 3, 4 \\ k & x=5 \\ 0 & \text{otherwise} \end{cases}$$

a   **i** Show that $k = \dfrac{1}{20}$. [2]

    ii Calculate the value of $E(X)$. [2]

    iii Show that $\text{Var}(X) = 1.5275$. [3]

    iv Find the probability that Isaac collects more than 2 treasures. [2]

b The number of points that Isaac scores for collecting treasures is Y, where $Y = 100X - 50$.

   Calculate the mean and standard deviation of Y. [4]

AQA MS2B June 2013

5 A discrete random variable X has the probability distribution

$$P(X=x) = \begin{cases} \dfrac{x}{20} & x=1, 2, 3, 4, 5 \\ \dfrac{x}{24} & x=6 \\ 0 & \text{otherwise} \end{cases}$$

a Calculate $P(X \geq 5)$. [2]

b   **i** Show that $E\left(\dfrac{1}{X}\right) = \dfrac{7}{24}$. [2]

    ii Hence, or otherwise, show that $\text{Var}\left(\dfrac{1}{X}\right) = 0.036$, correct to three decimal places. [3]

c Calculate the mean and variance of A, the area of rectangles having sides of length $X+3$ and $\dfrac{1}{X}$. [5]

AQA MS2B January 2008

6 The number of text messages, N, sent by Peter each month on his mobile never exceeds 40.

 When $0 \le N \le 10$, he is charged for 5 messages.

 When $10 < N \le 20$, he is charged for 15 messages.

 When $20 < N \le 30$, he is charged for 25 messages.

 When $30 < N \le 40$, he is charged for 35 messages.

a The number of text messages, Y, that Peter is charged for each month has the following probability distribution:

y	5	15	25	35
$P(Y=y)$	0.1	0.2	0.3	0.4

i Calculate the mean and standard deviation of Y. [4]

ii The Goodtime phone company makes a total charge for text messages, C pence, each month given by

 $C = 10Y + 5$

 Calculate E(C). [1]

b The number of text messages, X, sent by Joanne each month on her mobile phone is such that

 $E(X) = 8.35$ and $E(X^2) = 75.25$

 The Newtime phone company makes a total charge for text messages, T pence, each month given by

 $T = 0.4X + 250$

 Calculate Var(T). [4]

AQA MS2B June 2008

7 X and Y are independent random variables such that X has mean 10 and standard deviation 3 and Y has mean 15 and standard deviation 2.

a Find E($3X + 2Y$). [1]

b Find Var($Y - 4X$). [2]

c Find the mean and standard deviation of the random variable W, where W is the sum of 10 independent observations of X. [3]

8 A small hotel has a microwave oven and a combination oven.

The random variable X is the number of breakdowns in a year of the microwave oven and the random variable Y is the number of breakdowns in a year of the combination oven.

The mean of X is 4 and the variance is 4. The mean of Y is 5 and the variance is 5.

Find the mean and standard deviation of T, the total number of breakdowns of the two ovens in a year. [5]

15 Bernoulli and Binomial Distributions

Introduction

The Bernoulli and binomial distributions are two special discrete probability distributions.

The Bernoulli distribution provides a model for the number of successes in a single trial of an experiment, say the number of sixes in one throw of a die.

This has great importance in the development of the binomial distribution which is the model for the number of successes when you carry out the same trial a given number of times, say the number of sixes in 10 throws of the die.

James (or Jacques) Bernoulli (1654–1705) was a member of a very talented Swiss family. He graduated in theology from Basel University when he was 21 and returned as a physics lecturer when he was 29, becoming Professor of Mathematics at the age of 33. His main work was The Art of Conjecture, a study on probability.

Objectives

By the end of this chapter, you should know how to…

▶ Use the conditions for the application of a Bernoulli distribution.

▶ Find the mean and variance of a Bernoulli distribution.

▶ Use the conditions for a binomial distribution.

▶ Calculate binomial probabilities using a formula.

▶ Find binomial probabilities using cumulative probabilities in a table.

▶ Find the mean, variance and standard deviation of a binomial distribution.

Recap

You need to remember that for a discrete random variable X

▶ $\sum P(X = x) = 1$

▶ $E(X) = \sum x P(X = x)$

▶ $Var(X) = E(X^2) - \mu^2$

▶ standard deviation $= \sqrt{Var(X)}$

15.1 The Bernoulli distribution

The discrete random variable Y with probability distribution function

$$\begin{cases} P(Y=0) = 1 - p \\ P(Y=1) = p \end{cases} \quad \text{for } 0 \le p \le 1$$

y	0	1
$P(Y=y)$	$1-p$	p

is said to have a **Bernoulli distribution** with **parameter** p.

You can write

$Y \sim \text{Bernoulli}(p)$

This special discrete distribution arises from an experiment known as a **Bernoulli trial** in which there are only two possible outcomes: success or failure. The probability of success is p.

Note

Only the value of p is needed to describe a Bernoulli distribution fully.

Suppose you are playing a board game and you want to throw a six on a die. If the die is fair, $P(6) = \frac{1}{6}$. The act of throwing the die is a Bernoulli trial with two possible outcomes: success (throwing a six) and failure (not throwing a six). Let Y be the number of sixes when you throw the die.

y	0	1
$P(Y = y)$	$\frac{5}{6}$	$\frac{1}{6}$

Y can take the values 0 or 1.

$$P(Y = 0) = P(\text{failure}) = P(\text{not a six}) = \frac{5}{6}$$

$$P(Y = 1) = P(\text{success}) = P(\text{six}) = \frac{1}{6}$$

So Y has a Bernoulli distribution with $p = \frac{1}{6}$.

This is written $Y \sim \text{Bernoulli}\left(\frac{1}{6}\right)$

> If the random variable Y is the *number of successful outcomes* in a Bernoulli trial with probability p of success, then
> $$P(Y = 0) = P(\text{failure}) = 1 - p$$
> $$P(Y = 1) = P(\text{success}) = p$$
> and Y has a Bernoulli distribution, where $Y \sim \text{Bernoulli}(p)$

Example 1

A coin is biased so that it is three times as likely to show tails as heads. Write out the probability distribution for Y, the number of tails when the coin is tossed.

$P(\text{success}) = P(\text{tail}) = 0.75$.

If Y is the number of tails in a single toss of the coin,

then $Y \sim \text{Bernoulli}(0.75)$.

$$P(Y = 0) = 1 - p = 0.25$$

$$P(Y = 1) = p = 0.75$$

y	0	1
$P(Y = y)$	0.25	0.75

Mean and variance of a Bernoulli distribution

Often q is used for $1 - p$, so $q = 1 - p$.

So, if $Y \sim \text{Bernoulli}(p)$,

y	0	1
$P(Y = y)$	q	p

$$E(Y) = \sum y P(Y = y)$$
$$= (0 \times q) + 1 \times p$$
$$= p$$

You studied the mean of a discrete random variable in section 14.3

$$E(Y^2) = \sum y^2 P(Y = y)$$
$$= (0^2 \times q) + (1^2 \times p)$$
$$= p$$

$$Var(Y) = E(Y^2) - [E(Y)]^2$$
$$= p - p^2$$
$$= p(1 - p)$$
$$= pq$$

Variance of a discrete random variable was covered in section 14.4

If Y has a Bernoulli distribution such that $Y \sim \text{Bernoulli}(p)$, then

mean $= \text{E}(Y) = p$

variance $= \text{Var}(Y) = pq$, where $q = 1 - p$

standard deviation $= \sqrt{pq}$

Example 2

Question

A bag contains 6 red counters, 3 blue counters and 1 yellow counter.

A counter is picked at random from the bag. Find

a the mean number of red counters picked

b the standard deviation of the number of yellow counters picked.

Answer

a There appear to be three outcomes: red, blue, yellow, but you are only interested in whether the counter is red or not.

Define 'success' as picking a red counter and 'failure' as picking a counter that is not red.

This is a Bernoulli trial with $p = \text{P}(\text{red}) = 0.6$.

If R is the number of red counters picked, then $R \sim \text{Bernoulli}(0.6)$.

Mean number of red counters picked $= p = 0.6$.

b Now you are only interested in whether the counter is yellow or not.

Define 'success' as picking a yellow counter and 'failure' as picking a counter that is not yellow.

This is a Bernoulli trial with $p = \text{P}(\text{yellow}) = 0.1$.

If Y is the number of yellow counters picked, then $Y \sim \text{Bernoulli}(0.1)$.

Variance $= pq = 0.1 \times 0.9 = 0.09$.

Standard deviation $= \sqrt{0.09} = 0.3$.

Exercise 1

1 The discrete random variable Y has probability distribution

y	0	1
$\text{P}(Y = y)$	0.64	a

Find the mean and standard deviation of Y.

2 The discrete random variable Y has a Bernoulli distribution with $\text{P}(Y = 0) = \dfrac{5}{12}$. Find the mean and standard deviation of Y.

3 A bag contains 15 discs, numbered 1, 2, 3, ..., 15. A disc is randomly selected from the bag.

a Find the mean number of discs that show a multiple of 5.

b Find the standard deviation of the number of discs that show an odd number.

4 A fair cubical die has faces numbered 1, 1, 2, 3, 3, 4. The score is the face uppermost when the die is rolled. Find the mean and variance of the number of even scores obtained on one roll of the die.

5 Given that $W \sim \text{Bernoulli}(p)$ and $\text{Var}(W) = 0.16$, find the two possible values for p.

15.2 The binomial distribution

The **binomial distribution** is another special discrete distribution which has many important applications in statistics. The following example illustrates how a binomial situation arises.

Example 3

A coin is biased so that the probability of obtaining a head when it is tossed is 0.7.

Find the probability of obtaining exactly two heads when the coin is tossed

a   3 times

b   6 times.

a   A tree diagram shows clearly the possible outcomes in 3 tosses.

▶ Draw the tree with three sets of branches, one for each toss of the coin.

▶ Locate the end results that give exactly 2 heads, so 2 heads and 1 tail.

▶ Find the sum of the probabilities of these end results.

> **Note**
>
> See more about tree diagrams in section 13.4.

The trials are independent so you multiply the probabilities.

$P(\text{HHT}) = 0.7 \times 0.7 \times 0.3 = 0.7^2 \times 0.3$

$P(\text{HTH}) = 0.7 \times 0.3 \times 0.7 = 0.7^2 \times 0.3$

$P(\text{THH}) = 0.3 \times 0.7 \times 0.7 = 0.7^2 \times 0.3$

P(exactly 2 heads in 3 tosses)

$= P(\text{HHT}) + P(\text{HTH}) + P(\text{THH})$

$= \mathbf{3} \times 0.7^2 \times 0.3$

$= 0.441$

There are **3** ways of arranging 2 heads and 1 tail.

b   You could show the outcomes for 6 tosses by extending the tree to 6 layers but this would be difficult to draw and very time-consuming. Instead, just focus on the end results giving exactly 2 heads, that is the different arrangements of 2 heads and 4 tails.

One such arrangement is in the order HHTTTT.

Now $P(\text{HHTTTT}) = 0.7 \times 0.7 \times 0.3 \times 0.3 \times 0.3 \times 0.3$

$= 0.7^2 \times 0.3^4$

(continued)

But there are several other arrangements, each with probability $0.7^2 \times 0.3^4$, such as

HTHTTT, TTTHTH, TTHTHT, and so on …

The number of arrangements is the number of different ways to choose 2 spaces from 6 spaces for H and then fill the remaining spaces with T.

The number of ways to choose 2 from 6

$$= \binom{6}{2} = {_6}C_2 = \frac{6!}{2!4!} = 15$$

so there are 15 ways of arranging 2 heads and 4 tails.

P(exactly 2 heads in 6 tosses)

$$= \binom{6}{2} \times 0.7^2 \times 0.3^4 = 15 \times 0.7^2 \times 0.3^4 = 0.0595 \text{ (3 sf)}$$

6 choose 2   2 heads   4 tails

> **Note**
>
> You can find $_6C_2$ directly on your calculator: $\boxed{6}\,\boxed{\text{nCr}}\,\boxed{2}\,\boxed{=}$

Conditions for a binomial distribution

A binomial situation arises from repeating a Bernoulli trial a fixed number of times.

If Y is the number of successful outcomes in a **single** Bernoulli trial, where $Y \sim \text{Bernoulli}(p)$, then Y can take the values 0 and 1.

Now consider $X = Y_1 + Y_2 + Y_r + \ldots + Y_n$, where

$Y_r = 0$ if the outcome of the rth trial is a failure

$Y_r = 1$ if the outcome of the rth trial is a success.

> **The random variable X, the number of successful outcomes in n independent Bernoulli trials, is said to follow a binomial distribution with parameters n and p.**

You can write

$$X \sim \text{B}(n, p)$$

> **Note**
>
> Only the values of n and p are needed to describe the binomial distribution fully.

So, a discrete random variable X follows a binomial distribution when all of the following conditions are satisfied:

▶ the number of trials, n, is fixed
▶ the trials are independent
▶ each trial results in one of two outcomes: success or failure
▶ the probability of success, p, is constant for each trial.

Deciding whether a binomial distribution is appropriate

You must check that *all* the conditions for a binomial distribution are satisfied *in the context of the situation*. If any one is not satisfied, then X does not follow a binomial distribution.

Example 4

In a bag there are 8 green counters and 7 red counters. Leyla and Waleed take part in an experiment in which 6 counters are to be selected at random from the bag. Leyla is told to put the counter back into the bag after each trial and Waleed is told to put the counter into his pocket after each trial.

The random variable X is the number of green counters selected.

Explain why X follows a binomial distribution in Leyla's experiment but not in Waleed's experiment.

In both experiments the process is carried out 6 times, so the number of trials is fixed. Also each trial results in a successful outcome (the counter is green) or an unsuccessful outcome (the counter is red).

Leyla's experiment

Since the counter is put back into the bag after each selection, the outcome of a trial does not depend on the outcomes of preceding trials, so the trials are independent.

Also, since $P(\text{green}) = \dfrac{8}{15}$ for each trial, the probability of success is constant.

So X follows a binomial distribution in Leyla's experiment.

> **Note**
>
> Sampling *with* replacement.

Waleed's experiment

Since the counter is not put back into the bag after each selection, each outcome depends on the outcomes of preceding trials, so the trials are not independent. X does not follow a binomial distribution.

> **Note**
>
> Sampling *without* replacement.

Note that if you are sampling without replacement from a *large* population, the variation in the probability of success for each trial would be negligible, so it is usually considered to be a binomial situation.

Example 5

Kwasi is playing a board game where he has to throw a six on the die in order to start. The random variable X is the number of times he throws the die until he throws a six. Explain why X does not follow a binomial distribution.

Kwasi might get a six on his first throw, but he might need to throw the die many times to get a six.

Since there is no limit to the number of trials required, the number of trials is not fixed and so X does not follow a binomial distribution.

If you know only that a trial is repeated a fixed number of times, you will need to make certain assumptions *related to the context* in order to use a binomial distribution.

Example 6

Fred takes 12 shots at a goal. State any assumptions necessary for the number of goals he scores to follow a binomial distribution.

The following assumptions must be made:

▶ Scoring a goal in one shot is independent of scoring a goal in all other shots.

▶ The probability of scoring a goal on any shot is the same for all 12 shots, that is, Fred's ability to score a goal does not improve with practice.

Formula for calculating binomial probabilities

The probability of x successes in n trials is given by

$$P(X=x)=\binom{n}{x}p^x q^{n-x} \quad \text{for } x=0,1,2,...,n \quad \text{where } q=1-p$$

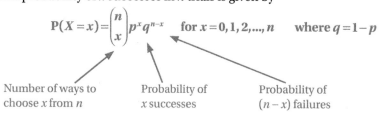

Number of ways to choose x from n

Probability of x successes

Probability of $(n-x)$ failures

Note

$$\binom{n}{x}={}_nC_x=\frac{n!}{x!(n-x)!}$$

Link between binomial probabilities and the binomial expansion of $(q+p)^n$

Writing $q=1-p$, it is easy to see the link between the probabilities in the binomial distribution and the terms in the binomial expansion of $(q+p)^n$, studied in P1.

For example, if you calculate the probabilities for $X \sim B(6, p)$ using the above formula, you will find that X has the following probability distribution:

x	0	1	2	3	4	5	6
$P(X=x)$	q^6	$6p^1q^5$	$15p^2q^4$	$20p^3q^3$	$15p^4q^2$	$6p^5q^1$	p^6

Note

The coefficients
1 6 15 20 15 6 1
appear in Pascal's Triangle on page 116.

Now expand $(q+p)^6$ using the **binomial theorem**.

You saw the binomial theorem in section 8.5, where

$$(q+p)^6=q^6+6q^5p^1+15q^4p^2+20\ q^3p^3+15q^2p^4+6q^1p^5+p^6$$

In statistics, the convention is to write the power of p before the power of q in each term. If you do this and then compare the expansion of $(q+p)^6$ with the probabilities in the table for $X \sim B(6, p)$, you will see that they correspond exactly.

$$(q+p)^6 = \quad q^6 \quad + \quad 6p^1q^5 \quad + \quad 15p^2q^4 \quad + \quad 20p^3q^3 \quad + \quad 15p^4q^2 \quad + \quad 6p^5q^1 \quad + \quad p^6$$

$\quad\quad\quad\quad\quad P(X=0) \quad P(X=1) \quad P(X=2) \quad P(X=3) \quad P(X=4) \quad P(X=5) \quad P(X=6)$

Check the sum of the probabilities:

$$\sum P(X=x)=(q+p)^6=(1-p+p)^6=1^6=1, \text{ as expected.}$$

Example 7

X follows a binomial distribution, where $X \sim B\left(11, \frac{1}{3}\right)$. Calculate

a P($X=4$)

b P($X=9$)

Answer

Question

$$X \sim B\left(11, \frac{1}{3}\right) \qquad\qquad n = 11, p = \frac{1}{3}, q = 1 - p = \frac{2}{3}$$

a $\quad P(X = 4) = \binom{11}{4} p^4 q^7 \qquad\qquad$ The indices add up to 11.

$$= \binom{11}{4} \times \left(\frac{1}{3}\right)^4 \times \left(\frac{2}{3}\right)^7$$

$$= 0.238 \text{ (3 sf)}$$

b $\quad P(X = 9) = \binom{11}{9} p^9 q^2$

$$= \binom{11}{9} \times \left(\frac{1}{3}\right)^9 \times \left(\frac{2}{3}\right)^2$$

$$= 0.00124 \text{ (3 sf)}$$

You can write down the following probabilities straight away

$$P(\text{no successes}) = P(X = 0) = \binom{n}{0} \times p^0 q^n \;\; = q^n \qquad\qquad \binom{n}{0} = 1 \text{ and } p^0 = 1$$

$$P(\text{all successes}) = P(X = n) = \binom{n}{n} \times p^n q^0 \;\; = p^n \qquad\qquad \binom{n}{n} = 1 \text{ and } q^0 = 1$$

Example 8

At a certain supermarket 49% of customers pay by debit card. Find the probability that in a random sample of 12 customers,

a exactly 7 pay by debit card

b at most 2 pay by debit card

c at least 3 but fewer than 5 pay by debit card

d all pay by debit card.

Let D be the number of customers in a sample of 12 who pay by debit card.

$$D \sim B(12, 0.49) \qquad\qquad n = 12, p = 0.49, q = 1 - p = 0.51$$

Note

Define the variable in words and say how it is distributed.

a $\quad P(\text{exactly 7})$

$$= P(D = 7)$$

$$= \binom{12}{7} p^7 q^5$$

$$= \binom{12}{7} \times 0.49^7 \times 0.51^5$$

$$= 0.1853\ldots = 0.185 \text{ (3 sf)}$$

b $\quad P(\text{at most 2})$

$$= P(D \leq 2)$$

$$= P(D = 0) + P(D = 1) + P(D = 2)$$

$$= q^{12} + \binom{12}{1} p^1 q^{11} + \binom{12}{2} p^2 q^{10}$$

$$= 0.51^{12} + \binom{12}{1} \times 0.49^1 \times 0.51^{11} + \binom{12}{2} \times 0.49^2 \times 0.51^{10}$$

$$= 0.02274\ldots = 0.0227 \text{ (3 sf)}$$

(continued)

(continued)

c P(at least 3 but fewer than 5)

$$= P(3 \le D < 5)$$

$$= P(D = 3) + P(D = 4)$$

$$= \binom{12}{3}p^3 q^9 + \binom{12}{4}p^4 q^8$$

$$= \binom{12}{3} \times 0.49^3 \times 0.51^9 + \binom{12}{4} \times 0.49^4 \times 0.51^8$$

$$= 0.1910... = 0.191 \text{ (3 sf)}$$

d P(all pay by debit card)

$$= P(D = 12)$$

$$= p^{12}$$

$$= 0.49^{12}$$

$$= 0.0001915... = 0.000192 \text{ (3 sf)}$$

It is also useful to recall that, for any discrete random variable,

$$\sum P(X = x) = 1,$$

so the sum of all possible probabilities is 1.

Example 9

On average, three quarters of the patients who have a check-up at a particular dental practice do not need follow-up treatment.

Find the probability that, in a random sample of 9 patients from the practice, the number that do not need follow-up treatment is

a at most 7   **b** at least 3.

Let X be the number of patients in a sample of 9 who do not need follow-up treatment.

$$X \sim B(9, 0.75) \qquad\qquad n = 9, p = 0.75, q = 0.25$$

a P(at most 7)

$$= P(X \le 7)$$

$$= 1 - P(X > 7)$$

$$= 1 - (P(X = 8) + P(X = 9))$$

$$= 1 - \left(\binom{9}{8}p^8 q^1 + p^9 \right)$$

$$= 1 - \left(\binom{9}{8} \times 0.75^8 \times 0.25^1 + 0.75^9 \right)$$

$$= 1 - 0.3003...$$

$$= 0.6996...$$

$$= 0.700 \text{ (3 sf)}$$

b P(at least 3)

$$= P(X \ge 3)$$

$$= 1 - (P(X = 0) + P(X = 1) + P(X = 2))$$

(continued)

274 **Bernoulli and Binomial Distributions**

(continued)

$$= 1 - \left(q^9 + \binom{9}{1} p^1 q^8 + \binom{9}{2} p^2 q^7 \right)$$

$$= 1 - \left(0.25^9 + \binom{9}{1} \times 0.75^1 \times 0.25^8 + \binom{9}{2} \times 0.75^2 \times 0.25^7 \right)$$

$$= 0.9986... = 0.999 \text{ (3 sf)}$$

Exercise 2

1 If $X \sim B(10, 0.3)$, find

  a $P(X=2)$ **b** $P(X=6)$ **c** $P(X=10)$ **d** $P(X=0)$

2 If $X \sim B(8, 0.25)$ find

  a $P(X=5)$ **b** $P(X=3)$ **c** $P(X \le 3)$ **d** $P(X \ge 7)$

3 If $X \sim B\left(9, \dfrac{1}{3} \right)$ find

  a $P(X=5)$ **b** $P(X<2)$ **c** $P(X>7)$

4 If $X \sim B(9, 0.45)$ find

  a $P(2 \le X \le 4)$ **b** $P(5 < X < 8)$

5 If $X \sim B(13, 0.7)$ find

  a $P(8 < X \le 10)$ **b** $P(8 \le X < 10)$

6 If $X \sim B(11, 0.64)$, find

  a $P(X \le 8)$ **b** $P(X>2)$

7 If $M \sim B(8, 0.73)$, find

  a $P(M \le 7)$ **b** $P(M>6)$

8 A coin is biased so that the probability of obtaining a tail is 0.42. The coin is tossed 12 times. Find the probability that the number of tails obtained is

  a exactly 6   **b** at least 10   **c** no more than 2.

9 The probability that a patient attending a clinic has a particular health condition is 0.4. Find the probability that in a randomly chosen group of 7 patients the number with the health condition is

  a exactly 3   **b** more than 5

  c fewer than 2   **d** at least 2 but no more than 4.

10 A 5-sided spinner is equally likely to stop on any of the numbers 1, 2, 3, 4 or 5. Aaliyah spins the spinner 10 times. Find the probability that the spinner stops on

  a an even number on exactly 7 spins

  b an odd number on more than 7 spins.

11 In an experiment 5 fair coins are tossed together. Find the probability that they land showing

  a exactly 3 tails

  b fewer than 3 tails

  c more than 3 heads

  d at least 2 heads but no more than 4 heads.

12 On average, 1 in 8 people living in a particular country were not born in that country. Determine the probability that, in a random sample of 20 people living in the country:

   a   exactly 5 were not born in the country

   b   at least 3 were not born in the country

   c   more than 18 were born in the country.

13 The discrete random variable X is such that $X \sim B(4, p)$.

Complete the probability distribution table for X in terms of p and q, where $q = 1 - p$.

x	0	1	2	3	4
$P(X = x)$		$4p^1q^3$			

15.3 The cumulative binomial distribution function

When $X \sim B(n, p)$, the **cumulative binomial distribution function** gives *cumulative* probabilities $P(X \leq x)$, where

$$P(X \leq x) = P(X = 0) + P(X = 1) + P(X = 2) + \cdots + P(X = x)$$

In the examination you will have access to a table giving cumulative probabilities. This is **Table 1** in the examination booklet: *Formulae and Statistical Tables.*

Different sections of the table give cumulative probabilities for $n = 2$ to $n = 15$, then for $n = 20, 25, 30, 40$ and 50.

In the extract printed below, you will see that values of p from 0.01 to 0.5 are written across the top and x-values are written down the side.

An extract from Table 1: Cumulative Binomial Distribution Function

The tabulated values give $P(X \leq x)$ when $n = 7$, so when $X \sim B(7, p)$.

p	0.01	0.02	0.03	0.04	0.05	0.06	0.07	0.08	0.09	0.10	0.15	0.20	0.25	0.30	0.35	0.40	0.45	0.50	p
x	$n = 7$																		x
0	0.9321	0.8681	0.8080	0.7514	0.6983	0.6485	0.6017	0.5578	0.5168	0.4783	0.3206	0.2097	0.1335	0.0824	0.0490	0.0280	0.0152	0.0078	0
1	0.9999	0.9921	0.9829	0.9706	0.9556	0.9382	0.9187	0.8974	0.8745	0.8503	0.7166	0.5767	0.4449	0.3294	0.2338	0.1586	0.1024	0.0625	1
2	1.0000	0.9997	0.9991	0.9980	0.9962	0.9937	0.9903	0.9860	0.9807	0.9743	0.9262	0.8520	0.7564	0.6471	0.5323	0.4199	0.3164	0.2266	2
3		1.0000	1.0000	0.9999	0.9998	0.9996	0.9993	0.9988	0.9982	0.9973	0.9879	0.9667	0.9294	0.7840	0.8002	0.7102	0.6083	0.5000	3
4				1.0000	1.0000	1.0000	1.0000	0.9999	0.9999	0.9998	0.9988	0.9953	0.9871	0.9712	0.9444	0.9037	0.8471	0.7734	4
5								1.0000	1.0000	1.0000	0.9999	0.9996	0.9987	0.9962	0.9910	0.9812	0.9643	0.9375	5
6											1.0000	1.0000	0.9999	0.9998	0.9994	0.9984	0.9963	0.9922	6
7													1.0000	1.0000	1.0000	1.0000	1.0000	1.0000	7

The cumulative probabilities in Table 1 are given to 4 decimal places. Notice that, in any particular column, after the value 1.0000 (4 dp) is first reached, subsequent entries (also 1.0000) are left blank.

Using Table 1 when $p \leq 0.5$

Suppose you want probabilities for $X \sim B(7, 0.2)$. You need the column for p headed 0.20. The relevant values are printed again below.

$n = 7$	$p = 0.20$	
x		
0	0.2097	$\longleftarrow$ P($X \leq 0$)
1	0.5767	$\longleftarrow$ P($X \leq 1$)
2	0.8520	$\longleftarrow$ P($X \leq 2$)
3	0.9667	$\longleftarrow$ P($X \leq 3$)
4	0.9953	$\longleftarrow$ P($X \leq 4$)
5	0.9996	$\longleftarrow$ P($X \leq 5$)
6	1.0000	$\longleftarrow$ P($X \leq 6$)
7		$\longleftarrow$ P($X \leq 7$)

> **Note**
>
> P($X \leq 0$) = P($X = 0$)

So, when $X \sim B(7, 0.2)$,   **Possible values of X**

a P($X \leq 5$) = 0.9996   0 1 2 3 4 5 6 7

b P($X \leq 3$) = 0.9667   0 1 2 3 4 5 6 7

c P($X = 0$) = P($X \leq 0$) = 0.2097   0 1 2 3 4 5 6 7

You will need to relate other probabilities to P($X \leq x$) and care must be taken to distinguish between $\leq$ and $<$ and between $\geq$ and $>$.

d P($X < 3$) = P($X \leq 2$) = 0.8520   0 1 2 3 4 5 6 7

e P($X > 4$) = 1 − P($X \leq 4$)   0 1 2 3 4 5 6 7
$\quad\quad\quad = 1 - 0.9953$
$\quad\quad\quad = 0.0047$

f P($X \geq 2$) = 1 − P($X \leq 1$)   0 1 2 3 4 5 6 7
$\quad\quad\quad = 1 - 0.5767$
$\quad\quad\quad = 0.4233$

g P($X = 4$) = P($X \leq 4$) − P($X \leq 3$)   0 1 2 3 4 5 6 7
$\quad\quad\quad = 0.9953 - 0.9667$
$\quad\quad\quad = 0.0286$

h P($1 \leq X \leq 5$) = P($X \leq 5$) − P($X \leq 0$)   0 1 2 3 4 5 6 7
$\quad\quad\quad = 0.9996 - 0.2097$
$\quad\quad\quad = 0.7899$

i P($1 < X \leq 5$) = P($X \leq 5$) − P($X \leq 1$)   0 1 2 3 4 5 6 7
$\quad\quad\quad = 0.9996 - 0.5767$
$\quad\quad\quad = 0.4229$

j P($1 \leq X < 5$) = P($X \leq 4$) − P($X \leq 0$)   0 1 2 3 4 5 6 7
$\quad\quad\quad = 0.9953 - 0.2097$
$\quad\quad\quad = 0.7856$

k P($1 < X < 5$) = P($X \leq 4$) − P($X \leq 1$)   0 1 2 3 4 5 6 7
$\quad\quad\quad = 0.9953 - 0.5767$
$\quad\quad\quad = 0.4186$

Example 10

The proportion of members of a health club who play tennis is 0.35.

A sample of 40 members of the health club is selected at random.

Determine the probability that the number of these members who play tennis is:

a   at most 12;

b   at least 10 but at most 15;

c   exactly 15.

AQA MS1A June 2010

Let X be the number of members in a sample of 40 who play tennis.

$n = 40$ and $p = 0.35$, so $X \sim B(40, 0.35)$.

Use cumulative binomial probability tables for $n = 40$ and $p = 0.35$

> **Note**
>
> See Table 1 in *Formulae and Statistical Tables*.

a $P(\text{at most } 12) = P(X \leq 12)$

$= 0.3143 = 0.314 \text{ (3 sf)}$

b $P(\text{at least 10 but at most 15})$

$= P(10 \leq X \leq 15)$ ... 8 9 **10 11 12 13 14 15** 16 17 ...

$= P(X \leq 15) - P(X \leq 9)$

$= 0.6946 - 0.0644$

$= 0.6302 = 0.630 \text{ (3 sf)}$

c $P(\text{exactly 15}) = P(X = 15)$

$= P(X \leq 15) - P(X \leq 14)$ ... 13 14 **15** 16 17 ...

$= 0.6946 - 0.5721$

$= 0.1225$

$= 0.123 \text{ (3 sf)}$

Alternatively, it may be quicker to use the formula as follows:

$$P(X = 15) = \binom{40}{15} \times 0.35^{15} \times 0.65^{25}$$

$$= 0.12256... = 0.123 \text{ (3 sf)}$$

In this example, the answers agree to 3 significant figures.

Example 11

In a particular production process, the probability that an item is faulty is 0.08. In a quality control test, a random sample of 50 items is taken. Find the probability that

a   exactly 4 are faulty

b   more than 5 but fewer than 9 are faulty

c   at least 40 are *not faulty*.

Let X be the number of faulty items in the sample of 50.

Then $X \sim B(50, 0.08)$.

Use cumulative binomial probability tables for $n = 50$ and $p = 0.08$

a P(exactly 4 are faulty)

$= P(X = 4)$

$= P(X \leq 4) - P(X \leq 3)$

$= 0.6290 - 0.4253$

$= 0.2037 = 0.204 \text{ (3 sf)}$

b P(more than 5 but fewer than 9 are faulty)

$= P(5 < X < 9)$ \qquad ... 4 5 **6 7 8** 9 ...

$= P(X \leq 8) - P(X \leq 5)$

$= 0.9833 - 0.7919$

$= 0.1914 = 0.191 \text{ (3 sf)}$

c P(at least 40 are not faulty)

$= $ P(at most 10 are faulty) \qquad Not faulty \quad ... **40 41 42 ... 49 50**

$= P(X \leq 10)$ \qquad\qquad\qquad Faulty \qquad ... **10 9 8 ... 1 0**

$= 0.9983 = 0.998 \text{ (3 sf)}$

Note that this technique is helpful when using the tables when $p > 0.5$.

$n = 50$	$p = 0.08$
$x = 0$	0.0155
1	0.0827
2	0.2260
3	0.4253
4	0.6290
5	0.7919
6	0.8981
7	0.9562
8	0.9833
9	0.9944
10	0.9983
11	0.9995
12	0.9999
13	1.0000

Exercise 3

Use cumulative binomial probability tables where appropriate.

1 If $Y \sim B(14, 0.4)$, find

  a $P(Y \leq 7)$ **b** $P(Y < 5)$ **c** $P(Y > 9)$

  d $P(Y \geq 6)$ **e** $P(Y = 8)$

2 If $X \sim B(10, 0.45)$, find

  a $P(2 \leq X \leq 7)$ **b** $P(2 < X \leq 7)$

  c $P(2 \leq X < 7)$ **d** $P(2 < X < 7)$

3 The random variable X follows a binomial distribution with $n = 25$ and $p = 0.35$. Find

  a $P(10 \leq X \leq 15)$ **b** $P(X = 12)$ **c** $P(X < 7)$

  d $P(X \geq 9)$ **e** $P(4 < X < 14)$

4 On average 1 in 10 of the chocolates produced in a factory are mis-shaped. In a random sample of 40 chocolates, find the probability that the number of mis-shaped chocolates is

  a fewer than 9   **b** at least 3 but at most 10

  c 6 or more.

5 A certain tribe is distinguished by the fact that 45% of the males have six toes on their right foot. Stating any necessary assumptions, find the probability that, in a selected group of 25 males from the tribe, more than 13 have six toes on their right foot.

6. One-fifth of a certain population has a minor eye defect. Find the probability that the number of people with this eye defect is

   a more than 2 in a random sample of 10 people

   b exactly 2 in a random sample of 10 people

   c at least 10 in a random sample of 40 people.

7. A manufacturer makes two sizes of screws: 40% are large and 60% are small. They are packed into packs of 20. Assuming that each pack of these screws contains a random selection, calculate the probability that, in a pack of 20 screws, there are

   a equal numbers of large and small screws

   b more than 15 large screws.

8. In a particular region 10% of people have blood type B.

   a Find the probability that exactly 3 have blood type B in a random sample of 5 people from the region.

   b Find the probability that at most 2 have blood type B in a random sample of 9 people from the region.

   c Find the probability that exactly 13 do **not** have blood type B in a random sample of 15 people from the region.

Using Table 1 when $p > 0.5$

When $p > 0.5$, use the symmetry property of the binomial distribution.

This is illustrated below in the probability distributions of $X \sim B(5, 0.3)$ and $Y \sim B(5, 0.7)$. Notice that $n = 5$ in both distributions, and the values of p add up to 1.

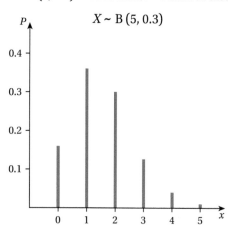
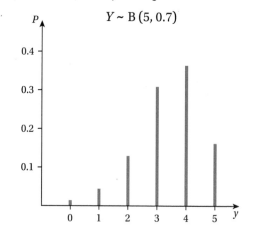

By symmetry, you can see that

$$P(X=0) = P(Y=5)$$
$$P(X=1) = P(Y=4)$$
$$P(X=2) = P(Y=3) \text{ and so on.}$$

Also note, for example, that

$$P(X<2) = P(Y>3) \qquad P(X \le 2) = P(Y \ge 3)$$
$$P(X<4) = P(Y<1) \qquad P(X \ge 4) = P(Y \le 1)$$

Example 12

The random variable X is distributed B(8, 0.6). Use cumulative binomial tables to find

a $P(X \geq 3)$

b $P(X \leq 2)$

c $P(X = 5)$

These two probabilities add up to 1.

$X \sim$ B(8, 0.6), so consider $Y \sim$ B(8, 0.4) and use the table for $n = 8$ and $p = 0.4$.

$Y \sim B(8, 0.4)$

$n = 8$	$p = 0.4$
$y = 0$	0.0618
1	0.1064
2	0.3154
3	0.5941
4	0.8263
5	0.9502
6	0.9915
7	0.9993
8	1.0000

a $P(X \geq 3) = P(Y \leq 5) = 0.9502$

These two numbers add up to 8.

b $P(X \leq 2) = P(Y \geq 6)$

$\qquad = 1 - P(Y \leq 5)$

$\qquad = 1 - 0.9502$

$\qquad = 0.0498$

c $P(X = 5) = P(Y = 3)$

$\qquad = P(Y \leq 3) - P(Y \leq 2)$

$\qquad = 0.5941 - 0.3154$

$\qquad = 0.2787$

Example 13

In a particular region, 85% of adults own a cell phone. A random sample of 30 adults is selected from the region. Determine the probability that the number of adults in the sample that own a cell phone is:

a   more than 25

b   more than 22 but fewer than 28

c   at least 20 but fewer than 28.

P(owns a cell phone) = 0.85

Let X be the number of people in 30 who own a cell phone.

Then $X \sim$ B(30, 0.85).

Since $p > 0.5$, consider Y, the number of people in 30 who *do not* own a cell phone.

P(does not own a cell phone) = $1 - 0.85 = 0.15$, so $Y \sim$ B(30, 0.15).

Using cumulative binomial tables for Y with $n = 30$, $p = 0.15$

> **Note**
>
> See Table 1 in *Formulae and Statistical Tables.*

a $P(X > 25) = P(Y < 5)$

$\qquad = P(Y \leq 4)$

$\qquad = 0.5245 = 0.525$ (3 sf)

X: Own a cell phone   ... 23 24 25 **26 27 28 29 30**

Y: Do not own a cell phone   ... 7 6 5 **4 3 2 1 0**

(continued)

(continued)

b
$$P(22 < X < 28) = P(23 \le X \le 27)$$

X: Own a cell phone   ... 22 **23 24 25 26 27** 28 29 30

Y: Do not own a cell phone   ... 8 **7 6 5 4 3** 2 1 0

$$= P(3 \le Y \le 7)$$
$$= P(Y \le 7) - P(Y \le 2)$$
$$= 0.9302 - 0.1514$$
$$= 0.7788 = 0.779 \ (3 \ \text{sf})$$

c
$$P(20 \le X < 28) = P(20 \le X \le 27)$$

X: Own a cell phone   ... 19 **20 21 22 23 24 25 26 27** 28 29 30

Y: Do not own a cell phone   ... 11 **10 9 8 7 6 5 4 3** 2 1 0

$$= P(3 \le Y \le 10)$$
$$= P(Y \le 10) - P(Y \le 2)$$
$$= 0.9971 - 0.1514$$
$$= 0.8457 = 0.846 \ (3 \ \text{sf})$$

Exercise 4

Use cumulative binomial probability tables where appropriate.

1 If $X \sim B(11, 0.65)$, find

  a $P(X = 6)$ **b** $P(X < 3)$ **c** $P(X \ge 9)$

2 The random variable X is distributed as $B(50, 0.7)$. Find

  a $P(X \ge 40)$ **b** $P(32 \le X \le 41)$

  c $P(X < 29)$ **d** $P(X = 35)$

3 When Alex tries to send a fax, the probability that he can successfully send it is 0.85. Each attempt is independent of all other attempts. He tries to send 50 faxes.

Find the probability that he can successfully send at least 46 faxes.

4 An experiment consists of taking shots at a target and counting the number of hits. The probability of hitting the target with a single shot is 0.8. Stating any necessary assumptions, find the probability that in 13 consecutive attempts the target is hit at most 11 times.

5 In a survey it is found that 65% of shoppers choose Soapy Suds washing powder.

Stating any necessary assumptions, determine the probability that, in a group of 10 shoppers, the number who buy Soapy Suds washing powder is

  a at most 8   **b** at least 4   **c** fewer than 3.

6 At a particular hospital, records show that each day, on average, 80% of people keep their appointment at the outpatients' clinic.

  a Find the probability that in a random sample of 10 patients, more than 7 keep their appointment.

  b Find the probability that in a day when 30 appointments have been booked the number that keep their appointments is

    i at most 21

    ii at least 23

    iii more than 20 but fewer than 27

    iv exactly 25.

> **Note**
>
> ► It is often quicker to use cumulative probability tables but sometimes you may have to use the formula as the tables are not available for all possible values of p or of n.
>
> ► The values given in the tables usually agree, to 3 decimal places, with calculated values.

7 A certain variety of flower seed is sold in packets containing a large number of seeds. It is claimed on the packet that 40% will bloom white and 60% will bloom red and this may be assumed to be accurate.

  a 15 seeds are planted. Find the probability that

    i more than 10 will bloom red

    ii fewer than 7 will bloom red.

  b 50 seeds are planted. Find the probability that exactly half will bloom red.

8 There are 40 boxes of balloons in a storeroom. Each box contains 3 red balloons and 7 green balloons. A balloon is taken at random from each box. Find the probability that

  a no more than 25 green balloons are taken

  b at least 10 red balloons are taken

  c more green balloons than red balloons are taken.

15.4 Mean, variance and standard deviation of a binomial distribution

To find the mean and variance of a binomial variable $X \sim \mathrm{B}(n, p)$, use the fact that X is the sum of n independent observations of a Bernoulli variable Y, where $Y \sim \mathrm{Bernoulli}(p)$,

so $X = Y_1 + Y_2 + \cdots + Y_n$

Now, when $Y \sim \mathrm{Bernoulli}(p)$,

    $\mathrm{E}(Y) = p,$ and $\mathrm{Var}(Y) = pq$ where $q = 1 - p$

So, $\mathrm{E}(X) = \mathrm{E}(Y_1 + Y_2 + \cdots + Y_n)$

        $= \mathrm{E}(Y_1) + \mathrm{E}(Y_2) + \cdots \mathrm{E}(Y_n)$

        $= p + p + \cdots + p$

        $= np$

$\mathrm{Var}(X) = \mathrm{Var}(Y_1 + Y_2 + \cdots + Y_n)$

        $= \mathrm{Var}(Y_1) + \mathrm{Var}(Y_2) + \cdots + \mathrm{Var}(Y_n)$

        $= pq + pq + \cdots + pq$

        $= npq$

  If $X \sim \mathrm{B}(n, p)$,

    mean $= \mathrm{E}(X) = np$

    variance $= \mathrm{Var}(X) = npq$ where $q = 1 - p$

    standard deviation $= \sqrt{npq}$

Example 14

In a large consignment of apples, 15% are rejected for being too small.
A random sample of 50 apples is taken from the consignment.

Determine the mean and standard deviation of the number of rejected apples
in the sample.

Let X be the number of rejected apples in a consignment of 50.

$X \sim B(50, 0.15)$ $n = 50, p = 0.15, q = 0.85$

mean $= np = 50 \times 0.15 = 7.5$

standard deviation $= \sqrt{npq} = \sqrt{50 \times 0.15 \times 0.85} = 2.524... = 2.52 \text{ (3 sf)}$

Example 15

The random variable X has distribution $B(20, 0.25)$. The mean of X is μ and the
standard deviation is σ.

a   Find the values of μ and σ.

b   Calculate the percentage of the distribution that lies within one standard
deviation of the mean. Give your answer to 2 significant figures.

$X \sim B(20, 0.25)$ $n = 20, p = 0.25, q = 0.75$

a $\mu = np$

$= 20 \times 0.25$

$= 5$

$\sigma = \sqrt{npq}$

$= \sqrt{20 \times 0.25 \times 0.75} = 1.936... = 1.94 \text{ (3 sf)}$

b $\mu - \sigma = 5 - 1.94 = 3.06$ and $\mu + \sigma = 5 + 1.94 = 6.94$

P(X lies within one standard deviation of the mean)

$= P(3.06 < X < 6.94)$

$= P(X = 4, 5, 6)$ *X takes integer values only*

$= P(X \leq 6) - P(X \leq 3)$ 0 1 2 3 4 5 6 7 8 ....

$= 0.7858 - 0.2252$

$= 0.5606$

$= 56\% \text{ (2 sf)}$

$n = 20$	$p = 0.25$
$x = 0$	0.0032
1	0.0243
2	0.0913
3	0.2252
4	0.4148
5	0.6172
6	0.7858
etc	

Example 16

The random variable X is distributed $B(n, p)$. The mean of X is 5 and the
standard deviation is 2.

Find the values of n and p.

Bernoulli and Binomial Distributions

Mean $= np = 5$ (1)

Variance $= npq = 2^2 = 4$ (2)

Substitute for np from (1) into (2)

$\quad 5q = 4$

$\quad q = 0.8$

$\quad p = 1 - 0.8 = 0.2$

Substitute for p in (1)

$\quad n \times 0.2 = 5$

$\quad n = \dfrac{5}{0.2} = 25$

So $n = 25$, $p = 0.2$ and $X \sim B(25, 0.2)$.

Example 17

A fair coin is tossed 10 times and the number of tails recorded.

a   The random variable X is the number of tails.

    State the mean and variance of X.

b   The number of tails is multiplied by 3 and denoted by the random variable Y.

    State the mean and standard deviation of Y.

a X is the number of tails in 10 tosses.

    $X \sim B(10, 0.5)$ $n = 10$, $p = 0.5$, $q = 1 - p = 0.5$

    Mean $= np = 10 \times 0.5 = 5$

    Variance $= npq = 10 \times 0.5 \times 0.5 = 2.5$

b $Y = 3X$

    $E(Y) = 3E(X) = 3 \times 5 = 15$

    $\text{Var}(Y) = 3^2\,\text{Var}(X) = 9\,\text{Var}(X) = 9 \times 2.5 = 22.5$

    s.d. of $Y = \sqrt{22.5} = 4.743\ldots = 4.74\,(3\,\text{sf})$

Exercise 5

1 The random variable X has distribution $B(14, 0.36)$. Find

   a   the mean   **b**   the variance

   c   the standard deviation.

2 The random variable Y has distribution $B(20, 0.4)$. Find the probability that Y is equal to the mean of Y.

3 The probability that an item produced by a machine is satisfactory is 0.92.

   a   Find the expected number of satisfactory items in a random sample of 25 items produced by the machine.

   b   Find the standard deviation of the number of unsatisfactory items in a random sample of 50 items produced by the machine.

4 The random variable Y has distribution $B(n, 0.3)$ and $E(Y) = 2.4$.

   a   Find n.   **b**   Find $P(Y = 5)$.

   c   Find the standard deviation of Y.

5 X is the number of tails obtained when a fair coin is tossed 10 times. The mean of X is μ and the standard deviation of X is σ.

  a Find μ and σ.

  b **i** Find $P(X < \mu - 2\sigma)$.   **ii** Find $P(X > \mu + 2\sigma)$.

  c Find the probability that X is more than two standard deviations away from the mean.

6 In a multiple choice test, for each question students have to choose the correct answer from a choice of four answers. There are 20 questions in the test. Jack decides not to read any of the questions, but to select an answer at random each time.

  a What is the probability that Jack answers a question correctly?

  b Find the mean and standard deviation of the number of correct answers that Jack gets.

  c Find the probability that the number of correct answers that Jack gets is within one standard deviation of the mean.

7 In a bag there are 6 red counters, 8 yellow counters and 6 green counters.

  Ronami selects a counter at random from the bag, notes its colour and then puts it back into the bag. She does this four times in all. Find

  a the probability that she selects 4 red counters

  b the expected number of yellow counters she selects

  c the variance of the number of green counters she selects.

8 The probability that a person chosen at random wears glasses is p.

  A random sample of n people is chosen and the number of people in the sample who wear glasses is denoted by X.

  It is given that $E(X) = 2.4$ and $Var(X) = 1.68$.

  a Find the value of p.

  b Find the value of n.

  c Find the probability that exactly 6 people in the sample wear glasses.

9 It is given that $X \sim B(n, p)$. The mean of X is 3.6 and the variance of X is 2.16. Find

  a n and p **b** $P(X \le 2)$.

10 A calculator randomly generates the digits 0, 1, 2, 3, 4, 5, 6, 7, 8 and 9, so that each digit has an equal chance of occurring.

  a What is the probability that a zero is generated?

  b Six digits are generated. Find the probability that there is exactly one zero.

  c Twenty digits are generated. Find the expected number of digits that are multiples of 3.

  d One hundred digits are generated. Find the variance of the number of odd digits.

11. Lilia travels to work by bus or by car. The probability that she travels by bus on any day is 0.7. If she travels by bus, there is a probability of 0.1 that she is late for work. If she travels by car there is a probability of 0.2 that she is late for work.

    a By drawing a tree diagram, or otherwise, find the probability that she is late for work on a particular day.

    b Find the expected number of days she is late for work in 20 working days.

    c Find the variance of the number of days she travels by car in 10 working days.

12. The discrete random variable X is such that $X \sim B(n, p)$ with $n = 3$ and $p = 0.4$.

    a Complete the probability distribution table for X.

x	0	1	2	3
$P(X=x)$		0.432		0.064

    b Use the values in the table to calculate $E(X)$ and $Var(X)$.

    c Calculate np and npq where $q = 1 - p$ and confirm that $E(X) = np$ and $Var(X) = npq$.

15.5 Further applications

Example 18

At a large college, an analysis of the number of distinctions each student was awarded in assignments resulted in the following information. For example, 26% of students achieved exactly 3 distinctions in their assignments.

Number of distinctions	0	1	2	3	4	5	6
Percentage of students	2	8	35	26	17	5	7

A random sample of 18 students was selected.

a Find the probability that the sample contains

   i fewer than 4 students with 2 distinctions

   ii at most 16 students with fewer than 4 distinctions.

b Find the mean and variance of the number of students in the sample with at least 3 distinctions.

a i From the table, 35% of students had 2 distinctions, so P(student has 2 distinctions) = 0.35.

Let X be the number of students in the sample of 18 with 2 distinctions.

$X \sim B(18, 0.35)$ $n = 18, p = 0.35, q = 1 - p = 0.65$

$P(X < 4) = P(X = 0) + P(X = 1) + P(X = 2) + P(X = 3)$

$$= 0.65^{18} + \binom{18}{1} \times 0.35^1 \times 0.65^{17} + \binom{18}{2} \times 0.35^2 \times 0.65^{16}$$

$$+ \binom{18}{3} \times 0.35^3 \times 0.65^{15}$$

$$= 0.07826\ldots = 0.0783 \text{ (3 sf)}$$

(continued)

ii P(student has fewer than 4 distinctions) $= 0.02 + 0.08 + 0.35 + 0.26 = 0.71$

Let Y be the number of students in the sample of 18 with fewer than 4 distinctions.

$Y \sim \text{B}(18, 0.71)$ $\qquad\qquad n = 18, p = 0.71, q = 0.29$

$P(Y \leq 16) = 1 - [P(Y = 17) + P(Y = 18)]$

$$= 1 - \left(\binom{18}{17} \times 0.71^{17} \times 0.29^{1} + 0.71^{18} \right)$$

$$= 0.9824\ldots = 0.982 \text{ (3 sf)}$$

b P(student has at least 3 distinctions) $= 0.26 + 0.17 + 0.05 + 0.07 = 0.55$

Let A be the number of students in the sample of 18 with at least 3 distinctions.

$A \sim \text{B}(18, 0.55)$ $\qquad\qquad n = 18, p = 0.55, q = 0.45$

Mean $= np = 18 \times 0.55 = 9.9$

Variance $= npq = 18 \times 0.55 \times 0.45 = 4.455$

Example 19

An amateur tennis club purchases tennis balls that have been used previously in professional tournaments.

The probability that each such ball fails a standard bounce test is 0.15.

The club purchases boxes each containing 10 of these tennis balls. Assume that the 10 balls in any box represent a random sample.

a Determine the probability that the number of balls in a box which fail the bounce test is:

i at most 2; $\qquad\qquad$ **ii** at least 2;

iii more than 1 but fewer than 5.

b Determine the probability that, in **5 boxes**, the total number of balls which fail the bounce test is:

i more than 5; $\qquad\qquad$ **ii** at least 5 but at most 10.

c Calculate the mean and variance for the total number of balls in **50 boxes** which fail the bounce test.

AQA MS1A June 2011

$n = 10$	$p = 0.15$
$x = 0$	0.1969
1	0.5443
2	0.8202
3	0.9500
4	0.9901
5	0.9986
6	0.9999
7	1.0000
8	
9	
10	

a Let X be the number of balls in a box of 10 which fail the bounce test. Then $X \sim \text{B}(n, p)$ where $n = 10, p = 0.15$, so $X \sim \text{B}(10, 0.15)$.

i P(at most 2) $= P(X \leq 2) = 0.8202 = 0.820 \text{ (3 sf)}$

ii P(at least 2) $= P(X \geq 2)$

$$= 1 - P(X \leq 1)$$

$$= 1 - 0.5443$$

$$= 0.4557 = 0.456 \text{ (3 sf)}$$

(continued)

(continued)

iii P(more than 1 but fewer than 5)

$$= P(X = 2, 3, 4)$$
$$= P(X \leq 4) - P(X \leq 1)$$
$$= 0.9901 - 0.5443$$
$$= 0.4458 = 0.446 \text{ (3 sf)}$$

b Let Y be the number of balls in 5 boxes which fail the bounce test.

Then $Y \sim B(n, p)$ where $n = 50$, $p = 0.15$, so $Y \sim B(50, 0.15)$.

i P(more than 5) = P($Y > 5$)

$$= 1 - P(Y \leq 5)$$
$$= 1 - 0.2194$$
$$= 0.7086 = 0.709 \text{ (3 sf)}$$

ii P(at least 5 but at most 10)

$$= P(5 \leq Y \leq 10)$$
$$= P(Y \leq 10) - P(Y \leq 4)$$
$$= 0.8801 - 0.1121$$
$$= 0.7680 = 0.768 \text{ (3 sf)}$$

c Let T be the total number of balls in 50 boxes which fail the bounce test.

$T \sim B(n, p)$ where $n = 500$, $p = 0.15$, so $T \sim B(500, 0.15)$

Mean $= E(T) = np = 500 \times 0.15 = 75$

Variance $= \text{Var}(T) = np(1 - p) = 500 \times 0.15 \times 0.85 = 63.75$

Example 20

A bank issues three versions of its credit card: classic, gold and platinum. A customer's application for a credit card may be refused because of the customer's poor credit rating.

a The proportion of customers who are refused the classic version of the bank's credit card is 0.275.

Calculate the probability that, from a random sample of 10 customers applying for the classic version, exactly 2 applications are refused.

b The proportion of customers who are refused the gold version of the bank's credit card is 0.65.

Determine the probability that, from a random sample of 40 customers applying for the gold version, the number of applications that are accepted is:

i no more than 15; **ii** at least 10 but at most 20.

c The proportion of customers who are refused the platinum version of the bank's credit card is 0.85.

Determine the probability that, from a random sample of 50 customers applying for the platinum version, more than 40 applications are refused.

AQA MS1A June 2012

a Let C be the number of customers in 10 who are refused the classic version. $\quad$ $p = 0.275$ is not given in the tables,

Then $C \sim B(n, p)$ where $n = 10$, $p = 0.275$ and $q = 1 - 0.275 = 0.725$, $\quad$ so the formula must be used.

so $C \sim B(10, 0.275)$.

$$P(C = 2) = \binom{10}{2} \times 0.275^2 \times 0.725^8 = 0.2597\ldots = 0.260 \text{ (3 sf)}$$

b P(refused gold version) = 0.65,

so P(accepted for gold version) = $1 - 0.65 = 0.35$

Let G be the number in 40 who are accepted for the gold version, where $G \sim B(40, 0.35)$.

Use cumulative binomial probability tables with $n = 40$, $p = 0.35$

i P(no more than 15 are accepted for the gold version)

$= P(G \leq 15)$

$= 0.6946 = 0.695 \text{ (3 sf)}$

ii P(at least 10 but at most 20 are accepted for the gold version)

$= P(10 \leq G \leq 20)$

$= P(G \leq 20) - P(G \leq 9)$

$= 0.9827 - 0.0644$

$= 0.9183 = 0.918 \text{ (3 sf)}$

c Let R be the number in 50 who are refused the platinum version, where $R \sim B(50, 0.85)$.

Since $p > 0.5$, consider A, the number in 50 who are accepted for the platinum version.

P(accepted) = $1 - 0.85 = 0.15$,

so $A \sim B(50, 0.15)$.

Since you want $P(R > 40)$, you need to find $P(A < 10)$.

$P(R > 40) = P(A < 10)$

$\quad\quad\quad = P(A \leq 9) \quad\quad$ Use Table 1 with $n = 50$ and $p = 0.15$.

$\quad\quad\quad = 0.7911 = 0.791 \text{ (3 sf)}$

Summary

Bernoulli distribution: $Y \sim$ Bernoulli(p)

Mean, variance and standard deviation

A discrete random variable Y with probability distribution function $\quad\quad$ Probability distribution for

$$\begin{cases} P(Y = 0) = 1 - p \\ P(Y = 1) = p \end{cases} \quad \text{for } 0 \leq p \leq 1$$

$Y \sim$ Bernoulli(p):

has a Bernoulli distribution with parameter p.

If Y is the number of successes in a single Bernoulli trial with probability of success p then $Y \sim$ Bernoulli(p).

y	0	1
$P(Y=y)$	$1 - p$	p

Mean, variance and standard deviation

mean $= E(Y) = p$

variance $= \text{Var}(Y) = pq$ where $q = 1 - p$

standard deviation $= \sqrt{pq}$

Binomial distribution: $X \sim B(n, p)$

The binomial distribution results from repeating a Bernoulli trial a fixed number of times.

For a discrete random variable to follow a binomial distribution, all of the following conditions must be satisfied:

- the number of trials, n, is fixed
- the trials are independent
- each trial results in one of two outcomes: success or failure
- the probability of success, p, is constant for each trial.

If X is the number of successes in n independent Bernoulli trials, each with probability of success p, then X follows a binomial distribution with parameters n and p, so $X \sim B(n, p)$.

Binomial probabilities

The probability of x successes in n trials is given by

$$P(X = x) = \binom{n}{x} p^x q^{n-x} \quad \text{for } x = 0, 1, 2, ..., n \qquad \text{where } q = 1 - p \text{ and } \binom{n}{x} = {}_nC_x = \frac{n!}{x!(n-x)!}$$

Number of ways to choose x from n

Probability of x successes

Probability of $(n - x)$ failures

Mean, variance and standard deviation

mean $= E(X) = np$

variance $= \text{Var}(Y) = npq$

standard deviation $= \sqrt{npq}$

Review

1. The probability that a component produced by a particular machine is unsatisfactory is 0.05.

   A random sample of 12 components is selected from the production line of the machine.

   Find the probability that the number of unsatisfactory components in the sample is

   a exactly 3   b at least 2.

2. The faces on a special fair cubical die are numbered 1, 2, 2, 3, 3, 3.

   Find the probability of obtaining fewer than 6 odd numbers in 7 throws of the die.

3 The random variable X has a binomial distribution with parameters $n = 16$ and $p = 0.15$. Find

   a $E(X)$

   b the probability that X is greater than $E(X)$

   c $Var(X)$.

4 Crocus and tulip bulbs are sold in mixed packs of 36 bulbs. On average, a pack contains three times as many crocus bulbs as tulip bulbs.

   a A pack is selected at random. Find the probability that two-thirds of the bulbs in the pack are crocus bulbs.

   b Find the mean and variance of the number of tulip bulbs in a pack.

5 The discrete random variable X has a binomial distribution with mean 2.4 and variance 1.92. Find the probability that $X = 4$.

6 The probability that Barry's cat, Sylvester, chooses to stay outside all night is 0.35, and the cat's choice is independent from night to night.

   a Determine the probability that, during a period of 2 weeks (14 nights), Sylvester chooses to stay outside:

     i on at most 7 nights;   **ii** on at least 11 nights;

     iii on more than 5 nights but fewer than 10 nights.

   b Calculate the probability that, during a period of **3 weeks**, Sylvester chooses to stay outside on exactly 4 nights.

   c Barry claims that, during the summer, the number of weeks, S, on which Sylvester chooses to stay outside can be modelled by a binomial distribution with $n = 7$ and $p = \dfrac{5}{7}$.

     i Assuming that Barry's claim is correct, find the mean and variance of S.

     ii For a period of 13 weeks during the summer, the number of nights per week on which Sylvester chose to stay outside had a mean of 5 and a variance of 1.5.

     Comment on Barry's claim.

AQA MS1B January 2010

7 In the holiday period, the probability that Grégoire plays football on any particular day is 0.2.

   a Find the probability that Grégoire plays football on exactly 5 days in a holiday period of 14 days.

   b Find the mean number of days on which Grégoire plays football in a holiday period of 21 days.

   c If the standard deviation of the number of days Grégoire plays football is 2, how many days are there in the holiday period?

8 Balloons are packaged in party bags. Each bag contains 20 balloons. The colours of the balloons in a party bag are shown in the table.

Colour	Red	Blue	Green	Yellow
Frequency	8	5	4	3

Serene buys 10 party bags of balloons and selects a balloon at random from each bag.

a Find the probability that she selects at least 2 green balloons.

b Find the mean number of blue balloons that she selects.

c Find the variance of the number of yellow balloons she selects.

9 In a certain city 35% of all shops advertise in the local newspaper.

a A random sample of 12 shops is taken. Find the probability that more than 9 advertise in the local newspaper.

b A random sample of 50 shops is taken. Find the probability that

  i at least 16 advertise in the local newspaper

  ii fewer than 30 do **not** advertise in the local newspaper.

10 The records at a passport office show that, on average, 15% of photographs that accompany applications for passport renewals are unusable.

Assume that exactly one photograph accompanies each application.

a Determine the probability that in a random sample of 40 applications:

  i exactly 6 photographs are unusable;

  ii at most 5 photographs are unusable;

  iii more than 5 but fewer than 10 photographs are unusable.

b Calculate the mean and the standard deviation for the number of photographs that are unusable in a random sample of **32** applications.

AQA MS1A January 2012

Assessment

1 The random variable X has a binomial distribution with parameters $n = 30$ and $p = 0.5$. Find

a $P(12 < X < 15)$ [2]

b $P(X = E(X))$.   [3]

2 The random variable X has the distribution B(9, 0.45). Find

a $E(X)$ [2]

b the standard deviation of X.   [3]

3 Ghyslaine passes through 6 sets of traffic lights when she walks to work each morning. She estimates that, for each set of traffic lights, the probability that the lights show red is 0.3.

Stating any necessary assumptions, find the probability that, as she walks to work on a particular morning,

a 3 or fewer sets show red   [3]

b more than 4 sets show red   [2]

c all 6 sets show red.   [2]

4 *Stopoff* owns a chain of hotels. Guests are presented with the bills for their stays when they check out.

The number of bills that contain errors may be modelled by a binomial distribution with parameter n and p, where $p = 0.30$.

Determine the probability that, in a random sample of 40 bills:

a at most 10 bills contain errors; [2]

b at least 15 bills contain errors; [2]

c at least 6 bills but at most 18 bills contain errors. [3]

AQA MS1A January 2013

5 The probability that Shayna receives at least one telephone call on any day is 0.8. The number of telephone calls she receives is independent from day to day.

a Calculate the probability that, during a particular fortnight, Shayna receives at least one telephone call on exactly 9 days. [3]

b Calculate the mean number of days in April on which she does not receive any telephone calls. [3]

6 The post office in a market town is located within a small supermarket.

The probability that an individual customer entering the supermarket requires a service from:

the post office only is 0.48;
the supermarket only is 0.30;
both the post office and the supermarket is 0.22.

It may be assumed that the service required is independent from customer to customer.

a For a random sample of 12 individual customers, calculate the probability that exactly 5 of them require a service from the post office only. [3]

b For a random sample of 40 individual customers, determine the probability that more than 10 but fewer than 15 of them require a service from the supermarket only. [3]

c For a random sample of 100 individual customers, calculate the mean and standard deviation for the number of them requiring a service from both the post office and the supermarket. [3]

AQA MS1A January 2007

7 Plastic clothes pegs are made in various colours.

The number of red pegs may be modelled by a binomial distribution with parameter p equal to 0.2.

The contents of packets of 50 pegs of mixed colours may be considered to be random samples.

a Determine the probability that a packet contains:

   i less than or equal to 15 red pegs; [2]

   ii exactly 10 red pegs; [2]

   iii more than 5 but fewer than 15 red pegs. [3]

b Sly, a student, claims to have counted the number of red pegs in each of 100 packets of 50 pegs. From his results the following values are calculated.

     Mean number of red pegs per packet = 10.5

     Variance of number of red pegs per packet = 20.41.

   Comment on the validity of Sly's claim. [4]

AQA MS1B January 2006

8 The table shows, for a particular population, the proportion of people in each of the four main blood groups.

Blood group	O	A	B	AB
Proportion	0.40	0.28	0.20	0.12

a A random sample of 20 people is selected from the population.

   Determine the probability that the sample contains:

   i at most 10 people with blood group O; [2]

   ii exactly 3 people with blood group A; [3]

   iii more than 4 but fewer than 8 people with blood group B. [3]

b A random sample of 500 people is selected from this population.

   Find values for the mean and variance of the number of people in the sample with blood group AB. [2]

AQA MS1A January 2006

Introduction

Mechanics is the study of how and why objects move in various ways, or do not move at all.

This chapter defines the terms used to describe motion.

Objectives

By the end of this chapter, you should know how to...

▶ Define displacement, velocity and acceleration.

▶ Sketch and use displacement-time graphs and velocity-time graphs to solve problems of motion in a straight line.

Recap

You need to remember how to...

▶ Use the relationship between distance, speed and time:

$$\text{distance} = \text{speed} \times \text{time} \quad \text{and} \quad \text{average speed} = \frac{\text{total distance}}{\text{total time}}$$

▶ Find the gradient of a straight line.

▶ Sketch a parabola from its equation.

▶ Use trapeziums to estimate the area under a curve.

▶ Differentiate a polynomial.

Applications

When a plane lands it decelerates. Designers of new planes need to calculate the time taken and distance traveled when the plane brakes so they know the minimum length of runway that a plane would need to stop.

. . . .

16.1 Displacement, speed, velocity and acceleration

Displacement

When a particle is moving in a straight line it can be moving in either direction along the line.

The difference between the two different directions is identified by giving a positive sign to one direction and a negative sign to the other direction.

The diagram shows a model engine that starts from a point O and moves in the direction OA along a straight line.

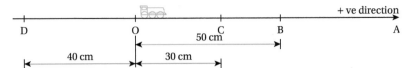

When the engine reaches B, it has travelled a distance of 50 cm and it is also 50 cm from O. The engine then reverses direction and moves 20 cm back towards O to point C, then

the total distance that the engine has travelled is 70 cm

but the distance of the engine from O is 30 cm (to the right).

The engine now continues moving towards O and carries on to point D, beyond O,

the total distance that it has travelled is 140 cm

but the distance of the engine from O is 40 cm (to the left).

Calling the direction from O to A the positive direction then

the **distance** from O in the direction from O to A is called the **displacement**.

Therefore, from O,   the displacement of B is 50 cm

the displacement of C is 30 cm

the displacement of D is −40 cm.

Displacement is a quantity which has *both* magnitude (size) *and* direction. Quantities of this type are called vectors.

Distance only has size – the direction doesn't matter.
Distance is a scalar quantity.

Example 1

Starting from floor 4, a lift stops first at floor 11, then at floor 1 and finally at floor 6. The distance between floors is 4 m. Taking the upward direction as positive write down, for each of the stops,

a   the displacement, of the lift floor from floor 4

b   the distance the lift floor has travelled since first leaving floor 4.

Taking the level of floor 4 as the start and the upward direction as positive:

a   At floor 11   the displacement is $+7 \times 4$ m $= 28$ m

the distance travelled is 28 m.

b   At floor 1   the displacement is -3×4 m $= -12$ m

the distance travelled is $(10+7) \times 4$ m $= 17 \times 4$ m $= 68$ m.

At floor 6   the displacement is $+2 \times 4$ m $= 8$ m

the distance travelled is $(10+7+5) \times 4$ m $= 22 \times 4$ m $= 88$ m.

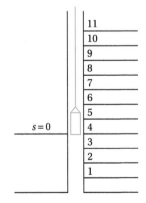

Exercise 1

1 Which of the following quantities are vectors and which are scalars?

    a 5 km due south   **b** 6 miles

    c A speed of 4 ms^{-1} **d** 200 miles north-east

    e A temperature of 25°C   **f** A force of 8 units vertically downwards

    g A mass of 6 kg   **h** A time of 7 seconds.

2 A particle starts at a point A and moves along a straight line. The point A is 1.2 m from a point O on the line.

This graph shows the motion of the particle.

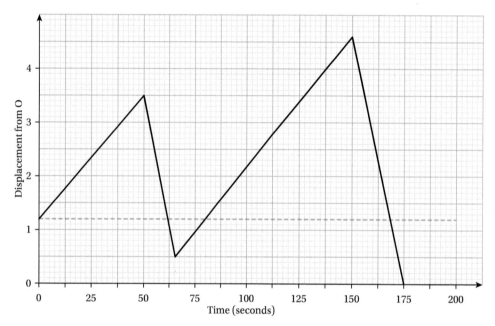

Take the upward direction as positive.

    a State the displacement of the particle from A after

        i 50 seconds   **ii** 80 seconds   **iii** 175 seconds.

    b Find **i** the maximum displacement of the particle from A

            ii the total distance travelled by the particle.

    c Find the average speed over the 175 seconds.

Velocity and speed

The **velocity** of an object is its speed *in a particular direction*.

Therefore velocity is a vector quantity. **Speed** is the magnitude of velocity.

When an object moves with a uniform velocity,(uniform means constant) then the object has a constant speed *and* its direction is constant, so it is moving one way along a straight line.

To define which direction the object is moves, a velocity in one direction along the line is taken as positive; a velocity in the opposite direction is then negative.

For example a particle P moving with a constant speed of 5 ms^{-1} along the line shown in the diagram.

When P starts from O and moves to the right, the positive direction, the velocity of P is +5 ms^{-1}.

After 1 second the displacement of P from O is +5 m, after 2 seconds it is +10 m, and so on.

The displacement is increasing at a rate of +5 ms^{-1}.

When P starts from A and moves to the left with the same speed, its velocity is −5 ms^{-1}.

The initial displacement of P from O is +11 m and after 1 second the displacement is 6 m.

The displacement has decreased by 5 m, so it has increased by −5 m.

Also, after 2 seconds the displacement is 1 m and has increased by −10 m.

The displacement is increasing at a rate of −5 ms^{-1}

therefore velocity is the rate at which the displacement increases.

Acceleration

Acceleration is the rate at which *velocity* is increasing.

Therefore when the velocity of a particle moving in a straight line increases steadily from 3 ms^{-1} to 11 ms^{-1} in 4 seconds, the acceleration is +2 ms^{-2}.

When the velocity decreases from 14 ms^{-1} to 5 ms^{-1} in 3 seconds (the velocity has increased by −9 ms^{-1}) the acceleration is −3 ms^{-2}.

Exercise 2

1 State whether each of the following statements is correct or incorrect. Give a reason for the statements that you think are incorrect.

a A car driving due north at 40 mph has a constant velocity.

b A toy train runs round a circular track with constant velocity 2 ms^{-1}.

c A plane flies in a straight line from London to Newcastle so its velocity is constant.

2 A particle moves along the straight line shown in the diagram. It passes through A, moves to B, then moves from B to C, from C to D and finally from D to E.

C A   B   E   D
|--|--|--|--|--|--|--|--|--|--|--|--|--|--|--→ +ve
-2 -1 0 1 2 3 4 5 6 7 8 9 10 11 12 13

Distance (metres)

This table gives the value of t at each point, where t is the number of seconds that have elapsed since the particle first passed through A.

	B	C	D	E
t	5	8	15	19

Find the velocity of the particle, constant in each section, in travelling from

a A to B   **b** B to C   **c** C to D   **d** D to E.

3 The velocity of a particle changes steadily from $-5\ \text{ms}^{-1}$ to $-21\ \text{ms}^{-1}$ in 4 seconds. What is the acceleration?

4 A particle moving in a straight line with a constant acceleration has a velocity $u\ \text{ms}^{-1}$ at one instant and t seconds later the velocity is $v\ \text{ms}^{-1}$. Find the acceleration of the particle when

a $u = 8, v = 2, t = 3$ **b** $u = 4, v = -11, t = 5$.

5 A body moving initially at $5\ \text{ms}^{-1}$ has a constant acceleration of $a\ \text{ms}^{-2}$. After 6 seconds its velocity is $v\ \text{ms}^{-1}$. Find v when

a $a = 3$ **b** $a = -2$ **c** $a = 0$.

16.2 Displacement-time and velocity-time graphs

Displacement-time graphs

When an object P moves in a straight line, a displacement–time graph shows how the distance of P *in a specified direction from a fixed point* varies with time.

For example, an object P moves in a straight line and moves through points O, A, B and C on the line. P covers each section at a constant speed.

This table gives the displacement, s metres, of each of these points from O, and the time, t seconds after leaving O, when P is at each point.

	O	A	B	C
s	0	10	6	-9
t	0	5	7	12

This displacement-time graph shows the information in the table.

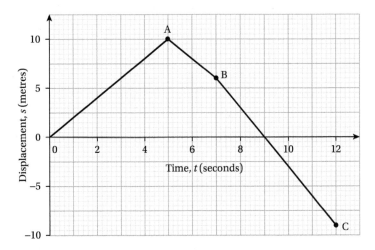

For the section from O to A,

P has travelled a distance of 10 m in the positive direction in 5 seconds, so the velocity is 2 ms⁻¹ and the gradient of the graph is 2.

For the section from A to B,

P has travelled a distance of 4 m in the negative direction in 2 seconds, so the velocity is −2 ms⁻¹ and the gradient of the graph is −2.

For the section from B to C,

P has travelled a distance of 15 m in the negative direction in 5 seconds, so the velocity is −3 ms⁻¹ and the gradient of the graph is −3.

In each section the gradient of the displacement-time graph represents the velocity.

The average velocity is the constant velocity that would produce the final increase in displacement in the total time interval.

For example the average velocity from O to B is $\frac{6-0}{7-0}$ ms⁻¹ = $\frac{6}{7}$ ms⁻¹ (this is equal to the gradient of the chord OB)

the average velocity from A to C is $\frac{-9-10}{12-5}$ ms⁻¹. = $-\frac{19}{7}$ ms⁻¹ (this is equal to the gradient of the chord AC)

the average velocity from O to C is $\frac{-9-0}{12-0}$ ms⁻¹ = $-\frac{3}{4}$ ms⁻¹ (this is equal to the gradient of the chord OC).

When P moves from O to C, the total *distance* that P has moved is
$(10+4+15)$ m = 29 m.

So the average *speed* of P is $\left(\frac{29}{12}\right)$ ms⁻¹ = 2.4 ms⁻¹ (correct 2 significant figures).

This shows that

the average speed is *different* from the average velocity.

For a curved displacement–time graph also, the average velocity over a time interval is given by the gradient of the chord corresponding to that interval.

The gradient of the tangent to the curve at a particular value of t, represents the velocity at that instant.

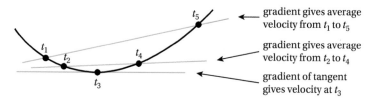

gradient gives average velocity from t_1 to t_5

gradient gives average velocity from t_2 to t_4

gradient of tangent gives velocity at t_3

For any motion over a given interval of time,

$$\text{average speed} = \frac{\text{distance covered in the time interval}}{\text{the time interval}}$$

$$\text{average velocity} = \frac{\text{increase in displacement over the time interval}}{\text{the time interval}}$$

The velocity at an instant is given by the gradient of the displacement–time graph at the corresponding point.

Example 2

Question

A and O are two fixed points on a straight line. A particle P moves on the line so that, at time t seconds its displacement, s metres, from O is given by $s = (t-1)(t-5)$.

When $t = 0$, the particle is at point A.

Sketch a displacement–time graph for values of t from 0 to 6.

a   At what times does the particle pass through O?

b   What is the average speed over the 6-second time interval?

c   What is the average velocity over the 6-second time interval?

d   At what time is the velocity zero?

e   Find the velocity when $t = 5$.

Answer

The curve whose equation is $s = (t-1)(t-5)$ is a parabola. The curve crosses the t-axis where $t = 1$ and $t = 5$ so is symmetrical about the line $t = 3$.

The curve has a minimum value of -4 where $t = 3$.

a   When the particle is at O its displacement from O is zero, that is when $s = 0$. The particle passes through O when $t = 1$ and $t = 5$.

b   P starts at A where $s = +5$. P then covers 5 m to O and continues beyond O for a further 4 m to B where $s = -4$.

Then P goes back to O, that is 4 m back, and a further 5 m to A.

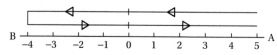

$$\text{average speed} = \frac{\text{total } distance \text{ covered}}{\text{time interval}} = \frac{5 + 4 + 4 + 5}{6} = 3 \text{ ms}^{-1}.$$

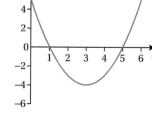

(continued)

(continued)

c The average velocity is given by the gradient of the chord joining the points where $t = 0$ and $t = 6$.

Therefore the average velocity is $\dfrac{5 - 5}{6 - 0} = 0$.

d The velocity is zero when the gradient of the tangent is zero, that is when the tangent is parallel to the time axis.

The velocity is zero when $t = 3$.

e The velocity when $t = 5$ is the gradient of the curve at the point where $t = 5$. The gradient of a curve can be found using differentiation.

$$s = (t - 1)(t - 5) \implies s = t^2 - 6t + 5$$

$$\frac{ds}{dt} = 2t - 6 \text{ and when } t = 5, \ \frac{ds}{dt} = 4$$

Therefore the velocity is 4 ms^{-1} when $t = 5$.

Exercise 3

1 A boy is kicking a ball straight towards a wall. For each kick he sees how far the ball rebounds. This graph shows, for one kick, the displacement of the ball from the boy in the direction towards the wall.

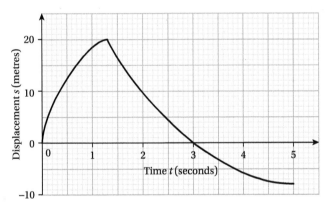

a Find the average velocity of the ball between $t = 0$ and

   i $t = 0.5$ **ii** $t = 3$ **iii** $t = 5$.

b **i** State the velocity when $t = 1.3$.

   ii Explain why the graph has a 'point' at this time.

c At what time does the ball pass the boy when rebounding?

2 A particle P is moving along a straight line. The displacement of P from O, a fixed point on the line, after t seconds is s metres where $s = t(4 - t)$.

Sketch a displacement-time graph and use it to answer the following questions.

a Find the average velocity from

   i $t = 0$ to $t = 2$ **ii** $t = 2$ to $t = 4$ **iii** $t = 0$ to $t = 4$ **iv** $t = 2$ to $t = 6$.

b Find the average speed for each of the time intervals in part **a**.

c Find the velocity when $t = 0$.

d At what time is the velocity zero?

3 A fixed point A is on a straight line and a particle P is moving on the line.
The displacement, s metres, of P from A after t seconds is given by $s = 5t - t^2$.

    a Sketch a displacement-time graph and use it to answer the following
        questions.

    b At what time is the velocity zero ?

    c Find the velocity when $t = 5$.

    d Find, for the 6-second journey,

        i the average velocity   **ii** the average speed.

4 This graph shows the motion of a ball bouncing vertically.

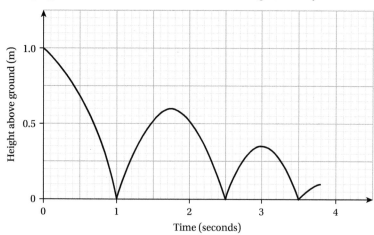

    a State the times when the velocity of the ball is zero.

    b Find the average speed during

        i the first second

        ii the first full bounce (from $t = 1$ to $t = 2.5$)

        iii the second full bounce.

    c Find the average velocity during each of the time intervals in part **b**.

    d What is the average velocity over the first 3 seconds.

Velocity-time graphs

For an object P travelling in a straight line, a velocity-time graph shows how the
speed of P *in a particular direction* varies with time.

The graph on the next page shows the velocity of P during a 20-second time
interval, starting from a point O on the straight line.

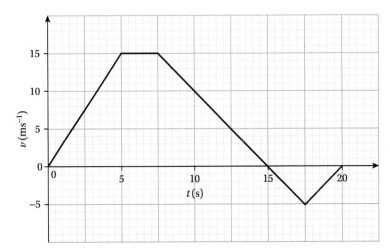

In the first 5 seconds the velocity increases steadily from zero to 15 ms^{-1} so the acceleration in this section is 3 ms^{-2} and the gradient of this section of the graph is 3.

In the next 2.5 seconds the velocity is constant so the acceleration is zero, and the gradient the graph is also zero.

Then the velocity decreases until after another 7.5 seconds the graph crosses the time axis. This shows that the velocity has reduced to zero and that P has momentarily come to rest.

For this section the acceleration is $-\dfrac{15}{7.5}$ ms^{-2} and the gradient is -2. (The velocity is decreasing but it is still positive, so P is still moving forward.)

When $t = 15$ the velocity is zero and immediately after that the velocity becomes negative, so P *stops* going forward and begins to move in the opposite direction with an acceleration of $-\dfrac{5}{2.5}$ ms^{-2} (the gradient is -2).

For the last 2.5 seconds the velocity is becoming less negative, so. it is increasing. The acceleration is $\dfrac{0-(-15)}{2.5}$ ms^{-2} = 2 ms^{-2} and the gradient is 2.

The velocity is still negative so P is still moving backwards until, when $t = 20$, P comes to rest.

Each section demonstrates that, for motion with constant acceleration, the gradient of the velocity-time graph represents the acceleration.

Now look at the distance moved by P in each section, remembering that P moves forward for 15 seconds and then moves in the opposite direction.

Using average velocity × time gives the following results.

Time interval (s)	0–5	5–7.5	7.5–15	15–17.5	17.5–20
Distance moved (m)	37.5	37.5	56.25	6.25	6.25
Direction moved	forward	forward	forward	backward	backward

P moves 131.25 m forward and then 12.5 m back so the *displacement* of P from O at the end of the 20 seconds is 131.25 − 12.5 m = 118.75 m.

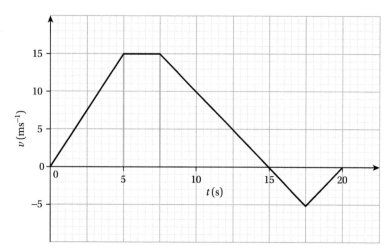

This shows that the distance moved in each section is represented by the area between that section of the graph and the time axis. Take any area that is below this axis as negative, the displacement of an object moving with constant acceleration is represented by the area between the velocity-time graph and the time axis.

When the acceleration of a moving object is not constant the velocity-time graph is curved. Similar arguments show that, for any velocity-time graph, straight or curved:

> The average acceleration over a time interval is represented by the gradient of the corresponding chord. A negative gradient gives a negative acceleration, showing that the velocity is decreasing.

> The acceleration at a given instant is represented by the gradient of the tangent to the curve at that particular point on the curve.

> The displacement is represented by the area between the curve and the time axis, regions below that axis are negative. An approximation to this area can be found using the area of trapeziums.

Exercise 4

1

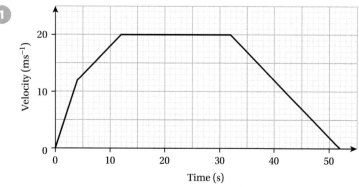

The graph shows the velocity of a car as it moves along a straight road, starting from a point A.

Find

a the acceleration during the first 4 seconds

b the deceleration during the final 10 seconds

c the distance travelled

   i while accelerating

   ii at constant speed

   iii while decelerating.

Explain why the displacement of the car from A at the end of the 52-second journey is equal to the total *distance* travelled from A.

2 A train is brought to rest from a velocity of 24 ms⁻¹ with a constant acceleration of −0.8 ms⁻².

Sketch a velocity-time graph and find the distance covered by the train while it is decelerating.

3 A particle is moving in a straight line with a velocity of 10 ms⁻¹ when it is given an acceleration of −2 ms⁻² for 8 seconds. Draw a velocity-time graph for the eight-second time interval and use it to find

a the time when the direction of motion of the particle is reversed

b the increase in displacement during the 8 seconds

c the total distance travelled in this time.

4 The velocity of a runner was recorded at different times and is illustrated by the velocity-time graph.

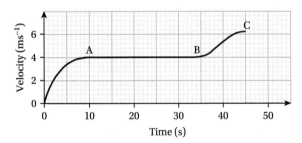

a Explain what is happening between

   i O and A

   ii A and B

   iii B and C.

b Estimate the length of the race explaining whether your answer is more or less than the actual length.

Note

Use a triangle between O and A, a rectangle between A and B and a trapezium between B and C.

5 A girl runs forwards to a point A and then runs back to a point B which is at a distance behind A.

This is the graph of her velocity plotted against time.

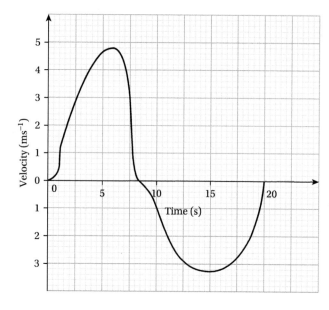

a At what time does the girl

   i reach point A   **ii** reach point B?

b At what times is the acceleration zero?

c Explain what happens after about 6 seconds.

d Use ordinates at $t = 5$, $t = 8.5$ and $t = 15$ to estimate the distance that the girl runs.

6 An aircraft starts from rest at the end of a straight runway. The aircraft accelerates uniformly for 30 seconds until it reaches its take-off speed of 90 ms^{-1}.

a Find the acceleration of the aircraft.

b Sketch a velocity-time graph.

c Find the length of the runway used by the aircraft to reach its takeoff speed.

7 A lift travels vertically up from rest at A to rest at B, which is 20 m above A. The lift accelerates from A at 1.5 ms^{-2} for 2 s then it travels at a constant speed before decelerating uniformly at 4 ms^{-2} to stop at B.

a Sketch a velocity-time graph.

b Find the time the lift takes to decelerate.

c Hence find the time that the lift takes to move from A to B.

Summary

For motion with constant speed:

$$\text{distance} = \text{speed} \times \text{time} \qquad \text{average speed} = \frac{\text{total distance}}{\text{total time}}$$

Velocity is the rate of increase of displacement.

Acceleration is the rate of increase of velocity.

In a displacement-time graph the velocity at a particular time, t, is given by the gradient of the tangent to the graph at that value of t.

In a velocity-time graph the acceleration at a particular time, t, is given by the gradient of the tangent to the graph at that value of t,

the displacement after time t is given by the area under the graph for that time interval.

Review

For Questions 1 to 4 state the letter that gives the correct answer.

1 The diagram shows the displacement-time graph for a particle moving in a straight line.

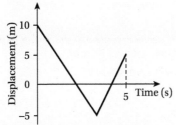

The average velocity for the interval from $t = 0$ to $t = 5$ is

a 0   **b** $6\,\text{ms}^{-1}$ **c** $-1\,\text{ms}^{-1}$ **d** $2\,\text{ms}^{-1}$.

2 The diagram shows the displacement–time graph for a particle moving in a straight line.

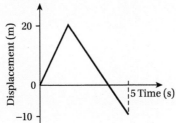

The distance covered by the particle in the interval from $t = 0$ to $t = 5$ is

a 20 m   **b** 50 m   **c** 15 m   **d** 5 m.

3

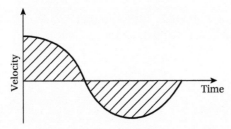

The diagram shows the velocity-time graph for a particle moving in a straight line. The sum of the two shaded areas represents

a the increase in displacement of the particle

b the average velocity of the particle

c the distance moved by the particle

d the average speed of the particle.

4 A particle moves in a straight line with a constant acceleration of $3\,\text{ms}^{-2}$ with an initial velocity of $-1\,\text{ms}^{-1}$. Its velocity 2 seconds later is

a $5\,\text{ms}^{-1}$ **b** $6\,\text{ms}^{-1}$ **c** $4\,\text{ms}^{-1}$ **d** $-7\,\text{ms}^{-1}$.

5 A test car starts from rest at time $t = 0$ seconds and moves with a uniform acceleration of $3\,\text{ms}^{-2}$ along a straight horizontal path. After T seconds, when its speed is $V\,\text{ms}^{-1}$, it stops accelerating and maintains this steady speed until it hits a wall when it comes instantly to rest. The car has then travelled a distance of 800 m in 30 s.

a   Sketch a speed-time graph to illustrate this information.

b   show $V = \dfrac{1600}{60} - T$.

Assessment

1 A car travels along a straight horizontal road with constant acceleration from a speed of 10 ms^{-1} to a speed of 20 ms^{-1} and covers a distance of 200 m.

a   Sketch a velocity-time graph showing this motion.

b   Find the time taken by the car to travel the 200 m.

c   Find the time taken by the car to travel the first 100 m.

2 The graph shows how the velocity of a train changes as it travels along a straight horizontal track.

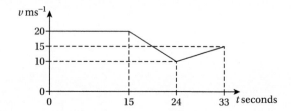

a   Find the distance travelled by the train in this 33 seconds.

b   Find the average velocity of the train in this 33 seconds.

3 The graph shows the journey of a train as it travels along a horizontal track.

The train accelerates from rest at a station A to a speed of 27 ms^{-1} and then decelerates to rest again at a station B. The stations are 5400 m apart.

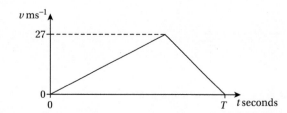

a   Find the time taken for the journey between the two stations.

   The acceleration of the train is 0.3 ms^{-2}.

b   Find the distance the train travels while it is accelerating.

4 The diagram shows the displacement–time graph for a particle moving in a straight line.

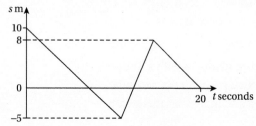

   a Find the distance travelled by the particle.

   b Find the average speed of the particle.

   c Find the average velocity of the particle.

5 A particle P is moving on a straight line. After t seconds the displacement, s metres, of P from A, a fixed point on the line, is given by $s = (1 - t)(3 - t)$.

   a Sketch a displacement-time graph for $0 \leq t \leq 6$.

   b At what time is the velocity zero?

   c Find the velocity when $t = 5$.

   d Find, for the first 6 seconds,

       i the average velocity   **ii** the average speed.

6 Two trains, A and B, are moving on straight horizontal tracks which run alongside each other and are parallel. The trains both move with constant acceleration. At time $t = 0$, the fronts of the trains pass a signal. The velocities of the trains are shown in the graph below.

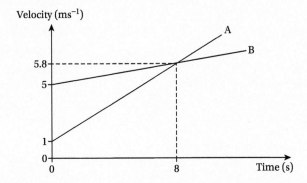

   a Find the distance between the fronts of the two trains when they have the same velocity and state which train has travelled further from the signal.

   b Find the time when A has travelled 9 metres further than B.

AQA MM1B June 2015

7 The graph shows how the speed of a cyclist, Hannah, varies as she travels for 21 seconds along a straight horizontal road.

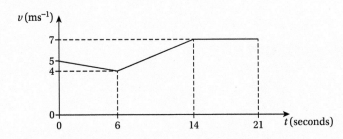

   a Find the distance travelled by Hannah in the 21 seconds.

   b Find Hannah's average speed during the 21 seconds.

AQA MM1B June 2013

17 Motion in a Straight Line

Introduction

When an object moves with constant acceleration, the relationships between displacement, velocity and acceleration give some simple formulae. These formulae together with velocity-time graphs can be used to solve problems when the acceleration is constant. This chapter also covers motion in a straight line with variable acceleration.

Objectives

By the end of this chapter, you should know how to...

▶ Derive and use the equations of motion with constant acceleration.

▶ Solve problems involving objects falling freely under gravity.

▶ Work with equations of motion with variable acceleration.

Recap

You need to remember...

▶ The gradient of a velocity-time graph gives the acceleration.

▶ The area under a velocity-time graph gives the displacement.

▶ The formula for solving a quadratic equation.

▶ How to solve a pair of simultaneous equations when one of them is quadratic.

▶ How to differentiate and integrate x^n.

▶ That $\dfrac{\mathrm{d}y}{\mathrm{d}x}$ means the rate at which y is changing with respect to x.

Applications

When a firework rocket is launched, it accelerates rapidly but gravity will act to slow it down as the fuel runs out.

Designers of fireworks need to know the distance it travels upwards before the explosion is triggered and the time taken to reach that point.

17.1 Equations of motion with constant acceleration

The velocity-time graph illustrates an object moving for t seconds with constant acceleration a units.

At the start of the time interval the velocity is u units and the end of the time interval it is v units.

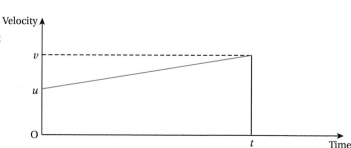

The velocity increases by a units each second so after t seconds the increase in velocity is at units. Therefore

$$v = u + at \qquad\qquad [1]$$

The area under the velocity-time graph represents the displacement. The area of this trapezium is $\frac{1}{2}(u+v) \times t$. Therefore

$$s = \frac{1}{2}(u+v)t \qquad\qquad [2]$$

Eliminating v from [1] and [2] by substituting $u + at$ for v in [2] gives $s = \frac{1}{2}(u+u+at)t$. Therefore

$$s = ut + \frac{1}{2}at^2 \qquad\qquad [3]$$

Eliminating u from [1] and [2] gives

$$s = vt - \frac{1}{2}at^2 \qquad\qquad [4]$$

Eliminating t from [1] and [2] gives

From [1] $\qquad\qquad t = \dfrac{v-u}{a}$

Substituting in [2] gives $s = \frac{1}{2}(v+u)\dfrac{(v-u)}{a} = \dfrac{1}{2a}(v^2-u^2) \quad \Rightarrow \quad 2as = v^2 - u^2$

Rearranging gives

$$v^2 = u^2 + 2as \qquad\qquad [5]$$

These five formulae can now be used to solve by calculation any problem on motion with constant acceleration.

Each formula contains four quantities from u, v, a, s and t but not the fifth quantity. The formula to use can be found by looking for the quantity that is *not* involved.

However, many problems can be solved using a sketch of a velocity-time graph. Also a velocity-time sketch graph often makes the solution clearer when using the formulae.

When acceleration is constant (acceleration is a vector so its direction is constant as well as its magnitude), it follows that all the motion takes place along a straight line but the object can move in either direction along the line,

Therefore start a problem by deciding which direction is positive; the opposite direction is then negative.

Then make a list of the given information, and what is needed, using u for the initial velocity and v for the final velocity. Many motion problems are clearer when a simple diagram is drawn, using different arrow heads to indicate different quantities.

This book uses:

for velocity,   for acceleration,   for a length

Example 1

A particle starts from a point A with velocity 3 ms⁻¹ and moves with a constant acceleration of $\frac{1}{2}$ ms⁻² along a straight line AB. It reaches B with a velocity of 5 ms⁻¹. Find

a the displacement of B from A

b the time taken from A to B.

Given: $u = 3$, $v = 5$, $a = \frac{1}{2}$

a Wanted: s

(t is not involved so use formula $v^2 - u^2 = 2as$)

$$5^2 - 3^2 = 2\left(\frac{1}{2}\right)s$$

$$\Rightarrow \quad s = 16$$

The displacement of B from A is 16 m.

b Wanted : t

(s is not involved so use the formula $v = u + at$)

$$5 = 3 + \frac{1}{2}t \quad \Rightarrow \quad t = 4$$

The time taken from A to B is 4 seconds.

Alternatively, use a velocity–time graph sketched from the given information.

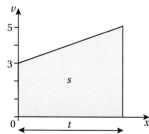

The gradient gives the acceleration.

$$\text{Gradient} = \frac{5 - 3}{t} = \frac{1}{2}$$

Therefore $t = 4$ so the time is 4 seconds.

The area gives the displacement.

$$\text{Area} = \frac{1}{2}(3 + 5) \times t = \frac{1}{2}(8)(4) = 16$$

Therefore $s = 16$ so the displacement is 16 m.

Example 2

A cyclist starts to ride up a straight steep hill with a velocity of 8 ms⁻¹. At the top of the hill, which is 96 m long, the velocity is 4 ms⁻¹. Find the value of the constant acceleration.

Given: $u = 8$, $v = 4$, $s = 96$

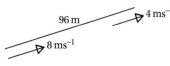

Wanted: a, so use the formula without t in it,

$$v^2 - u^2 = 2as$$

$$4^2 - 8^2 = 2a(96)$$

$$\Rightarrow a = \frac{-48}{192} = \frac{-1}{4}$$

The acceleration is $-\frac{1}{4}$ ms⁻².

Example 3

The driver of a train decelerates at steady rate of 0.2 ms^{-2} and brings the train to rest in 1 minute 30 seconds.

Find a   the speed of the train in kmh^{-1} at the instant when the brakes are applied,

   b   the distance the train then travels before it stops.

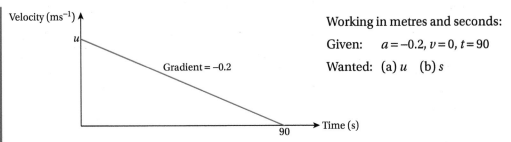

Working in metres and seconds:

Given: $a = -0.2$, $v = 0$, $t = 90$

Wanted: (a) u (b) s

a   Using $v = u + at$ gives

$$0 = u + (-0.2)(90)$$

Therefore $u = 18$.

The speed of the train was $18 \text{ ms}^{-1} = \dfrac{18}{1000} \times 60 \times 60 \text{ kmh}^{-1} = 64.8 \text{ kmh}^{-1}$.

b   Using $s = vt - \dfrac{1}{2}at^2$ gives

$$s = (0)(90) - \frac{1}{2}(-0.2)(90)^2 = 810$$

This is the displacement of the point where the train stops from the point where the brakes were applied. Therefore the distance the train travels is 810 m.

Example 4

A particle A starts from rest at a point O and moves on a straight line with constant acceleration 2 ms^{-2}. At the same instant another particle B, 12 m behind O, is moving with velocity 5 ms^{-1} and has a constant acceleration of 3 ms^{-2}. How far from O are the particles when B overtakes A?

A and B travel different distances. Using s to represent the shorter distance, that is the distance that A travels; B then travels a distance $(s + 12)$ m. A and B move for the same time.

For A

Given: $u = 0$, $a = 2$ wanted: distance s, time t is the same for both particles

$$s = ut + \frac{1}{2}at^2 \implies s = \frac{1}{2}(2)t^2 \tag{1}$$

For B

Given: $u = 5$, $a = 3$ wanted: distance $(s + 12)$ m, time t is the same for both particles

$$s = ut + \frac{1}{2}at^2 \implies s + 12 = 5t + \frac{1}{2}(3)t^2 \tag{2}$$

(continued)

(continued)

These two equations can be solved simultaneously. *s* is wanted but it is easier to first find *t*.

$$[2] - [1] \text{ gives } 12 = 5t + \frac{1}{2}t^2 \implies t^2 + 10t - 24 = 0$$
$$\implies (t+12)(t-2) = 0$$
$$\implies t = 2 \text{ or } -12$$

Using the positive value of *t* in [1] gives *s* = 4.

B overtakes A at a distance 4 m from O.

Exercise 1

Give answers that are not exact correct to 3 significant figures.

In Questions 1 to 10 an object is moving with constant acceleration a ms^{-2} along a straight line. The velocity at the initial point O is u ms^{-1} and t seconds after passing O the velocity is v ms^{-1} and the displacement from O is s m.

1. $u = 0$, $a = 3$, $v = 15$;   find t.

2. $t = 10$, $s = 24$, $u = 6$;   find v.

3. $a = 5$, $u = 4$, $s = 2$;   find v.

4. $u = 16$, $v = 8$, $t = 5$;   find s.

5. $t = 7$, $u = 3$, $v = 17$;   find a.

6. $t = 7$, $u = 17$, $v = 3$;   find a.

7. $u = 5$, $t = 3$, $a = -2$;   find s.

8. $a = -4$, $u = 6$, $v = 0$;   find t.

9. $v = 3$, $t = 9$, $a = 2$;   find s.

10. $v = 7$, $t = 5$, $a = 3$;   find u.

11. A particle starts from rest and moves in a straight line with a constant acceleration of 2 ms^{-2}. Find the distance covered

    a in the first four seconds

    b in the fourth second of its motion.

12. A particle moves in a straight line with a constant acceleration of −3 ms^{-2}. The particle has an initial velocity at point A of 10.5 ms^{-1}.

    a Show that the times when the displacement from A is 15 m are given by $t^2 - 7t + 10 = 0$.

    b Find the time when the displacement from A is −15 m.

13. A car is travelling at 10 ms^{-1} when the driver sees a broken-down car on the road 100 m ahead. The driver brakes, decelerating at 18 ms^{-2}. How far from the broken-down car does the driver stop?

14. A body moving in a straight line with constant acceleration takes 3 seconds and 5 seconds to cover two successive distances of 1 m. Find the acceleration.

> **Note**
>
> Hint: use distances of 1 m and 2 m from the start of the motion.

15 The displacements from a fixed point O, of an object moving in a straight line with constant acceleration, are 10 m and 14 m at times of 2 seconds and 4 seconds respectively after leaving O. Find

  a the initial velocity

  b the acceleration

  c the time interval between leaving O and returning to O.

16 A particle accelerates from rest from a point A to a velocity of 12 ms⁻¹. It then moves with constant velocity for 42 seconds before decelerating to rest at a point B. The time taken to travel from A to B is 60 seconds. The acceleration and deceleration are equal. Find the distance between A and B.

17.2 Free fall motion under gravity

Experiments show that, however heavy, all objects with negligible air resistance have the same acceleration vertically downward when falling freely.

This acceleration due to gravitational attraction is represented by the letter g. Its value on the surface of the earth is approximately 9.8 ms⁻².

An object falling freely travels in a vertical line, so problems on its motion can be solved by using the equations for uniform acceleration in a straight line.

For objects that are dropped, the downward direction is usually taken as positive, so g is positive.

For objects that are thrown vertically upwards, the upward direction is usually taken as positive so the acceleration is $-g$.

Example 5

An object is dropped from a point A and hits the ground 3 seconds later. Find the height of A above ground.

Anything that is 'dropped' from a stationary point is not thrown but released from rest, so its initial velocity is zero.

Taking the downward direction as positive:

Known: $u = 0$ wanted: s

$a = 9.8$

$t = 3$

Using $s = ut + \frac{1}{2}at^2$ gives

$s = 0 + \frac{1}{2}(9.8)(3^2)$

Therefore $s = 44.1$.

The height of point A is 44.1 m.

Example 6

A ball is thrown vertically upwards from a point A, 7 m above the ground.

a   The ball passes A on its way down 2 seconds later. What speed was the ball thrown at?

b   Find the velocity of the ball when it passes A on the way down.

c   At what speed does the ball hit the ground?

Taking the upward direction as positive:

a

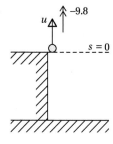

Known: $t = 2$ Required: u

          $a = -9.8$

          $s = 0$

Using $s = ut + \dfrac{1}{2}at^2$ gives

        $0 = 2u + \dfrac{1}{2}(-9.8)(4)$

Therefore $u = 9.8$.

The ball was thrown with a speed of 9.8 ms^{-1}.

b

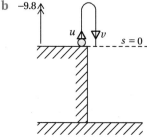

Known: $t = 2$ Required: v

          $a = -9.8$.

          $s = 0$

Using $s = vt - \dfrac{1}{2}at^2$ gives

        $0 = 2v - \dfrac{1}{2}(-9.8)(4)$

Therefore $v = -9.8$.

The ball is *falling* at 9.8 ms^{-1}.

> ### Note
>
> Parts **a** and **b** show that a particle thrown upwards with a velocity u, returns to the same level with a velocity $-u$, that is with equal speed in the downward direction.

c   The ground is 7 m below A so the final displacement of the ball is −7 m.

Working from the time when the ball was thrown we have:

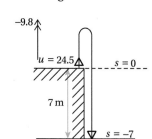

Known: $u = 9.8$ wanted: v

          $s = -7$

          $a = -9.8$

Using $v^2 - u^2 = 2as$ gives

      $v^2 - (9.8)^2 = 2(-9.8)(-7)$

$\Rightarrow$ $v^2 = 96.04 + 137.2 = 233.24$

Therefore $v = 15.27$.

The ball hits the ground at 15.3 ms^{-1} correct to 3 significant figures.

Example 7

An object A is fired vertically upwards with a speed of 40 ms⁻¹ and 1 second later another object B is fired from the same point with the same velocity. In this question take g as 10.

a Find the time taken before they collide.

b Find the height above the point of projection at which they collide.

When B has been in the air for t seconds, A has been in the air for $(t+1)$ seconds.

Taking the upward direction as positive we have:

For A

Known: $u = 40$

$a = -10$

time $= (t+1)$

wanted: s

For B

Known: $u = 40$

$a = -10$

time $= t$

wanted: s

> **Note**
>
> Using t for the shorter time interval and $(t+1)$ for the longer one, rather than t for the first and $(t-1)$ for the second, avoids mistakes that can happen when minus signs are involved.

a Using $s = ut + \dfrac{1}{2}at^2$ gives:

For A $\quad s = 40(t+1) + \dfrac{1}{2}(-10)(t+1)^2$

For B $\quad s = 40t + \dfrac{1}{2}(-10)t^2$

Therefore $40(t+1) - 5(t+1)^2 = 40t - 5t^2$

$\Rightarrow \qquad\qquad 40 - 10t - 5 = 0$

$\Rightarrow \qquad\qquad\qquad t = 3.5$

The projectiles collide 3.5 s after B is fired.

b As A and B are at the same height when they collide we can find s for either of them.

For B,

$s = 40(3.5) - 5(3.5)^2 = 78.75$

A and B collide 78.8 m above the point of projection.

Example 8

A boy kicks a ball vertically up next to a wall that is 2.5 m high. He kicks the ball from a height of 0.4 m with a speed of 14 ms⁻¹. How long is the ball above the top of the wall?

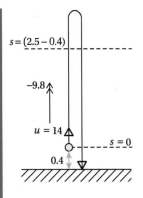

Known: $u = 14$ wanted: t

$a = -9.8$

$s = 2.1$ (measuring displacement upward from a point 0.4 m above the ground)

Using $\qquad s = ut + \dfrac{1}{2}at^2$ gives

$2.1 = 14t - 4.9t^2$

$\Rightarrow \qquad\qquad 0.3 = 2t - 0.7t^2$

$\Rightarrow \qquad 7t^2 - 20t + 3 = 0$

$\Rightarrow \qquad\qquad t = \dfrac{20 \pm \sqrt{400 - 84}}{14} = 2.698 \text{ or } 0.159$

(continued)

(continued)

The ball is at the height of the top of the wall at two different times. Therefore it takes 0.159 seconds to reach the top of the wall when going up, and returns to that height 2.698 seconds from the start.

So the ball is above the wall for $(2.698 − 0.159)\ s = 2.54\ s$, correct to 3 significant figures.

Exercise 2

Take g as 9.8 ms^{-2} and give answers correct to 2 significant figures.

1   **a**   A stone dropped from the top of a cliff takes 5 seconds to reach the ground below. Find the height of the cliff.

     b   A stone is thrown vertically down with from the top of the cliff. Find the velocity needed for the stone to land on the ground 4 seconds later.

2   A particle is projected vertically upward from ground level with a speed of 24.5 ms^{-1}. Find

     a   the greatest height reached

     b   the time taken for the particle to return to the ground.

3   A brick falls from the top of a building. Find how far it falls

     a   in the first second

     b   in the first two seconds

     c   during the third second.

4   A ball is thrown vertically upward and is caught at the same height 3 seconds later. Find

     a   the distance it rose

     b   the speed with which it was thrown.

5   A brick is dropped down a shaft, 50 m deep.

     a   For how long does it fall?

     b   With what speed does it hit the bottom?

6   A parachutist is descending vertically at a steady speed of 2 ms^{-1} when he drops a watch. If the watch hits the ground 3 seconds later at what height was the parachutist when he dropped it?

7   A flare is projected straight up from the bottom of a shaft that is 30 m deep. To be seen, the flare must reach at least 10 m above the ground. What is the least speed with which it must be fired?

8   A ball is thrown vertically, with a speed of 7 ms^{-1} from a balcony 14 m above the ground. Find how long it takes to reach the ground if it is thrown

     a   downwards   **b**   upwards.

     Find also the speed with which it reaches the ground in each of these cases.

9   A stone is dropped from the top of a building and at the same instant another stone is thrown vertically upward from the ground below at a speed of 15 ms^{-1}. The stones pass each other after 1.2 seconds. Find the height of the building.

10 A boy kicks a ball vertically upward with initial speed 15 ms^{-1} next to a wall that is 3 m high.

    a What is the greatest height reached by the ball?

    b For how long can it be seen by someone on the other side of the wall?

11 A ball is dropped from a point 1.6 m above the floor.

    a Find the speed of the ball when it hits the floor.

    b When it hits the floor its speed is halved by the impact. How high does it bounce?

12 A stone is dropped from the top of a building to the ground. During the last second of its fall it moves through a distance which is $\frac{1}{5}$ of the height of the building. How high is the building?

17.3 Motion in a straight line with variable acceleration

Velocity, v, is the rate at which displacement, s, varies with time, t, so $v = \dfrac{ds}{dt}$

and acceleration, a, is the rate at which velocity, v, varies with time, t, so $a = \dfrac{dv}{dt}$.

Also as $a = \dfrac{dv}{dt}$ and $v = \dfrac{ds}{dt}$ then $a = \dfrac{d}{dt}\left(\dfrac{ds}{dt}\right) = \dfrac{d^2s}{dt^2}$.

These relationships can be used to solve problems in which the motion varies with time.

Starting with the acceleration, then

$$a = \frac{dv}{dt} \Rightarrow v = \int a\, dt$$

and $v = \dfrac{ds}{dt} \Rightarrow s = \int v\, dt$

Therefore when the acceleration of a moving body is a function of time,

    velocity can be found by integrating a with respect to t

and

    displacement can be found by integrating v with respect to t.

Example 9

A body moves along a straight line so that its displacement, s metres, from a fixed point O on the line after t seconds, is given by $s = t^3 - 3t^2 - 9t$.

a Find the velocity after t seconds.

b Find the time(s) when the velocity is zero.

c Sketch the velocity-time graph.

a $s = t^3 - 3t^2 - 9t$

$v = \dfrac{ds}{dt} = 3t^2 - 6t - 9$

b When $v = 0$, $3t^2 - 6t - 9 = 0$

$\therefore \quad 3(t^2 - 2t - 3) = 0 \Rightarrow 3(t-3)(t+1) = 0 \Rightarrow t = 3 \text{ or } -1$

Therefore the velocity is zero *after 3* seconds; it was also zero 1 second *before* the body reached O.

(continued)

(continued)

c   The expression for the velocity is a quadratic function for which the graph is
a parabola crossing the *t*-axis where $t = 3$ and $t = -1$.

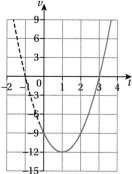

Example 10

Question

A particle P moves in a straight line and has an initial velocity of 2 ms⁻¹ at a point O
on the line. The acceleration of the particle *t* seconds later is given by $(2t - 6)$ ms⁻².

When $t = 5$ find expressions for

a   the velocity

b   the displacement of P from O.

Answer

a $v = \int a\,\mathrm{d}t = \int(2t - 6)\mathrm{d}t = t^2 - 6t + c_1$

   $v = 2$ when $t = 0$ therefore $c_1 = 2$

   Therefore $v = t^2 - 6t + 2$

   When $t = 5$, $v = 25 - 30 + 2 = -3$

   The velocity when $t = 5$ is -3 ms⁻¹

   so P is moving with speed 3 ms⁻¹ *towards* O.

b $s = \int v\,\mathrm{d}t = \int(t^2 - 6t + 2)\mathrm{d}t = \dfrac{1}{3}t^3 - 3t^2 + 2t + c_2$

   $s = 0$ when $t = 0$ therefore $c_2 = 0$

   $\therefore$ $s = \dfrac{1}{3}t^3 - 3t^2 + 2t$

   When $t = 5$, $s = \dfrac{125}{3} - 75 + 10 = -23\dfrac{1}{3}$

When $t = 5$ the displacement of P from O is $-23\dfrac{1}{3}$ m so P has changed direction,
moving back towards O, and then passes through O in the opposite direction.

Example 11

Question

A particle moves in a straight line. The velocity, *t* seconds after the particle
passes through a point O on the line, is given by $v = 3t^2 - 12t + 9$. Find the
times(s) when the particle changes direction.

Answer

The direction of motion is defined by the sign of the velocity (*not* by the sign of
the displacement). Hence, whenever the direction of motion is reversed, the
velocity is instantaneously zero.

   When $v = 0$, $3t^2 - 12t + 9 = 0$ $\Rightarrow$ $3(t^2 - 4t + 3) = 0$ $\Rightarrow$ $3(t - 1)(t - 3) = 0$

   Therefore $v = 0$ when $t = 1$ or 3.

(continued)

(continued)

Therefore the particle changes direction after 1 second and again after 3 seconds.

A sketch of the velocity-time graph confirms that v changes sign at $t = 1$ and $t = 3$.

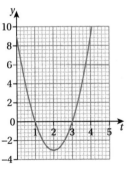

Example 12

A particle P moves in a straight line. The acceleration of P is given by $a = 2(t + 1)$ at any time t seconds. Initially P has a velocity of 2 m s^{-1}. Express the velocity of P as a function of t.

$$a = 2(t + 1) \quad \Rightarrow \quad v = \int a \, dt$$

Therefore $\quad v = \int 2(t + 1) dt = (t^2 + 2t) + c$

When $\quad t = 0, v = 2 \quad \Rightarrow \quad c = 2$

Therefore $\quad v = t^2 + 2t + 2$

Exercise 3

In each Question from 1 to 10, a particle P moves on a straight line and O is a fixed point on that line. After a time of t seconds the displacement of the particle from O is s metres, the velocity is $v \text{ ms}^{-1}$ and the acceleration is $a \text{ ms}^{-2}$.

1. Given that $s = 4t^3 - 5t^2 + 7t + 6$, find v when $t = 3$.

2. Given that $v = 9t^2 + 14t + 6$, find a when $t = 2$.

3. When $s = t^3 - 2t^2 + 9t$, find a when $t = 5$.

4. Given that P is at O when $t = 0$, and that $v = 2t^2 + 3t + 4$, find s when $t = 4$.

5. A particle P starts from O with velocity 3 m s^{-1} with acceleration given by $a = 12t - 5$. Find v and s when $t = 2$.

6. When $s = t^3 - 9t^2 + 24t - 11$, find the time(s) when $v = 0$.

7. Find the times when the direction of motion of P changes given that $s = 6t^3 - 9t^2 + 4t$.

8. A particle P starts from rest at O and moves with an acceleration given by $2t \text{ ms}^{-2}$. Find v and s in terms of t.

9. When $t = 0$, a particle P passes through a point with a displacement of 2 m from O with a velocity of 4 ms^{-1}. Given that $a = t^2 + 1$, find the velocity and displacement when $t = 4$.

10. When $t = 0$, P passes through O with velocity -4 ms^{-1}. Given that $a = 8 - 6t$, find

    a. the times when P is instantaneously at rest

    b. the displacement of P from O at these times.

11 A particle P moves in a straight line and O is a fixed point on the line. The displacement, s metres, of P from O at any time t seconds is given by $s = t^3 + t^2 + 12t - 23$. Show that the motion is always in the same direction.

12 A particle moves in a straight line with an acceleration at any time given by $(3t - 1)$ ms^{-2}.

The particle has a velocity of 3 ms^{-1} and is 7 m from a fixed point O on the line when $t = 2$. Find

  a its velocity when $t = 5$

  b its displacement from O when $t = 4$.

13 Given that $a = -\dfrac{1}{t^3}$ and that $v = 3$ when $t = 1$,

  a find the velocity when $t = 4$

  b show that, as the value of t becomes large, the velocity approaches a particular value (called the terminal velocity) and state this value.

14 A particle starts from rest at a fixed point A and moves in a straight line with an acceleration t seconds after leaving A, that is given by $a = 4t$.

After 2 seconds the particle reaches a point B and the acceleration is then zero. Find

  a the velocity when the particle reaches B

  b the distance AB.

15 The displacement, s metres, of a body from a point O after t seconds is given by $s = t^2 + \dfrac{1}{t}$. Find in terms of t an expression for the acceleration of the body.

Summary

Equations of motion in a straight line with constant acceleration

$v = u + at$

$s = \dfrac{1}{2}(u + v)t$

$s = ut + \dfrac{1}{2}at^2$ and $s = vt - \dfrac{1}{2}at^2$

$v^2 - u^2 = 2as$

where u = initial velocity, v = final velocity, a = acceleration, t = time and s = displacement.

The acceleration due to gravity is denoted by g and on the surface of the earth, $g = 9.8$ ms^{-2}.

Motion in a straight line with variable acceleration

$v = \dfrac{ds}{dt}$ and $a = \dfrac{dv}{dt}$

$v = \int a\, dt$ and $s = \int v\, dt$

Review

Give answers that are not exact correct to 3 significant figures.

1 A ball is dropped from a point 10 m above the ground.

   a Find the speed at which the ball hits the ground.

   b The ball rebounds upwards with a speed of 5 ms⁻¹. Find the time taken for the ball to hit the ground again.

2 A train has a maximum speed of 50 ms⁻¹ which it reaches with an acceleration of 0.25 ms⁻².

   a Find the time the train takes to reach its maximum speed and the distance travelled in this time.

   The train has a maximum deceleration of 0.5 ms⁻².

   b Find the shortest time it takes the train to stop from its maximum speed and the distance travelled in this time.

   The train stops at two stations that are 10 km apart.

   c What is the shortest time for the train to travel between these two stations?

3 A car is moving along a straight road with uniform acceleration. The car passes a point A with a speed of 12 ms⁻¹ and another point C with a speed of 32 ms⁻¹. The distance between A and C is 2000 m.

   Find the time taken by the car to move from A to C.

4 A particle is projected vertically upwards at time $t = 0$ from a point A with speed 20 ms⁻¹. Two seconds later a second particle is projected vertically upwards from A with the same speed.

   a Express the heights above A of both particles in terms of t.

   b Hence, find the value of t when the particles collide.

5 A particle P moves in a straight line and passes a fixed point O on the line at time $t = 0$. After t seconds, P is moving with a velocity v m s⁻¹ where $v = 2t^3 + 2t + 1$.

   a Find the acceleration of P when $t = 2$.

   b Find the distance covered by P between $t = 1$ and $t = 4$.

6 A particle P moves in a straight line and passes a fixed point O on the line at time $t = 0$. For values of $t > 2$, the velocity of the particle is given by $v = \dfrac{3}{t^2}$.

   a Explain why the particle is decelerating for all values of t greater than 2.

   b Find the distance covered by P in the interval between $t = 4$ and $t = 5$.

Assessment

1 A ball is thrown vertically downwards with an initial speed of 3 ms^{-1} from a point 10 m above horizontal ground. The ball moves freely under gravity.

   a Find the speed of the ball as it hits the ground.

   b The ball rebounds vertically upwards with an initial speed of 6 ms^{-1}.

      Find the time taken for the ball to reach the ground again.

2 A ball is dropped from rest at a point which is 3 m above horizontal ground and moves freely under gravity.

  When the ball hits the ground it rebounds vertically upwards to a height of 1.5 m above the ground.

  The ball then falls back to the ground.

   a Find the speed of the ball when it first hits the ground.

   b Find the speed of the ball the instant after it first rebounds from the ground.

   c Find the time between the first and second impacts of the ball with the ground.

3 A particle moves in a straight line. Initially the particle is at a fixed point P and t seconds after leaving P it has velocity $\left(4t^2 + t - 5\right)\,\text{ms}^{-1}$.

   a Find the acceleration of the particle when it comes instantaneously to rest.

   b Find the displacement of the particle from O when it comes instantaneously to rest.

4 A car travels on a straight horizontal race track. The car slows down with constant deceleration from a speed of 18 ms^{-1} to a speed of 10 ms^{-1} as it travels a distance of 700 metres.

   a Find the time that it takes the car to travel the first 640 metres.

   b Find the deceleration of the car during the first 640 metres.

5 A particle P moves in a straight line. The acceleration of P at time t seconds is $(t - 6)$ ms^{-2}. The velocity of P at time t seconds is v ms^{-1}. When $t = 0$, $v = 10$.

   a Find v in terms of t.

   b Find the values of t when P is instantaneously at rest.

   c Find the distance between the two points at which P is instantaneously at rest.

6 A car travels on a straight horizontal race track. The car decelerates uniformly from a speed of 20 ms^{-1} to a speed of 12 ms^{-1} as it travels a distance of 640 metres.

  The car then accelerates uniformly, travelling a further 1820 metres in 70 seconds.

   a **i** Find the time that it takes the car to travel the first 640 metres.

       ii Find the deceleration of the car during the first 640 metres.

b i Find the acceleration of the car as it travels the further 1820 metres.

  ii Find the speed of the car when it has completed the further 1820 metres.

c Find the average speed of the car as it travels the 2460 metres.

AQA MM1B January 2013

7 A car is travelling at a speed of 20 ms⁻¹ along a straight horizontal road. The driver applies the brakes and a constant braking force acts on the car until it comes to rest.

a Assume that no other horizontal forces act on the car.

  i After the car has travelled 75 metres, its speed has reduced to 10 ms⁻¹. Find the acceleration of the car.

  ii Find the time taken for the speed of the car to reduce from 20 ms⁻¹ to zero.

AQA MM1B June 2012 (part question)

18 Forces and Newton's Laws

Introduction

Force is a familiar quantity. To move a heavy object across a floor we need push it; to raise a heavy object from the floor onto a table we need to pull it up. Pushes and pulls are both forces and these simple examples show that force is needed to make an object start to move.

To stop a shopping trolley rolling down a hill we can get in front of it and push or hold it from behind and pull. So a force can also make a moving object slow down.

It was not until 1687, when Newton's Laws of Motion were published, that it was explained how force and motion are connected.

Recap

You need to remember...
▶ The equations of motion with constant acceleration.

Objectives

By the end of this chapter, you should know how to...
▶ Identify different types of force.
▶ Use Newton's Laws of Motion.
▶ Investigate the motion of connected particles.
▶ Understand and use the coefficient of friction.

Applications

When an object is moving some people think there must be a force acting on it to *keep* it moving.

A puck is struck with a stick and sent moving across an ice rink (a force *starts* the motion), but the puck continues to move although there is nothing to push it once it has left the stick. This shows that a force is *not* necessary to keep the keep the puck moving. However the puck will gradually slow down so a force is acting to slow it down but not to keep it moving.

18.1 Types of force

Weight

When a stone is dropped it begins to move down so a force must be *pulling it* down.

The force that attracts an object to the earth is the force due to gravity. It is called the **weight** of the object.

The effect of weight can be seen when an object falls and it can be felt when an object is held, so the weight of an object acts on it at all times whether the object is moving or stationary.

Normal reaction

A book at rest on a horizontal table does not fall down, so a force must be stopping it from falling. This is the force exerted by the table to hold the book up. As the book does not move then the upward force and the weight of the book down must balance. When an object does not move, the object, and the forces that act on it, are in **equilibrium**.

The force exerted upward on the book by the surface is called a **normal reaction**. Normal means perpendicular so a normal reaction acts perpendicular to the surface of contact and away from that surface. Therefore the normal reaction acts vertically upward on the book.

The two forces acting on the book, the normal reaction, R, and the weight, W, are shown in the diagram.

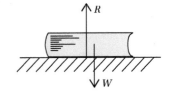

Friction

When a small push P is applied horizontally to the book then, if the book and the surface are not slippery, the book probably will not move. Therefore there must be another force equal and opposite to the push to balance it. This is a **friction** force. Friction happens only when objects are in rough contact.

A friction force acts on a body along the surface of contact and in a direction which opposes the potential movement of that body.

This diagram shows all the forces acting on the book.

It is unusual for there to be no friction at all between an object and a surface but there can be so little that its effect can be ignored. In this case the contact is called **smooth**.

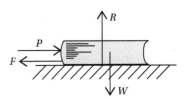

Tension

A stone at rest hangs by a string from a fixed point. The stone does not move so its weight acting down must be balanced by an upward force. This force is the **tension** in the string.

A string can never push and it can pull only if it is taut.

A string cannot be taut at one end and slack at the other, so a taut string exerts an *inward* pull at each end on the object which is attached at that end. The tension in a string acts along the string.

A rod can be in tension or in thrust.

For example when a rod is attached to a beam and a heavy object is suspended vertically attached to the rod at the other end, the rod is in tension.

When a rod is clamped at both ends and the clamps are tightened, the rod is in thrust.

In both cases the forces at each end of the rod are equal and opposite.

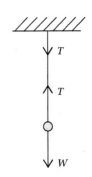

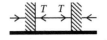

Drawing diagrams

To work on any problem involving the action of forces on a body the first (and necessary) step is to draw a clear diagram of the forces acting on the object.

Some points to remember are:

▶ Unless the object is light (this means it is taken as weightless), its weight acts vertically down.

▶ When the object is in contact with another surface, a normal reaction always acts on the body. Also, unless the contact is smooth, there may be a frictional force.

▶ When the object is attached to another by a string or rod, a force acts on the body at the point of attachment.

▶ Draw a diagram that is large enough and make the force lines long enough to be seen clearly.

Example 1

A truck is attached by a rope to an engine which is being driven along a horizontal smooth track.

Show the forces acting on

a   the truck

b   the engine.

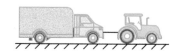

a   Looking at the truck, the only forces acting are the weight of the truck, the vertical normal reaction and the tension in the rope which acts away from the truck (inward along the rope).

The driving force of the engine *does not* act on the truck, it is the tension in the rope that pulls the truck forward.

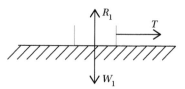

b   Acting on the engine alone is the weight, the normal reaction, the driving force and the tension in the rope which acts towards the centre of the rope (it is a drag on the engine).

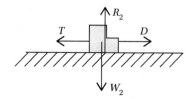

Example 2

A stone is projected vertically upwards. Draw a diagram to show the forces acting on the stone when it is

a   going up

b   falling back down.

a   The stone is not attached to anything so the only force acting is the weight of the stone.

b   The stone is not attached to anything so the only force acting is the weight of the stone.

> **Note**
>
> There is no force in the direction of motion.

Exercise 1

1 A block is at rest on a horizontal surface.

Draw a diagram to show the forces acting on the block.

2 A plank is resting on two supports, one at each end.

Draw a diagram to show the forces acting on the plank.

3 A particle is attached to one end of a light string whose other end is fixed to a point A.

The particle is hanging at rest.

Draw a diagram to show the forces acting on the particle.

4 A shelf AB is supported by two vertical strings, one at each end. A vase is placed on the shelf, a quarter of the way along the shelf from A.

Draw separate diagrams to show the forces acting on

a the shelf
b the vase.

5 Two bricks are placed, one on top of the other, on a horizontal surface.

Draw separate diagrams to show the forces acting on

a the top brick

b the lower brick.

6 A person standing at the edge of a flat roof lowers a package using a rope over the edge.

Draw diagrams to show

a the forces acting on the package

b the forces acting on the person on the roof.

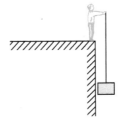

18.2 Newton's first law of motion

This law states that:

> An object will continue in its state of rest, or of uniform motion in a straight line, unless an external force is applied to it.

Newton's first law also implies that:

▶ When a body is at rest, or is moving with constant velocity, any forces that are acting must balance exactly, so they must be in equilibrium.

▶ When the speed of a moving object is changing, the forces acting on it are not in equilibrium.

▶ When the *direction* of motion of a moving object is changing, so it is not moving in a straight line, the forces acting on it are not in equilibrium. (So there is always a force acting on a body that is moving in a curve, even if the speed is constant.)

Resultant force

The diagram shows an object connected to a string.

The forces acting on it are its weight and the tension in the string.

If T is equal to W, the forces are in equilibrium and the object will not move.

If T is greater than W, the forces are not in equilibrium. There is an excess force equal to $T - W$ acting vertically upwards and the object will move.

$T - W$ is called the **resultant force**.

Newton's first law defines what force is: force is the quantity that, when acting on a body, changes the velocity of that body.

When the velocity of a body changes, there is an acceleration, so:

> **When a body has an acceleration there is a resultant force acting on it.**
> **When a body has no acceleration there is no resultant force acting on it.**

A body has no acceleration when it is at rest, or when it is moving with constant velocity so it is clear that no force is needed to keep a body moving with constant velocity.

Example 3

Question

An object is at rest under the action of the forces shown. All forces are measured in the same unit.

Find the values of F and R.

Answer

The object is at rest therefore there is no resultant force either horizontally or vertically.

Horizontally $\quad 5 - F = 0 \quad \Rightarrow \quad F = 5$

Vertically $\quad R - 60 = 0 \quad \Rightarrow \quad R = 60$

Example 4

Question

A particle of weight 7 units, hanging at the end of a vertical light string, is moving upward with constant velocity. Find the tension in the string.

Answer

The velocity of the particle is constant so it has no acceleration and therefore the resultant force is zero.

Vertically $\quad T - 7 = 0 \quad \Rightarrow \quad T = 7$

Therefore the tension in the string is 7 units.

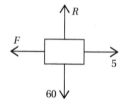

Exercise 2

In Questions 1 to 5, the diagram shows the forces, all in the same unit, acting on an object which is at rest.

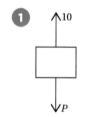
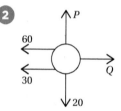
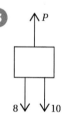
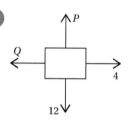
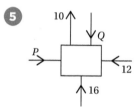

Find P.	Find P and Q.	Find P.	Find P and Q.	Find P and Q.

In each Question from 6 to 9 state whether or not the object shown in the diagram has an acceleration.

If there is an acceleration state whether it is horizontal or vertical.

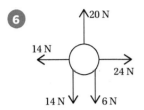
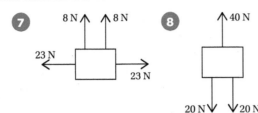
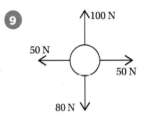
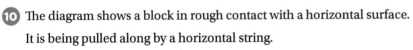
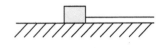

10 The diagram shows a block in rough contact with a horizontal surface.

It is being pulled along by a horizontal string.

a Make a copy of the diagram and on it mark all the forces acting on the block.

b Explain what it means for the tension in the string compared with the frictional force when the block

    i is accelerating   **ii** moves with constant velocity.

Newton's second law

This law defines the relationship between force, mass and acceleration.

Experience shows that

i for a body of a particular mass, the bigger is the force, the bigger will be the acceleration

ii the larger is the mass, the larger will be the force needed to produce a particular acceleration.

Experimental evidence shows that the force F is proportional both to the acceleration a and to the mass m.

Therefore $F \propto ma$

or $F = kma$ where k is a constant

When $m = 1$ and $a = 1$ then $F = k$, so the amount of force needed to give a mass of 1 kg an acceleration of 1 ms^{-2} is given by k.

When this amount of force is chosen as the unit of force then $k = 1$ and

    $F = ma$

    The unit of force is called the newton (N) and is defined as the amount of force that gives 1 kg an acceleration of 1 ms^{-2}.

When more than one force acts on a body, F is the resultant force.

When the force is constant the acceleration also is constant and when the force varies, so does the acceleration.

When the acceleration is zero, the resultant force is zero – so Newton's first law follows from the second.

Example 5

A resultant force of 12 N acts on a body of mass 5 kg. Find the acceleration of the body.

Using $F = ma$ gives $12 = 5a$ $\Rightarrow$ $a = 2.4$

The acceleration is 2.4 ms^{-2} in the direction of the force.

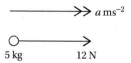

Example 6

A set of forces act on a mass of 3 kg and give it an acceleration of 11.4 ms^{-2}.

Find the magnitude of the resultant force.

When $m = 3$ and $a = 11.4$ using $F = ma$ gives

$F = 3 \times 11.4 = 34.2$

The resultant force is 34.2 N.

Example 7

The diagram shows the forces that act on a particle making it move vertically downward with an acceleration a ms^{-2}. Find the values of P and a.

There is no horizontal motion

Therefore $P - 7 = 0$ $\Rightarrow$ $P = 7$

The resultant force vertically down is $(6 - 3)$ N = 3 N

Using $F = ma$ gives

$$3 = 5a \quad \Rightarrow \quad a = \frac{3}{5}$$

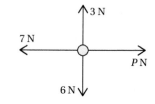

Some problems give some facts about how an object is *being made to move* with constant acceleration and other information about the *motion* of the object. So use both $F = ma$ and one of the equations of motion which contains a.

Example 8

A resultant force F newtons acts on a particle of mass 3 kg.

a   The particle accelerates uniformly from 2 ms^{-1} to 8 ms^{-1} in 2 seconds. Find the value of F.

b   If $F = 6$ find the displacement of the particle 4 seconds after starting from rest.

a For the motion of the particle:

$u = 2$, $v = 8$, $t = 2$ and a is required.

Using $v = u + at$ gives $8 = 2 + 2a$ $\Rightarrow$ $a = 3$

Now using $F = ma$ gives $F = 3 \times 3$

$\Rightarrow$ $F = 9$

The force is 9 N.

b The force is known so we use Newton's Law first.

Using $F = ma$ gives $6 = 3a$

$\Rightarrow$ $a = 2$

For the motion of the particle:

$u = 0$, $a = 2$, $t = 4$ and s is required.

Using $s = ut + \dfrac{1}{2}at^2$ gives $s = 0 + \dfrac{1}{2}(2)(4)^2$

$\Rightarrow$ $s = 16$

The displacement of the particle is 16 m.

Exercise 3

1 A resultant force of 12 N acts on a body of mass 8 kg. Find the acceleration of the body.

2 The acceleration of a particle of mass 2 kg is 14 ms⁻². Find the resultant force acting on the particle.

3 A force of 420 N acts on a block, giving it an acceleration of 10.5 ms⁻². Assuming that no other force acts on the block, find its mass.

4 In each diagram the forces shown (measured in newtons) make an object of mass 8 kg to move with the acceleration shown. Find P and/or Q in each case.

a

b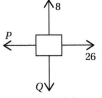

c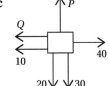

5 Each diagram shows the forces acting on a body of mass 3 kg. Find the magnitude and direction of the acceleration of the body in each case.

a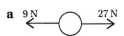

b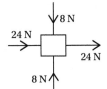

c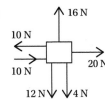

6 In each diagram the mass of the body is m kilograms. Find m and P.

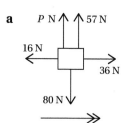

a

P N ↑ ↑ 57 N

16 N ←

→ 36 N

80 N ↓

⟹ 5 ms⁻²

b

↑ 18 N

30 N ←

→ P N

12 N ↓ ↓ 36 N

6 ms⁻² ↓

c

↑ 40 N

P N ←

→ 20 N

→ 20 N

20 N ↓

2 ms⁻² ↑↑

7 A body of mass 3 kg is accelerating vertically down at 5 ms⁻² under the action of the forces shown, all measured in newtons. Find the values of P and Q.

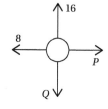

↑ 16

8 ←

→ P

Q ↓

8 A body of mass 2 kg accelerates uniformly from rest to 16 ms⁻¹ in 4 seconds. Find the resultant force acting on the body.

9 A force of 100 N acts on a particle of mass 8 kg. The particle is initially at rest. Find how far it travels in the first 5 seconds of its motion.

10 A block of mass 6 kg is pulled along a smooth horizontal surface by a horizontal string. The block accelerates uniformly and reaches a speed of 20 ms⁻¹ in 4 seconds from rest. Find

a the acceleration

b the tension in the string.

11 A constant force of 80 N acts for 7 seconds on a body, initially at rest, giving it a velocity of 35 ms⁻¹. Find

a the acceleration

b the mass of the body.

12 A force of 12 N acts on a particle of mass 60 kg making the velocity of the particle increase from 3 ms⁻¹ to 7 ms⁻¹. Find the distance that the particle travels in this time interval.

13 A body of mass 120 kg is moving in a straight line at 8 ms⁻¹ when a force of 40 N acts in the direction of motion for 18 seconds. What is the speed of the body at the end of this time?

Weight

An object of mass m kg, falls freely under gravity with an acceleration g ms⁻².

The force producing the acceleration is the weight, W newtons, of the object, so using $F = ma$ gives $W = mg$.

> **Therefore a body of mass m kilograms has a weight of mg newtons.**

For example, taking the value of g as 9.8,

the weight, W newtons, of a person whose mass is 55 kg is given by

$W = mg = 55 \times 9.8 = 539$

therefore the person's weight is 539 N.

The weight of a rockery stone is 1078 N, so its mass, m kg, is given by

$$W = mg \quad \Rightarrow \quad m = \frac{W}{g} = \frac{1078}{9.8} = 110$$

Therefore the mass of the stone is 110 kg.

Example 9

Question

A bucket attached to the end of a rope is used to raise water from a well. The mass of the empty bucket is 1.2 kg and it can hold 10 kg of water when full. Find the tension in the rope when

a the empty bucket is lowered with an acceleration of 2 ms^{-2}

b the full bucket is pulled up with an acceleration of 0.3 ms^{-2}.

Answer

a The acceleration of the bucket is downward so the resultant force acts downwards.

    The weight of the bucket is $1.2g$ N $= 1.2 \times 9.8$ N.

    The resultant force is vertically down and equal to $(1.2g - T)$ N.

    Using $F = ma$ gives

    $1.2 \times 9.8 - T = 1.2 \times 2$

    $T = 1.2 \times 9.8 - 1.2 \times 2 = 9.36$

    The tension in the rope is 9.36 N.

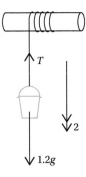

b The acceleration of the full bucket is upward so the resultant force acts upwards.

    The weight of the full bucket is 11.2×9.8 N.

    Resultant force up is $(T - 11.2g)$ N.

    Using $F = ma$ gives

    $T - 11.2 \times 9.8 = 11.2 \times 0.3$

    $T = 11.2 \times 0.3 + 11.2 \times 9.8$

    $= 113.12$

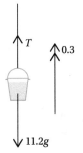

The tension in the rope is 113 N (correct to 3 significant figures).

Exercise 4

1 a Find the weight of a body of mass 5 kg.

   b Find the mass of a sack of potatoes of weight 147 N.

   c Find the weight of a ball of mass 60 g.

2 On the moon the acceleration due to gravity is 1.2 ms^{-2}.

   Repeat question 1 with this value of g.

3 A particle of mass 2 kg is attached to the end of a vertical light string.

   The particle is being pulled up with an acceleration of 5.8 ms^{-2} by the string.

   Find the tension in the string.

4 A mass of 6 kg moves vertically at the end of a light string.

   Find the tension in the string when the mass has an acceleration of

   a 5 ms^{-2} downwards   b 7 ms^{-2} upwards   c zero.

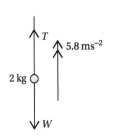

5 The tension in a string, which has a particle of mass
m kilograms attached to its lower end, is 70 N.

Find the value of m if the particle has

a an acceleration of 3 ms^{-2} upwards

b an acceleration of 9 ms^{-2} downwards

c a constant velocity of 4 ms^{-1} upwards

d a constant velocity of 4 ms^{-1} downwards.

6 A block with a mass of 750 kg can be raised and lowered using a rope. More
blocks can be added up to a maximum extra mass of 1000 kg.

a Find the tension in the rope when

i raising the fully loaded rope with an acceleration of $\dfrac{1}{2}$ ms^{-2}

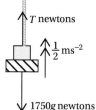

ii lowering the rope with just one block with an acceleration of $\dfrac{3}{4}$ ms^{-2}.

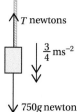

b The tension in the rope is 14 700 N when the rope, partly loaded, is being
raised at constant speed. Find the mass of the extra blocks.

7 A balloon of mass 1400 kg is descending vertically with an acceleration of
2 ms^{-2}. Find the upward force exerted on the balloon by the atmosphere.   This force is called air resistance.

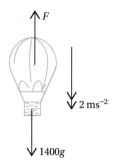

8 A block of mass 4 kg is lying on the floor of a lift that is accelerating at 5 ms^{-2}.

Find the normal reaction exerted on the block by the lift floor if the lift is

a going up   **b** going down.

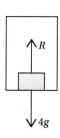

Newton's third law

Newton's third law states that:

Action and reaction are equal and opposite.

This means that if a body A exerts a force on a body B then B exerts an equal force in the opposite direction on A.

This is true whether the two bodies are in contact or are connected but some distance apart, whether they are moving or are stationary.

For example, a mass is resting in a scale pan.

The scale pan is exerting an upward force on the mass and the mass is exerting an equal force downward on the scale pan.

These two particles are connected by a taut string. The objects are not in direct contact but exert equal and opposite forces on each other because of the equal tensions in the string which act inwards at each end.

This is true even if the string passes round a *smooth* body, such as a pulley, which changes the direction of the string.

The tensions in the two portions of the string are the same and each portion exerts an inward pull at each end. So in each portion the tension at one end acts on the particle and at the other end the tension acts on the pulley; all these tensions are equal.

(If the string passes round a rough surface the tensions in the different portions of the string are not equal.)

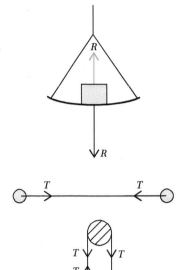

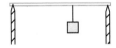

Exercise 5

Copy the diagram in each question. Make your copy twice as big and mark on it all the forces that are acting on each body (in Question 4 draw a small block to represent the person). Use either a different colour or a different type of line (for example broken and solid) for the forces that act on separate objects. Ignore forces that act on fixed surfaces – these are shown by hatching.

1 A load hangs from a beam which is supported at each end.

2 A mass on a table is linked by a string to a mass hanging over the smooth edge.

3 A mass B hangs by string from another mass A which hangs from a fixed point.

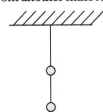

4 A passenger is in a lift that is being drawn up by a cable.

18.3 Connected particles

An inextensible (or inelastic) string is a string whose length does not alter.

Smooth pulleys have bearings and a rim that are completely without friction.

The diagram shows two particles A and B that are connected by a light inextensible string passing over a smooth fixed pulley.

The masses of A and B are m_A and m_B respectively and $m_A > m_B$. As the particles move, one of them moves up in the same way as the other moves down, so

the upward speed of B is equal to the downward speed of A,

the upward acceleration of B is equal to the downward acceleration of A,

the distance B moves up is equal to the distance A moves down.

The forces that act on *each particle alone* determine the way that each particle moves, so to find the motion look at the particles separately as shown:

Example 10

A light inextensible string passes over a smooth fixed pulley and particles of masses 5 kg and 7 kg, are attached, one at each end. The system is moving freely. Find in terms of g

a   the acceleration of each particle

b   the tension in the string

c   the force exerted on the pulley by the string.

The two particles have the same acceleration, a ms^{-2}, and the two parts of the string have the same tension, T newtons.

For each particle find the resultant force in the direction of motion and use the equation of motion, $F = ma$, in that direction.

(continued)

(continued)

For A: ↑

The resultant force is $T - 5g$.

Using $F = ma$ gives $T - 5g = 5a$. [1]

For B: ↓

The resultant force is $7g - T$.

Using $F = ma$ gives $7g - T = 7a$. [2]

a [1] + [2] gives $2g = 12a$ ⇒ $a = \dfrac{1}{6}g$

The acceleration of each particle is $\dfrac{g}{6}$ ms^{-2}.

b From [1], $T - 5g = 5\left(\dfrac{g}{6}\right)$ ⇒ $T = \dfrac{35g}{6}$

The tension in the string is $\dfrac{35g}{6}$ N.

c The string exerts a downward pull on each side of the pulley.

Therefore the resultant force exerted on the pulley by the string is $2T$ downwards,

⇒ $\dfrac{35g}{3}$ N downwards.

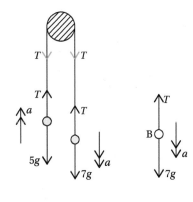

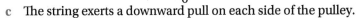

Example 11

A small block of mass 6 kg rests on a table top and is connected by a light inextensible string that passes over a smooth pulley, fixed on the edge of the table, to another small block of mass 5 kg which is hanging freely. Find, in terms of g, the acceleration of the system and the tension in the string when

a the table is smooth

b the table is rough and a frictional force of $2g$ N acts between the table and the block.

Use $F = ma$ for each block in its direction of motion.

a For the block A → $T = 6a$ [1]

For the block B ↓ $5g - T = 5a$ [2]

[1] + [2] ⇒ $5g = 11a$ ⇒ $a = \dfrac{5}{11}g$

and from [1] $T = 6a = \dfrac{30}{11}g$

The acceleration is $\dfrac{5}{11}g$ ms^{-2} and the tension is $\dfrac{30}{11}g$ N.

b For the block A → $T - 2g = 6a$ [3]

For the block B ↓ $5g - T = 5a$ [4]

[3] + [4] ⇒ $3g = 11a$ ⇒ $a = \dfrac{3}{11}g$

and $T = 6a + 2g = \dfrac{40}{11}g$

The acceleration is $\dfrac{3}{11}g$ ms^{-2} and the tension is $\dfrac{40}{11}g$ N.

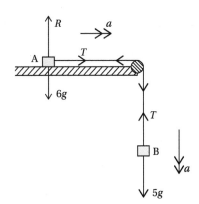

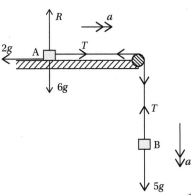

Example 12

The diagram shows the forces acting on a car and trailer at the instant when the driving force is 5000 N.

The car and trailer are connected by a light inextensible rope and are accelerating along a horizontal road against the resistive forces shown.

Find the acceleration of the system and the tension in the string at this instant.

For the car → $5000 - T - 100 = 1500a$ [1]

For the trailer → $T - 50 = 500a$ [2]

$[1] + [2] \Rightarrow 4850 = 2000a \Rightarrow a = 2.425$

and $T = 50 + (500)(2.425) = 1262.5$

The acceleration is 2.43 ms^{-2} and the tension in the rope is 1260 N to 3 significant figures.

Exercise 6

Give the answers in terms of g.

1 Each diagram shows the forces, in newtons, acting on two particles connected by a light inextensible string which passes over a fixed smooth pulley. In each case find the acceleration of the system and the tension in the string.

a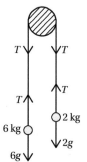

b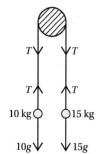

c

2 Two particles are connected by a light inextensible string which passes over a fixed smooth pulley. Find the acceleration of the system and the tension in the string when the masses of the particles are

a 5 kg and 10 kg

b 12 kg and 8 kg.

3 Two particles A and B of masses 8 kg and 4 kg respectively hang one at each end of a light inextensible string which passes over a fixed smooth pulley. Find

a the acceleration of the system when the particles are released from rest

b the distance that each particle moves during the first 5 seconds.

4 A particle of mass 5 kg rests on a smooth horizontal table and is attached to one end of a light inelastic string. The string passes over a fixed smooth pulley at the edge of the table and a particle of mass 3 kg hangs freely at the other end.

When the system is released from rest find

a the acceleration of the system

b the tension in the string.

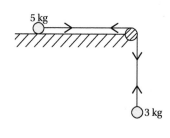

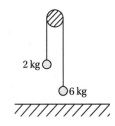

5 Two particles of masses 2 kg and 6 kg are attached, one to each end of a long light inextensible string which passes over a fixed smooth pulley. The system is released from rest and the heavier particle hits the ground after 2 seconds. Find

  a the acceleration of the system

  b the height of particle B above the ground when it was released

  c the speed at which B hits the ground.

6 Two particles, A of mass 4 kg and B of mass 5 kg, are connected by a light inextensible string passing over a smooth pulley.

Initially B is 1 m above a fixed horizontal plane. The system is released from rest in this position. Find

  a the acceleration of each particle

  b the speed of each particle when B hits the plane.

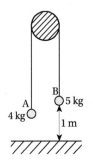

7 The diagram shows a particle P lying in contact with a smooth table top 1.5 m above the floor. A light inextensible string of length 1 m connects P to another particle Q hanging freely over a small smooth pulley at the edge of the table.

The mass of each particle is 2 kg, and P is held at a point distant 0.5 m from the edge of the table. The system is released from rest. Find

  a the acceleration of the system

  b the tension in the string

  c the speed of each particle when P reaches the edge of the table.

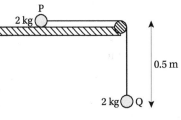

8 The diagram shows a trailer coupled to a car by a solid rod. The mass of the car is 800 kg and the mass of the trailer is 200 kg. The constant resistance to motion of the car is 70 N and the constant resistance to motion of the trailer is 30 N.

When the acceleration is 0.5 ms⁻² find

  a the tension in the rod

  b the driving force of the car.

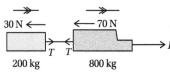

9 The car and trailer described in question 8 brakes.

Find the braking force on the car at the instant when the deceleration of the system is 1.5 ms⁻² and the resistive forces are as shown in Question 8.

18.4 Dynamic friction

A book lies on a table. If the table is rough then a force P applied to the book may not move it. This is because there is frictional resistance to motion.

The frictional force F, acting on the book, acts along the table in a direction opposite to the potential direction of motion.

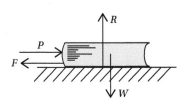

If P and F are the only forces acting horizontally on the book then, as long as it does not move, P and F must be equal.

When P is gradually increased the book will at some point be *just on the point of moving*.

> When the book is on the point of moving, F has reached its maximum value.
>
> When the book moves, F stays at its maximum value which is called dynamic friction.

A further increase in P will make the book move and then $P > F$.

Once an object begins to move, the frictional force opposing motion remains constant.

The coefficient of friction

For two particular surfaces in rough contact, experiments show that the limiting value of the frictional force is a fixed fraction of the normal reaction between the surfaces.

This fraction is called the **coefficient of friction** and it is denoted by the Greek letter μ.

> **For limiting and dynamic friction $F = \mu R$.**

As this is the maximum value of the frictional force, F can take any value from zero up to μR, therefore

> $0 \leq F \leq \mu R$

The value of μ depends upon the materials which make up the *two* surfaces in contact – it is *not* a property of *one* surface – so ideally we should always refer to **rough contact** but it is not always used. A shorter form of wording is often used such as 'a block moves on a rough plane' and it means that there is friction between the block and the plane.

In the context of mechanics, 'rough' means that there *is* friction at the contact; 'smooth' means that friction is ignored at the contact.

Example 13

A small block of weight 32 N is lying on a rough horizontal plane. A horizontal force of P newtons is applied to the block until it is on the point of moving the block.

a   When $P = 8$ find the coefficient of friction μ between the block and the plane.

b   If $\mu = 0.4$, find the value of P.

When the block is on the point of moving, it has no acceleration so the horizontal forces are balanced.

a   Horizontally $8 - \mu R = 0 \implies \mu R = 8$

Vertically $R - 32 = 0 \implies R = 32$

$$\mu = \frac{\mu R}{R} = \frac{8}{32}$$

The coefficient of friction is $\frac{1}{4}$.

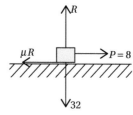

b   Horizontally $P - 0.4 \times R = 0$

Vertically again gives $R = 32$

Therefore $P - 0.4 \times 32 = 0 \implies P = 12.8$

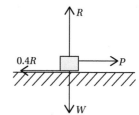

Example 14

The diagram shows a particle A lying on a rough table.

A light inelastic string, attached to A, passes over a smooth pulley at the edge of the table and is attached to another particle B hanging freely. Particles A and B have equal mass of 50 kg and the coefficient of friction between particle A and the table is $\frac{1}{2}$.

The particles are released from rest. Find

a the acceleration of the system

b the tension in the string.

a For A: $\uparrow$ $R = 50g$

Using $F = ma$: $\rightarrow$ $T - \dfrac{1}{2}R = 50a$ $\Rightarrow$ $T - 25g = 50a$ [1]

For B: Using $F = ma$: $\downarrow$ $50g - T = 50a$ [2]

[1] + [2] $\Rightarrow$ $25g = 100\,a$ $\Rightarrow$ $a = \dfrac{1}{4}g = 2.45$

The acceleration of the system is 2.45 ms^{-2}.

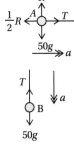

b Substituting for a in [1] gives $T = 25g + 50 \times \dfrac{1}{4}g = 37.5g = 367.5$.

The tension in the string is 367.5 N.

Exercise 7

Give answers that are not exact corrected 3 significant figures.

1 A particle of weight 24 N is on a rough horizontal plane and is being pulled by a horizontal string. If the particle is just on the point of moving when the tension in the string is 8 N, find the value of the coefficient of friction, μ.

2 A particle of weight 16 N is on a rough horizontal plane and is being pulled at a steady speed by a horizontal string. The coefficient of friction between the particle and the table is 0.4. Find the tension in the string.

3 A block is pulled along a rough horizontal table by a string and is accelerating at 5 ms^{-2}. The tension in the string is 80 N and the coefficient of friction between the block and the table is 0.3.

Find the weight of the block.

4 A horizontal force P newtons is applied to an object of weight 80 N, in rough contact with a horizontal plane.

The coefficient of friction between the object and the plane is $\dfrac{1}{2}$. What is the magnitude of the frictional force when

a $P = 10$ b $P = 40$ c $P = 50$?

State in each case whether or not the body moves. If it does move find its acceleration.

5 The diagram shows a particle A lying in rough contact with a table.

A light inelastic string attached to A, passes over a smooth pulley at the edge of the table and is attached to another particle B hanging freely.

The particles are of equal mass 40 kg and the coefficient of friction between A and the table is $\frac{2}{5}$. Find

a the acceleration of B   **b** the tension in the string.

6 Two particles A and B are connected by a light inextensible string. A is of mass 40 N and is resting on a smooth horizontal surface. B is of mass 20 N and is resting on a rough horizontal surface. The two surfaces are at the same level. The coefficient of friction between B and the surface is 0.5. B is pulled horizontally by a force of 40 N.

Find

a the acceleration of the system   **b** the tension in the string.

Summary

Normal reaction is the force acting on an object and is perpendicular to the surface of contact.

Tension in a taut string acts along the string.

Newton's Laws of Motion

First law: An object will continue in its state of rest, or of uniform motion in a straight line, unless an external force is applied to it.

Second law: The resultant force acting on a body of constant mass is equal to the mass of the body multiplied by its acceleration.

$$F\,(\text{N}) = m\,(\text{kg}) \times a\,(\text{ms}^{-2})$$

The resultant force and the acceleration are in the same direction.

Third law: Action and reaction are equal and opposite.

When a body has an acceleration there is a resultant force acting on it.

When a body has no acceleration there is no resultant force acting on it.

A body of mass m kilograms has a weight of mg newtons

Friction

Friction exists when two objects are in rough contact and have a tendency to move.

The frictional force F is just large enough to prevent motion, up to a limiting value.

When the limiting value is reached, $F = \mu R$ where R is the normal reaction and μ is the coefficient of friction.

For rough contact $0 \leq F \leq \mu R$.

When contact is smooth $\mu = 0$.

Review

1. Write down the letters of the statements that are correct.

   a A particle is moving with uniform velocity so there is a resultant force acting on the particle.

   b When a friction force acts on a body, it is not always of value μR where R is the normal contact force.

   c When a body has a resultant force acting on it the body will accelerate in the direction of the force.

   d A particle is hanging freely attached to a light inextensible string. The string is made to accelerate vertically upward. The tension in the string is greater than the weight of the particle.

   e A car is towing a trailer at a steady speed. The tension in the tow rope is greater than the resistance to the motion of the trailer.

   f When a particle has a constant acceleration it must be moving in a straight line.

2. A resultant force P acts on a particle of mass 2 kg for 5 seconds, increasing the velocity of the particle from 5 ms^{-1} to 10 ms^{-1}. Find

   a the acceleration of the particle.   **b** the value of P.

3. A block of mass 10 kg is pulled along a rough horizontal surface by a horizontal rope. The acceleration of the block is 0.2 ms^{-2} and the coefficient of friction between the block and the surface is 0.45. Find the tension in the rope.

4. A light inextensible string passes over a small fixed smooth pulley. A particle of mass 1 kg is attached at one end of the string and a particle of mass 1.5 kg is attached at the other end. Find the acceleration of the particles and the tension in the string.

5. A smooth pulley is fixed at a height 1 m above a horizontal table and a light inextensible string hangs over the pulley. A particle A of mass 2 kg is attached to one end of the string and a particle B of mass 4 kg is attached to the other end. The system is held at rest with the particles both hanging $\frac{1}{2}$ m above the table. The system is then released from rest. Find

   a the acceleration of the system

   b the speed with which the particle of mass 4 kg hits the table.

Assessment

1. A lorry, whose mass is 5000 kg is towing a trailer whose mass is 3500 kg along a horizontal road. The lorry and trailer are connected by a rod.

   As they move, there is a constant resistance force of 900 newtons acting on the lorry and a constant resistance force of R newtons acting on the trailer. The driving force of the lorry is 3500 newtons. The acceleration of the system is 0.3 ms^{-2}.

   a Find the resistive force R acting on the trailer.

   b Find the force that the rod exerts on the trailer.

2 Two particles A and B are connected by a light string that passes over a smooth fixed peg as shown in the diagram.

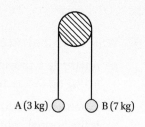

The mass of A is 3 kg and the mass of B is 7 kg.

The particles are released from rest when P and Q are at the same level.

a Form an equation of motion for each particle and hence find the acceleration of the system.

b Find the tension in the string.

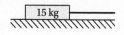

3 A block of mass 15 kg is pulled along a rough horizontal surface by a horizontal rope. The acceleration of the block is 0.2 ms⁻² and the coefficient of friction between the block and the surface is 0.45.

a On a copy of the diagram show all the forces acting on the block as it moves.

b Find the tension in the rope.

4 A block of mass 40 kg is pulled at a constant speed along a rough horizontal surface by a horizontal rope.

The tension in the rope is 220 N.

a Draw a diagram to show all the forces acting on the block as it moves.

b State the friction force acting on the block.

c Find the coefficient of friction between the block and the surface.

5 The particles A, B and C have masses of 2 kg, 3 kg and 1 kg respectively.

The particles A and B are connected by a light string that passes over a smooth fixed peg. The particle C is attached to the particle B. The particles are released from rest when P and Q are at the same level.

a Find the acceleration of the system when it is released from rest.

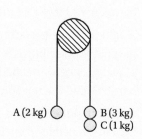

b The particle C falls away from B before A reaches the peg.

Find the acceleration of the system just after C falls away.

c Explain whether the tension in the string has increased or decreased after the particle C falls from B.

6 A block, of mass 3*m*, is placed on a horizontal surface at a point A. A light inextensible string is attached to the block and passes over a smooth peg. The string is horizontal between the block and the peg. A particle, of mass 2*m*, is attached to the other end of the string. The block is released from rest with the string taut and the string between the peg and the particle vertical, as shown in the diagram.

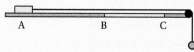

Assume that there is no air resistance acting on either the block or the particle, and that the size of the block is negligible.

The horizontal surface is smooth between the points A and B, but rough between the points B and C. Between B and C, the coefficient of friction between the block and the surface is 0.8.

a By forming equations of motion for both the block and the particle, find the acceleration of the block between A and B.

b Given that the distance between the points A and B is 1.2 metres, find the speed of the block when it reaches B.

c By forming equations of motion for both the block and the particle, find the acceleration of the block between B and C.

d Given that the distance between the points B and C is 0.9 metres, find the speed of the block when it reaches C.

e Explain why it is important to assume that the size of the block is negligible.

AQA MM1B June 2015

7 A wooden block, of mass 4 kg, is placed on a rough horizontal surface. The coefficient of friction between the block and the surface is 0.3. A horizontal force, of magnitude 30 newtons, acts on the block and causes it to accelerate.

30 N

a Draw a diagram to show all the forces acting on the block.

b Calculate the magnitude of the normal reaction force acting on the block.

c Find the magnitude of the friction force acting on the block.

d Find the acceleration of the block.

AQA MM1B June 2011

Momentum and Impulse

Objectives
By the end of this chapter, you should...
- ► Understand momentum and impulse.
- ► Know how to use the conservation of linear momentum

Recap
You need to remember...
- ► Newton's laws of motion.
- ► The equations of motion with constant acceleration.

Applications
When two snooker balls collide, they move away from each other with different velocities. Trying to predict where the balls will go and their speed after the collision is part of playing the game.

· ·

19.1 Momentum and impulse

Momentum

The momentum of an object is the product of its mass and its velocity.

> For an object of mass m, moving with velocity v
> momentum $= mv$

Because momentum is a multiple of velocity, momentum is a vector and has the same direction as the velocity.

When the velocity of a body is constant and its mass does not change, its momentum is constant.

Because a force is needed to change the velocity of an object its follows that a force must act on the object to change its momentum.

The relationship between a force and the change in momentum that it produces is found by combining Newton's Second Law with the equations of motion with constant acceleration:

> A constant force F acts for a time t on a body of mass m in the direction of its motion and causes the velocity to increase from u to v. As the force is constant, the acceleration a that it produces, is also constant.

Therefore using $F = ma$ and $v = u + at$ gives

$$v = u + t\left(\frac{F}{m}\right)$$

$$\Rightarrow \quad Ft = mv - mu$$

Therefore the change in momentum, that is final momentum minus initial momentum, is given by the product of the force and the time for which it acts.

Impulse

The product of the force and the time for which it acts is called the **impulse** of the force and is denoted by I,

so $\quad I = Ft \quad$ Therefore

impulse = change in momentum

giving $\quad I = mv - mu$

This relationship shows that impulse is also a vector quantity. Therefore, if a force exerts an impulse on an object in a direction opposite to that of motion, the impulse is negative. It follows that the change in momentum is negative, so the final momentum is less than the initial momentum.

For any problem involving direction along a straight line, it is important to define the positive direction and this applies to problems involving impulse and momentum.

Units of impulse and momentum

The unit of impulse is the product of a force unit and a time unit so, for a force in newtons acting for a time in seconds, the unit of impulse is the newton second, Ns. (This is *not* newtons *per* second.)

Momentum (mass × velocity) can be measured in kilogram metres per second, $(kgms^{-1})$ but usually the impulse unit, Ns, is used instead.

Example 1

Question

A hammer of mass 0.8 kg is moving at 12 ms^{-1} when it strikes a nail and is brought to rest. What is the magnitude of the impulse exerted on the hammer?

Answer

All the initial momentum of the hammer is lost when it hits the nail.

The change in the momentum of the hammer is 0.8×12 Ns = 9.6 Ns.

Therefore the impulse exerted on the hammer is 9.6 Ns.

Example 2

Question

A particle of mass 2 kg is moving in a straight line with a speed of 5 ms^{-1}. A force of 11 N acts on the particle for 6 seconds in the direction of motion. Find

a   the magnitude of the impulse exerted on the particle

b   the speed of the particle at the end of this time.

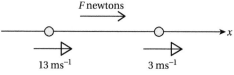

Take the positive direction as being to the right.

a The impulse of the force is I Ns where

$$I = Ft$$
$$= 11 \times 6$$

The magnitude of the impulse is 66 Ns.

b Initial momentum is 2×5 Ns.

Final momentum is $2v$ Ns.

Using $I = mv - mu$ gives

$$66 = 2v - 10 \quad \Rightarrow \quad v = 38$$

The velocity after 6 seconds is 38 ms^{-1}.

Example 3

The velocity of a particle of mass 7 kg, travelling along a straight line, changes from 13 ms^{-1} to 3 ms^{-1} in 5 seconds under the action of a constant force. Find the magnitude of the force.

Let the constant force be F N.

Initial momentum is 7×13 Ns; Final momentum is 7×3 Ns.

The impulse of the force is $5F$ Ns.

Using $\quad I = mv - mu$ gives

$$5F = 21 - 91 = -70$$
$$F = -70 \div 5 = -14$$

The magnitude of the force is 14 N.

Example 4

A truck of mass 1200 kg is travelling at 4 ms^{-1} when it hits a buffer and is brought to rest in 3 seconds. What is the average force exerted by the buffer?

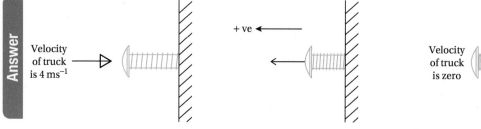

(continued)

(continued)

Taking the direction of the force as the positive direction, the initial velocity of the truck is -4 ms^{-1} and the final velocity is zero.

Using $Ft = mv - mu$

gives $F \times 3 = 1200 \times 0 - 1200 \times (-4)$

$\Rightarrow$ $F = 1600$

The average force exerted by the buffer is 1600 N.

Example 5

A particle of mass 3 kg is moving along a straight line with speed 6 ms^{-1} when a force is applied to it. After 4 seconds the particle is moving in the opposite direction with speed 2 ms^{-1}. Find the magnitude and direction of the force.

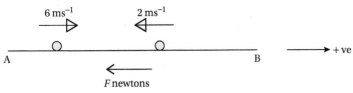

The initial momentum is 3×6 Ns, the final momentum is $3 \times (-2)$ Ns and the impulse of the force is $(-F) \times 4$ Ns.

Using $Ft = mv - mu$ gives

      $-4F = -6 - 18 \quad \Rightarrow \quad F = 6$

The force is 6 N acting in the direction opposite to the initial direction of movement.

Impulsive forces

When a large force acts for a very short time, it is not easy to find either the force or the time for which it acts, for example, a cricket bat hitting the ball, a shot being fired, a footballer kicking a ball.

These are examples of **impulsive forces** and in such cases the impulse of the force cannot be calculated using $I = Ft$. The change in momentum caused by the impulse can be used to find the impulse.

Example 6

A cricket ball of mass 0.2 kg has a speed of 20 ms^{-1} when the bat strikes it at right angles and reverses the direction of the ball. The speed of the ball immediately after being struck is 36 ms^{-1}.

Find the impulse given by the bat to the ball.

The final momentum of the ball is 0.2×36 N s $= 7.2$ Ns.

The initial momentum of the ball is $0.2 \times (-20)$ N s $= -4$Ns.

The impulse, I Ns, given by the bat is given by $I = 7.2 - (-4) = 11.2$.

Therefore the bat exerts an impulse of 11.2 Ns on the ball.

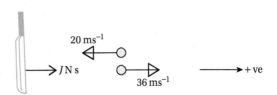

Exercise 1

1 Write down the momentum of

    a a child of mass 40 kg running with a speed of 3 ms⁻¹

    b a lorry of mass 1200 kg moving at 20 ms⁻¹

    c a missile of mass 92 kg travelling at 120 ms⁻¹

    d a train of mass 214 tonnes travelling at 55 ms⁻¹

    e a bullet of mass 100 g travelling at 40 ms⁻¹.

2 Find the magnitude of the impulse exerted by

    a a force of 14 N acting for 6 s

    b a force of 12 tonnes acting for 1 minute

    c a force that causes an increase in momentum of 88 Ns.

In Questions 3 to 7 a force of magnitude F newtons acts in the direction AB on a particle P of mass 2 kg. Initially the velocity of P is u ms⁻¹ and t seconds later it is v ms⁻¹, each in the direction AB.

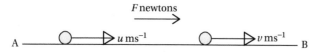

3 When $u = 4$, $v = 7$ and $t = 3$,   find F.

4 When $u = 7$, $v = 4$ and $t = 3$,   find F.

5 When $u = 5$, $t = 4$ and $F = 10$,   find v.

6 When $u = 8$, $t = 5$ and $F = -4$,   find v.

7 When $u = 10$, $v = 6$ and $F = -3$,   find t.

8 In what time will a force of 12 N reduce the speed of a particle of mass 1.5 kg from 36 ms⁻¹ to 12 ms⁻¹?

9 An object of mass 5 kg is moving in a straight line with a velocity of 10 ms⁻¹ when a force F is applied to it for 4 seconds. Find the velocity at the end of this time when

    a $F = 20$ **b** $F = -20$.

10 An object of mass 4 kg is moving with speed 7 ms⁻¹ when a force is applied to it for 8 seconds. Its speed then is again 7 ms⁻¹ but in the opposite direction. Find the magnitude of the force that has caused this change.

11 A dart of mass 40 g hits the dartboard at a speed of 16 ms⁻¹. The dart comes to rest in the board in 0.02 seconds, find the average force exerted by the board on the dart.

12 A particle of mass 5 kg moving in a straight line has a velocity 16 ms⁻¹ when a force −4 N begins to act on it.

Find the velocity of the particle when the force has been acting for

    a $\dfrac{1}{3}$ s   **b** 5 s.

13 A stationary truck is shunted into a siding by a locomotive that exerts a force of 2600 N on the truck for 12 seconds.

   a What is the momentum of the truck at the end of this time?

   The truck carries on without change of speed until it is brought to rest in 2 seconds when it hits the buffers at the end of the line.

   b What is the magnitude of the impulse exerted on the truck by the buffers?

   c What is the average force exerted on the truck by the buffers?

In Questions 14 to 18 calculate the impulse given.

14 A ball of mass 1.1 kg strikes a wall at right angles with a speed of 6 ms^{-1} and bounces off at 5 ms^{-1}.

15 The speed of a ball just before it is hit is 38 ms^{-1}.

   The bat hits the ball at right angles, giving it a return speed of 30 ms^{-1}. The mass of the ball is 0.15 kg.

16 A shot of mass 50 g, fired at 250 ms^{-1}, is stopped when it hits a steel barrier.

17 A bird of mass 60 g is stopped when it flies at 12 ms^{-1} directly into a window.

18 A stone, of weight 24 N, dropped from a window, hits the ground at 45 ms^{-1} and does not bounce.

19.2 Conservation of linear momentum

When two objects are in contact, they exert *equal and opposite* forces on each other.

Whether they are in contact for a measurable time, or just for a split second, then each is in contact with the other for the *same time*. Therefore they exert *equal and opposite impulses* on each other.

As long as neither object is fixed, these equal and opposite impulses produce equal and opposite changes in momentum, so the overall change in momentum of the two objects caused by the collision is zero.

Therefore, *as long as no external force acts on either object*, the total momentum of the two objects remains constant.

This property is known as *The Principle of Conservation of Linear Momentum* and is expressed formally as follows:

> When no external force affects the motion of a system, the total momentum of the system remains constant.

In some problems involving a collision the two colliding objects bounce and so have individual velocities after impact. Other objects collide and join together at impact, for example trucks which become coupled. (Such objects are said to **coalesce**.)

In either case there are different velocities before and after impact so draw separate 'before' and 'after' diagrams and define the chosen positive direction.

Example 7

A particle A of mass 3 kg, travelling at 5 ms^{-1} in a straight line collides with a particle B with mass 2 kg and travelling at 4 ms^{-1} in the same direction. After impact, B moves in the opposite direction at 2 ms^{-1}. Find the velocity of A.

Total momentum before impact is $3 \times 5 + 2 \times (-4)$ Ns.

Total momentum after impact is $3u + 2 \times 2$ Ns.

Using conservation of linear momentum gives

$$15 - 8 = 3u + 4 \quad \Rightarrow \quad u = 1$$

The velocity of A is 1 ms^{-1} in the same direction as before the impact.

In this example the direction of motion of A after collision is not given. We guessed that it was to the right and got a positive value for u showing that the guess was correct. If we had thought that A moved with speed u to the left, we would have found that $u = -1$; this also shows that A moves to the right. Therefore it does not matter in which direction we mark an unknown velocity; the sign of the answer gives the correct direction.

Example 8

A truck of mass 2 tonnes moves along a track at 8 ms^{-1} towards a truck of mass 5 tonnes moving at 5 ms^{-1} on the same track. The trucks become coupled at impact. Find the velocity at which they continue to move given that before impact they move

a   in the same direction

b   in opposite directions.

a

Before impact, the momentum of A is 3000×8 Ns $= 24\,000$ Ns

the momentum of B is 5000×5 Ns $= 25\,000$ Ns

the total momentum is $49\,000$ Ns.

After impact, the combined momentum is $8000 \times v$ Ns.

Using conservation of linear momentum gives

$$8000v = 49\,000 \quad \Rightarrow \quad v = 6.125$$

The velocity of the coupled trucks is 6.13 ms^{-1} correct to 3 significant figures.

(continued)

(continued)

b

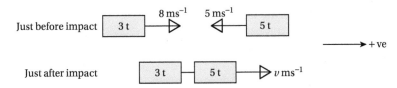

Before impact, the momentum of A is $\quad 3000 \times 8 \text{ Ns} = 24\,000 \text{ Ns}$

the momentum of B is $\quad 5000 \times (-5) \text{ Ns} = -25000 \text{ Ns}$

the total momentum is $\quad -1000 \text{ Ns}.$

After impact, the combined momentum is $8000 \times v \text{ Ns}.$

Using conservation of linear momentum gives

$$8000v = -1000 \quad \Rightarrow \quad v = -0.125$$

The velocity of the coupled trucks is 0.125 ms^{-1} in the direction of motion of the heavier truck before impact.

Exercise 2

In Questions 1 to 5 a body A of mass m_A travelling with velocity u_A, collides directly with a body B of mass m_B moving with velocity u_B. They coalesce at impact. The velocity with which the combined body moves on is v.

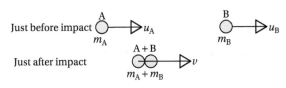

1 $m_A = 4 \text{ kg}, \quad u_A = 4 \text{ ms}^{-1}, \quad m_B = 2 \text{ kg}, \quad u_B = 1 \text{ ms}^{-1}. \quad$ Find v.

2 $m_A = 6 \text{ kg}, \quad u_A = 1 \text{ ms}^{-1}, \quad m_B = 2 \text{ kg}, \quad u_B = -3 \text{ ms}^{-1}. \quad$ Find v.

3 $m_A = 9 \text{ kg}, \quad u_A = 5 \text{ ms}^{-1}, \quad m_B = 4 \text{ kg}, \quad v = 3 \text{ ms}^{-1}. \quad$ Find u_B.

4 $m_A = 3 \text{ kg}, \quad u_A = 16 \text{ ms}^{-1}, \quad m_B = 5 \text{ kg}, \quad v = 6 \text{ ms}^{-1}. \quad$ Find u_B.

5 $u_A = 3 \text{ ms}^{-1}, \quad m_B = 6 \text{ kg}, \quad u_B = -5 \text{ ms}^{-1}, \quad v = -1 \text{ ms}^{-1}. \quad$ Find m_A.

6 A bullet of mass 0.1 kg is fired horizontally at 80 ms^{-1} into a stationary block of wood that is free to move on a smooth horizontal plane. The wooden block, with the bullet embedded in it, moves off with speed 5 ms^{-1}. Find the mass of the block.

7 A particle A of mass 5 kg travelling with speed 6 ms^{-1}, collides directly with a stationary particle B of mass 10 kg. A is brought to rest by the impact. Find the speed with which B begins to move.

8 The masses of two particles, P and Q, are 0.18 kg and 0.1 kg respectively. They are moving directly towards each other at speeds of 4 ms^{-1} and 12 ms^{-1} respectively. After they collide the direction of motion of each particle is reversed and the speed of Q is 6 ms^{-1}. Find the speed of P after impact.

9 A ball P of mass 2 kg is moving at 4 ms^{-1} when it collides with a ball Q of mass 1 kg moving in the same direction at 3 ms^{-1}. After the impact, both P and Q move on in the same direction as before, P at $u \text{ ms}^{-1}$ and Q at $v \text{ ms}^{-1}$.

Given that $7u = 2v$, find u and v.

10 Two skaters, a father and son, are standing at rest on the ice. The masses of the father and son are 70 kg and 50 kg respectively. The father slides a stone over the ice giving it a velocity of 8 ms⁻¹. The son catches it. The mass of the stone is 5 kg. Find

    a the speed with which the father starts to move backwards after releasing the stone

    b the common velocity of the son and the stone after he has caught it

    c the impulse exerted by the stone on the son. (Assume that there is no friction between the stone and the ice.)

Summary

An object of mass m, moving with velocity v, has momentum where momentum $= mv$.

impulse $=$ change in momentum so $\quad I = mv - mu$

The Principle of Conservation of Linear Momentum states that when in a given direction, no external force affects the motion of a system, the total momentum in that direction remains constant.

Review

In Questions 1 to 3, write down the letter that gives the correct answer.

1 A sphere A of mass $2m$ collides directly with a sphere B of mass m. Before impact the spheres are moving in opposite directions with speed u. A is brought to rest by the collision.

After impact the speed of B is

    a $\dfrac{1}{2}u$ **b** u **c** $2u$ **d** $-u$.

2 Two masses with the same mass collide and coalesce as shown in the diagram. What is the speed V of the combined mass just after impact?

    Just before impact $\bigcirc\!\!-\!\!\!\triangleright 3u$ $\bigcirc\!\!-\!\!\!\triangleright u$

    Just after impact $\infty\!\!-\!\!\!\triangleright V$

    a $2u$ **b** $3u$ **c** u **d** $5u$.

3 A particle of mass 2 kg moving with speed 4 ms⁻¹ is given a blow which changes the speed to 1 ms⁻¹ without deflecting the particle from a straight line. The impulse of the blow is

    a 10 Ns   **b** 6 Ns

    c it could be either 10 Ns or 6 Ns.

4   A railway truck, of mass 1900 kg, is moving at a speed of 5 ms^{-1} along a horizontal track. It collides with a stationary truck, of mass 600 kg.

The two trucks couple and move on together. Calculate

  a the speed, in ms^{-1} of the pair of trucks immediately after the collision

  b the magnitude of the impulse, in Ns, on the stationary truck due to the collision.

5   A brick of mass 3.5 kg falls from rest from a point vertically 10 m above firm horizontal ground. It does not rebound.

Find **a** the speed with which the brick hits the ground

      b the impulse of the force on the ground by the brick.

6   An object of mass 2 kg, moves along a straight line with speed 6 ms^{-1} and collides with an object of mass 0.5 kg moving with speed 3 ms^{-1} in the same direction along the same straight line. The bodies coalesce and move on together. Calculate their speed immediately after the collision.

7   A cricket ball, of mass 0.15 kg, is moving horizontally with speed 30 ms^{-1} when it hits a vertically held cricket bat.

The ball rebounds horizontally with speed 15 ms^{-1}.

Calculate the magnitude of the impulse, in Ns, of the force exerted by the ball on the bat.

Assessment

1   A ball of mass 1.5 kg falls from rest from a point vertically 4 m above firm horizontal ground.

  a Find the speed with which the ball hits the ground

The ball rebounds with speed 4 ms^{-1}.

  b Find the change in momentum of the ball during the rebound.

  c Hence find the impulse of the ground on the ball.

2   Two particles A and B have masses of 2 kg and 1 kg respectively. The particles are moving in the same direction along the same straight line on a smooth horizontal surface when they collide and coalesce.

Before they collide the velocities of A and B are 2 ms^{-1} and 1 ms^{-1} respectively.

Find their combined velocity after they collide.

3   Two balls A and B are moving towards each other along the same straight line on a smooth horizontal surface when they collide. The mass of a ball A is 0.4 kg.

Before they collide the velocities of A and B are 1.5 ms^{-1} and −0.4 ms^{-1} respectively.

After they collide their combined velocity is 1 ms^{-1}.

Find the mass of the particle B.

4 Two balls A and B are moving in the same direction along the same straight line on a smooth horizontal surface when they collide. The mass of ball A is 0.8 kg. Before the collision the speed of A is 3 ms^{-1} and the speed of B is 2 ms^{-1}. After the collision the speed of A is 1 ms^{-1} and both balls move in the same direction along the line.

a Find the impulse on the ball A in the collision.

b The mass of the ball B is 1.8 kg. Find the speed of B after the collision.

5 A ball of mass m is travelling horizontally with a velocity of 3 ms^{-1} at the instant before it hits a vertical wall. The wall gives an impulse of 6 Ns to the ball. Find an expression for the velocity of the ball in terms of m immediately after it hits the wall.

6 A ball A is moving along a horizontal line when it receives an impulse of 6 Ns in the direction of its movement.

The mass of A is 2 kg. The velocity of A before the impulse is 2.5 ms^{-1}.

a Find the change in momentum of the ball.

After receiving the impulse the ball collides with a stationary ball B. The ball A then reverses direction and moves with velocity 0.5 ms^{-1}. The ball B moves with velocity 0.3 ms^{-1} in the opposite direction to that of A.

b Find the mass of the ball B.

7 Two toy trains, A and B, are moving in the same direction on a straight horizontal track when they collide. As they collide, the speed of A is 4 ms^{-1} and the speed of B is 3 ms^{-1}. Immediately after the collision, they move together with a speed of 3.8 ms^{-1}.

The mass of A is 2 kg. Find the mass of B.

AQA MM1A June 2012

8 A trolley, of mass 5 kg, is moving in a straight line on a smooth horizontal surface.

It has a velocity of 6 ms^{-1} when it collides with a stationary trolley, of mass m kg.

Immediately after the collision, the trolleys move together with velocity 2.4 ms^{-1}.

Find m.

AQA MM1B January 2011

Glossary

A

acceleration the rate of increase of velocity

acute angle an angle between 0° and 90°

addition law for mutually exclusive events the relationship between P(A or B) and P(A and B) when A and B cannot occur at the same time

addition law of probability the general relationship between P(A or B) and P(A and B)

ambiguous two or more different meanings

angle of elevation the angle between a horizontal line and a line to a point above the horizontal

arc of a circle a section of a circle

arithmetic series a series where any term differs from the term before it by the same constant

asymptote a straight line a curve approaches but does not cross

average a typical or representative value of a set of data

B

Bernoulli distribution the distribution of the number of successes in a Bernoulli trial

Bernoulli trial a trial with two outcomes: success and failure

binomial coefficient the coefficient of a power of x in a binomial expansion

binomial distribution the distribution of the number of successes in n independent Bernoulli trials

binomial theorem the expansion of $(1 + x)^n$ as a series of ascending powers of x

C

Cartesian coordinates the directed distances (x, y) of a point from the x-axis and the y-axis

chord a straight line between two point on a curve

chord a straight line between two points on a circle

circumscribe a figure that passes through the vertices of another figure

coalesce join together to form one object

coefficient of friction the ratio of the friction force to the normal reaction when friction is maximum

coefficient the number multiplied by an expression containing just letters

collinear in a straight line

common difference the difference between consecutive terms in an arithmetic series

common factor a number or letter that is a factor of each term in an expression

common ratio the ratio between consecutive terms in a geometric series

complement of A event A does not occur, written A'

completing the square adding a number to an expression of the form $ax^2 + bx$ to form a perfect square

compound angle the sum or difference between two or more angles

compound area an area made up of two or more different shapes

conditional probability the probability of an event is conditional on the outcome of another event

constant of integration the unknown constant added to an integral

continuous variable a variable that cannot be given exactly, such as height, mass, time

convergent a sequence or a series is convergent when the nth term or the sum respectively approaches a finite value as n increases

cosine of an angle the ratio of the adjacent side to the hypotenuse of an angle in a right-angled triangle

cube root one of three equal factors whose product is a number

cubical die a die with 6 faces, usually marked 1, 2, 3, 4, 5, 6; the score on the die is the face uppermost when it is thrown or rolled

cubic function a polynomial where the highest power of x is 3

cumulative binomial distribution function a function giving cumulative binomial probabilities $P(X \leq x) = P(X = 0) + P(X = 1) + ... + P(X = x)$

cyclic a function with a repeating pattern

D

definite integration the process of integrating a function between limits

derivative the general expression for the gradient of a curve

derived function the general expression for the gradient of a curve

difference of two squares the difference between two terms that are perfect squares, for example $a^2 - b^2$

differentiation The process of finding a general expression for the gradient of a curve at any point

directed distance the distance in a given direction

discrete random variable X a variable whose value is determined by chance, taking individual values with a given probability

discrete variable a variable whose values can be given exactly

discriminant the expression $b^2 - 4ac$ where a, b and c are the coefficients of x^2, x and the constant in a quadratic expression

displacement a vector giving the length of a line and its direction of one point from another point

distance how far an object travels

divergent a sequence or a series is divergent when the nth term or the sum respectively does not approach a finite value as n increases

dividend the expression that is divided by another expression

divisor the expression that divides another expression

dynamic friction the maximum value of the friction force

E

equally likely outcomes each outcome has the same probability of occurring

equation of a curve the equation that gives the relationship between the x and y coordinates of the points on a curve or line

equilibrium an object is in equilibrium when it is not accelerating

event one or more possible outcomes of an experiment; a subset of the possibility space

event: A and B outcomes in $A \cap B$, so in both A and B

event: A only outcomes in A but not in B

event: A or B outcomes in $A \cup B$, so in A or B or both (in at least one of A and B)

event: neither A nor B outcomes that are not in $A \cup B$

expectation of a function of X, $E(g(X))$ the mean or expected value of $g(X)$

expectation of X, $E(X)$ the mean μ or expected value of the random variable X

expected value expectation, average value

experimental mean $\bar{x}$ the mean of a set of data

exponential function a function where the variable is part of a power

exponent index or power

F

factorial the product of the integers from one to a given number

factorising expressing as the product of factors

factor theorem $(x - a)$ is a factor of a polynomial in x when $f(a) = 0$

fair unbiased

finite series a series with a finite number of terms

force the quantity that, when acting on a body, changes the velocity of that body.

frequency distribution a table showing how many times each observation occurs

frequency the number of times an observation occurs

friction a force that opposes motion or potential motion

function a function is an expression involving one variable which gives a single answer, when a number is substituted for the variable

G

geometric series a series where each term is the same constant multiple of the term before it

gradient function the general expression for the gradient of a curve

gradient the gradient of a straight line is the increase in y divided by the increase in x between one point and another point on the line.

gravitational attraction the force that attracts an object to the earth

I

improper algebraic fraction a fraction where the highest power of the numerator is equal to or greater than the highest power of the denominator

impulse the product of a force and the time for which it acts

impulsive force a large force that acts for a very short time

indefinite integration integration where an unknown constant needs to added to the integral

independent events the outcome of an event is unaffected by the outcome of any other event

index (plural: indices) the index of a number, written as a superscript, gives how many of the same number are multiplied together, also called the power

inequality a comparison between two unequal quantities

infinite series a series whose terms continues indefinitely

integration the process of finding a function from its derivative

intercept the distance from the origin where a graph crosses the y or x axis

interquartile range difference between the quartiles: upper quartile – lower quartile

intersection of sets $A \cap B$, the overlap of sets

intersection the points where two graphs cut each other

irrational number a number that cannot be expressed as a/b where a and b are integers

L

like terms terms in an expression that contain the same combination of variables

limiting friction the maximum value of a friction force

limiting value value to which a quantity tends as n tends to infinity

linear combination of X and Y combination of the form $aX + bY$, where a and b are constants

line of symmetry a line that divides a figure into two that are the mirror image of each other

logarithm the power to which the base must be raised to equal a number

long-term relative frequency value to which r/n tends as n tends to infinity

lower quartile the median of all the observations before the median in an ordered set of data; the observation ¼ of the way through an ordered set of data

M

mass a measure of the quantity of matter in an object

maximum point a point on a curve where the gradient changes from positive to negative

maximum value the value of a function at a point on a curve where the gradient changes from positive to negative

mean of a random variable X μ, the expectation or expected value of X

mean of a set of data $\bar{x}$, the sum of the observations divided by the number of observations

measure of central tendency of data average or typical value measured by mean, mode, median

measures of spread of data the variability of data measured by range, interquartile range, standard deviation

median middle value in an ordered set of data

minimum point a point on a curve where the gradient changes from negative to positive

minimum value the value of a function at a point on a curve where the gradient changes from negative to positive

mode observation that occurs most often (most popular) in a set of data

momentum the product of mass and velocity

multiplication law for independent events relationship between P(A), P(B) and P(A and B) when the outcome of event A does not affect the outcome of event B

multiplication law of probability the general relationship between P(A and B) and conditional probabilities relating to A and B

mutually exclusive events events that cannot occur at the same time

N

newton the unit of force where 1 N is equal to the product of a mass of 1 kg and acceleration in ms^{-2}

normal reaction a force acting on a body on a surface perpendicular to the surface of contact and away from that surface

normal to a curve a straight line at right angles to a tangent to a curve

O

obtuse angle an angle between 90° and 180°

one way stretch a process that stretches a shape in one direction

ordinate the length of a line from the x-axis to a curve

origin the point where the x-axis and y-axis cross

outcome result of a trial in an experiment

P

parabola the shape of a curve whose equation is $y = ax^2 + bx + c$

perfect square an expression or number that can be expressed as the product of two equal factors

periodic a function with a repeating pattern

period of a function the length of a repeating pattern

perpendicular the angle between perpendicular lines is a right angle

polynomial a collection of terms containing powers of x which are positive integers, for example, $2x^3 - 5x + 2$

possibility space diagram a diagram showing equally likely outcomes as pairs on a grid, such as the outcomes when two dice are rolled

possibility space set of all distinct possible outcomes in an experiment

power of a number the power of a number, written as a superscript, gives how many of the same number are multiplied together, also called the index

probability (relative frequency) table a two-way table showing probabilities relating to events

probability distribution of X the probabilities of the values of the random variable X, P($X = x$)

probability function the probability distribution of X written as a function of x

probability the likelihood of an event happening

proper algebraic fraction a fraction where the highest power of the numerator is less than the highest power of the denominator

Q

quadratic equation an equation of the form $ax^2 + bx + c = 0$ where a. b and c are constants and $a \neq 0$

quadratic function a polynomial where the highest power of x is 2

quotient the result of a division

R

radian an angle unit, 1 radian is the angle subtended at the centre of a circle by an arc equal in length to the radius

random selection all outcomes are equally likely; there is no bias

range the spread of the distribution given by the difference between the maximum and minimum values

rational number a number that can be written as a/b where a and b are integers

real number the set of numbers that includes the rational and irrational numbers

recurrence relation a rule that gives the next term in a sequence in terms of the previous term

reflection reflection in a line creates a mirror image of a shape

relative frequency (probability) table   a two-way table showing frequencies written as proportions or percentages of the total

remainder   the expression or number left over after a division

remainder theorem   the value of a polynomial when a is substituted for x

resultant force   a single force equivalent to all the forces acting on a body

roots   the values that satisfy an equation

rough contact   contact between an object and a surface where friction opposes motion along the surface

S

sampling without replacement   selecting an item but not returning it to the population before another item is selected

sampling with replacement   selecting an item then returning it to the population before another item is selected

scalar   a quantity that has magnitude but no direction

secant of a circle   a straight line passing through a circle

second derivative   the rate at which the gradient function increases

sector of a circle   the part inside a circle bounded by two radii and the circle

segment   the part inside a circle bounded by a chord and the circle

sequence   a set of numbers in a given order

series   the sum of the terms of a sequence

simultaneous equations   equations that contain more than one variable and that can be solved to find values that satisfy all the equations

sine of an angle   the ratio of the opposite side to the hypotenuse of an angle in a right-angled triangle

sine wave   the shape of the curve $y = \sin x$

smooth contact   contact between an object and a surface where there is no frction force

speed   the rate of change of distance

square root   one of two equal factors whose product is a number

standard deviation of a set of data   measure of the spread of the observations about the mean $\bar{x}$

standard deviation of X σ, a measure of the spread of the random variable X about μ

stationary point   the coordinates of a point on a curve where the gradient is zero

stationary value   the value of a function at a point where the gradient is zero

statistical mode   calculator mode giving statistical values such as mean and standard deviation directly

subtends   a line or an arc forming an angle at a particular point when straight lines from its ends are drawn to that point

summary data   a summarised form of the data, such as Σx or Σx^2, rather than individual observations

surd   an irrational number left in square root form such as $\sqrt{2}$

T

tangent of an angle   the ratio of the opposite side to the adjacent side of an angle in a right-angled triangle

tangent to a curve   a line that touches a curve at a point, but does not cross the curve at that point

tension   the force in a taut string

terms   the parts of an expression separated by plus or minus signs

tetrahedral die   a die with 4 faces usually numbered 1, 2, 3, 4; the score is the value on which die lands when it is thrown or rolled

thrust   the force in a rod that is in compression

transformation   a process that moves or reflects in a line or stretches a curve

translation   a process that moves a curve

tree diagram   a diagrammatic way of showing outcomes of combined events

trigonometry   relationships involving lengths and angles in triangles

turning point   a point on a curve where the gradient changes sign

U

union of sets $A \cup B$; all of sets A and B, A or B or both

unlike terms   terms in an expression that do not contain the same combination of variables

upper quartile   the median of all the observations above the median in an ordered set of data; the observation ¾ way through an ordered set of data

V

variable   a quantity that can take more than one value

variance of a set of data   square of the standard deviation of a set of observations

variance of X $\sigma^2 = \text{Var}(X)$, the expectation of $(X - \mu)^2$

vector   a quantity that has magnitude and direction

velocity   the rate of increase of displacement

Venn diagram   a diagrammatic way of representing sets

vertex   the point at which a parabola turns or the point where two sides of a polygon meet

vertical line graph   a diagram consisting of vertical lines illustrating discrete data

W

weight   the force that attracts an object to the earth

Answers

1 Expanding Brackets, Surds and Indices

Exercise 1

1 $3x^2 - 4x$ **2** $a - 12$ **3** $2y - xy + y^2$

4 $5pq - 9p^2$ **5** $3xy + y^2$ **6** $x^3 - x^2 + x + 7$

7 $5 + t - t^2$ **8** $a^2 - ab - 2b$ **9** $7 - x$

10 $4x - 9$ **11** $3x^2 + 18x - 20$ **12** $ab - 2ac + cb$

13 $11ct - 2ct^2 - 55t^2$ **14** $-x^3 + 7x^2 - 7x$ **15** $-4y^2 + 24y - 10$

16 -7 **17** 2 **18 a** 1 **b** 0 **c** -3

Exercise 2

1 $x^2 + 6x + 8$ **2** $x^2 + 8x + 15$ **3** $a^2 + 13a + 42$

4 $t^2 + 15t + 56$ **5** $s^2 + 17s + 66$ **6** $2x^2 + 11x + 5$

7 $5y^2 + 28y + 15$ **8** $6a^2 + 17a + 12$ **9** $35t^2 + 86t + 48$

10 $99s^2 + 49s + 6$ **11** $x^2 - 5x + 6$ **12** $y^2 - 5y + 4$

13 $a^2 - 11a + 24$ **14** $b^2 - 17b + 72$ **15** $p^2 - 15p + 36$

16 $2y^2 - 13y + 15$ **17** $3x^2 - 13x + 4$ **18** $6r^2 - 25r + 14$

19 $20x^2 - 19x + 3$ **20** $6a^2 - 7ab + 2b^2$ **21** $x^2 - x - 6$

22 $a^2 + a - 56$ **23** $y^2 + 2y - 63$ **24** $s^2 + s - 30$

25 $q^2 + 8q - 65$ **26** $2t^2 + 3t - 20$ **27** $4x^2 + 11x - 3$

28 $6q^2 - q - 15$ **29** $x^2 - xy - 2y^2$ **30** $2s^2 + st - 6t^2$

Exercise 3

1 $x^2 - 4$ **2** $25 - x^2$ **3** $x^2 - 9$

4 $4x^2 - 1$ **5** $x^2 - 64$ **6** $x^2 - a^2$

7 $x^2 - 1$ **8** $9b^2 - 16$ **9** $4y^2 - 9$

10 $a^2b^2 - 36$ **11** $25x^2 - 1$ **12** $x^2y^2 - 16$

13 $x^2 + 8x + 16$ **14** $x^2 + 4x + 4$ **15** $4x^2 + 4x + 1$

16 $9x^2 + 30x + 25$ **17** $4x^2 + 28x + 49$ **18** $x^2 - 2x + 1$

19 $x^2 - 6x + 9$ **20** $4x^2 - 4x + 1$ **21** $16x^2 - 24x + 9$

22 $25x^2 - 20x + 4$ **23** $9t^2 - 42t + 49$ **24** $x^2 + 2xy + y^2$

25 $4p^2 + 36p + 81$ **26** $9q^2 - 66q + 121$ **27** $4x^2 - 20xy + 25y^2$

28 $3x^3 - 4x^2 + 4x - 4$

     coefficient of x^2 is -4 coefficient of x is 4

Exercise 4

1 $11x - 2x^2 - 12$ **2** $x^2 - 49$

3 $6 - 25x + 4x^2$ **4** $14p^2 - 3p - 2$

5 $9p^2 - 6p + 1$ **6** $15t^2 + t - 2$

7 $16 - 8p + p^2$ **8** $14t - 3 - 8t^2$

9 $x^2 + 4xy + 4y^2$ **10** $16x^2 - 9$

11 $9x^2 + 42x + 49$ **12** $15 - R - 2R^2$

13 $a^2 - 6ab + 9b^2$ **14** $4x^2 - 20x + 25$

15 $49a^2 - 4b^2$ **16** $9a^2 + 30ab + 25b^2$

17 $x^3 - x^2 - x - 2$ **18** $3x^3 - 5x^2 - x + 2$

19 $4x^3 - 8x^2 + 13x - 5$ **20** $x^3 - 2x^2 + 1$

21 $2x^3 - 9x^2 - 24x - 9$ **22** $x^3 + 6x^2 + 11x + 6$

23 $x^3 + 4x^2 - x - 4$ **24** $x^3 - 4x^2 + x + 6$

25 $2x^3 + 7x^2 + 7x + 2$ **26** $x^3 + 4x^2 + 5x + 2$

27 $4x^3 + 4x^2 - 7x + 2$ **28** $27x^3 - 27x^2 + 9x - 1$

29 $4x^3 - 9x^2 - 25x - 12$ **30** $4x^3 - 4x^2 - x + 1$

31 $6x^3 + 13x^2 + x - 2$ **32** $x^3 + 3x^2 + 3x + 1$

33 $x^3 + x^2 - 4x - 4$ **34** $2x^3 + 7x^2 - 9$

35 $24x^3 + 38x^2 - 51x + 10$ **36** $4x^3 - 42x^2 + 68x + 210$

37 $x^3 + 3x^2y + 3xy^2 + y^3$ **38** $x^4 + 4x^3y + 6x^2y^2 + 4xy^3 + y^4$

39 $x^6 - 15x^4 + 75x^2 - 125$ **40** $8 - 18x^2 + 18x^4 - 27x^6$

41 $6, -17$ **42** -36

Exercise 5

1 $2\sqrt{3}$ **2** $4\sqrt{2}$ **3** $3\sqrt{3}$

4 $5\sqrt{2}$ **5** $10\sqrt{2}$ **6** $6\sqrt{2}$

7 $9\sqrt{2}$ **8** $12\sqrt{2}$ **9** $5\sqrt{3}$

10 $4\sqrt{3}$ **11** $10\sqrt{5}$ **12** $2\sqrt{5}$

13 $2\sqrt{3} - 3$ **14** $5\sqrt{2} + 8$ **15** $2\sqrt{5} + 5\sqrt{15}$

16 4 **17** $\sqrt{6} + \sqrt{2} - \sqrt{3} - 1$

18 $13 + 7\sqrt{3}$ **19** 4 **20** $5 - 3\sqrt{2}$

21 $22 - 10\sqrt{5}$ **22** 9 **23** $10 - 4\sqrt{6}$

24 $31 + 12\sqrt{3}$ **25** $\left(4 + \sqrt{5}\right)\left(4 - \sqrt{5}\right) = 11$

26 $\left(\sqrt{11} - 3\right)\left(\sqrt{11} + 3\right) = 2$ **27** $\left(2\sqrt{3} + 4\right)\left(2\sqrt{3} - 4\right) = -4$

28 $\left(\sqrt{6} + \sqrt{5}\right)\left(\sqrt{6} - \sqrt{5}\right) = 1$ **29** $\left(3 + 2\sqrt{3}\right)\left(3 - 2\sqrt{3}\right) = -3$

30 $\left(2\sqrt{5} + \sqrt{2}\right)\left(2\sqrt{5} - \sqrt{2}\right) = 18$

Exercise 6

1 $\dfrac{3}{2}\sqrt{2}$ **2** $\dfrac{1}{7}\sqrt{7}$ **3** $\dfrac{2}{11}\sqrt{11}$

4 $\dfrac{3}{5}\sqrt{10}$ **5** $\dfrac{1}{9}\sqrt{3}$ **6** $\dfrac{1}{2}\sqrt{2}$

7 $\sqrt{2} + 1$ **8** $\dfrac{1}{23}\left(15\sqrt{2} - 6\right)$ **9** $\dfrac{1}{3}\left(4\sqrt{3} + 6\right)$

10 $-5\left(2 + \sqrt{5}\right)$ **11** $\dfrac{1}{4}\left(\sqrt{7} + \sqrt{3}\right)$ **12** $4\left(2 + \sqrt{3}\right)$

13 $\sqrt{5} - 2$ **14** $\dfrac{1}{13}\left(7\sqrt{3} + 2\right)$ **15** $3 + \sqrt{5}$

16 $3\left(\sqrt{3} + \sqrt{2}\right)$ **17** $\dfrac{3}{19}\left(10 - \sqrt{5}\right)$ **18** $3 + 2\sqrt{2}$

19 $\dfrac{2}{3}\left(7 - 2\sqrt{7}\right)$ **20** $\dfrac{1}{2}\left(1 + \sqrt{5}\right)$ **21** $\dfrac{1}{4}\left(\sqrt{11} + \sqrt{7}\right)$

22 $\dfrac{1}{6}\left(9 + \sqrt{3}\right)$ **23** $\dfrac{1}{14}\left(9\sqrt{2} - 20\right)$ **24** $\dfrac{1}{6}\left(3\sqrt{2} + 2\sqrt{3}\right)$

25 $\dfrac{1}{2}\left(2 + \sqrt{2}\right)$ **26** $\dfrac{1}{42}\left(3\sqrt{7} - \sqrt{21}\right)$ **27** $\dfrac{1}{9}\left(\sqrt{30} + 2\sqrt{3}\right)$

Exercise 7

1 $\dfrac{1}{2^4} = \dfrac{1}{16}$ **2** $\dfrac{1}{2^2} = \dfrac{1}{4}$ **3** $3^2 = 9$

4 x^2 **5** 1 **6** t^4

7 1 **8** 2 **9** $y^{\frac{3}{2}}$

10 x^5 **11** $\dfrac{1}{y^{\frac{3}{4}}}$ **12** p

13 3 **14** $\dfrac{1}{32}$ **15** $\dfrac{1}{2}$

16 2 **17** 27 **18** $\dfrac{9}{4}$

19 1 **20** 16 **21** $\dfrac{5}{4}$

22 -5 **23** 1331 **24** $\dfrac{3}{5}$

25 6 **26** $\dfrac{16}{27}$ **27** 8

28 5 **29** 1 **30** 1

31 2 **32** -1 **33** 0

34 -1

35 -5

36 $-\dfrac{1}{2}$

37 5

38 $-\dfrac{1}{2}$

39 3

40 $\dfrac{3}{2}$

Assessment

1 a -3 **b** $\dfrac{1}{2}$

2 a 21 **b** $p=18, q=108$

3 a 6 **b** 5 **c** $p=1, q=\dfrac{3}{2}$

4 a $2^{-\frac{1}{2}}, r=\dfrac{1}{2}$ **b** $\dfrac{3}{2}x^1, a=\dfrac{3}{2}, b=1$

5 a $2x^3-x^2-4x+3$ **b** $3, -8$

6 a i 18 **ii** 30 **b** $3-\sqrt{10}$ **7** $5-2\sqrt{3}$

Review

1 $3a^2-a$

2 $2x^2+3x-35$

3 $16x^2-24x+9$

4 $4x^2-9$

5 $3+x-10x^2$

6 $2x^3+x^2y-4xy^2-3y^3$

7 -54

8 $5\sqrt{6}$

9 $43-24\sqrt{3}$

10 d

11 a

12 B

2 Quadratic Polynomials and Equations

Exercise 1

1 $(x+5)(x+3)$ **2** $(x+7)(x+4)$ **3** $(x+6)(x+1)$

4 $(x+4)(x+3)$ **5** $(x-1)(x-9)$ **6** $(x-3)^2$

7 $(x+6)(x+2)$ **8** $(x-8)(x-1)$ **9** $(x+7)(x-2)$

10 $(x+4)(x-3)$ **11** $(x-5)(x+1)$ **12** $(x-12)(x+2)$

13 $(x+7)(x+2)$ **14** $(x-1)^2$ **15** $(x-3)(x+3)$

16 $(x+8)(x-3)$ **17** $(x+2)^2$ **18** $(x-1)(x+1)$

19 $(x-6)(x+3)$ **20** $(x+5)^2$ **21** $(x-4)(x+4)$

22 $(4+x)(1+x)$ **23** $(2x-1)(x-1)$ **24** $(3x+1)(x+1)$

25 $(3x-1)^2$ **26** $(3x+1)(2x-1)$ **27** $(3+x)^2$

28 $(2x-3)(2x+3)$ **29** $(x+a)^2$ **30** $(xy-1)^2$

Exercise 2

1 $(3x-4)(2x+3)$ **2** $(4x-3)(x-2)$ **3** $(4x-1)(x+1)$

4 $(3x-2)(x-5)$ **5** $(2x-3)^2$ **6** $(1-2x)(3+x)$

7 $(5x-4)(5x+4)$ **8** $(3+x)(1-x)$ **9** $(5x-1)(x-12)$

10 $(3x+5)^2$ **11** $(3-x)(1+x)$ **12** $(3+4x)(4-3x)$

13 $(1+x)(1-x)$ **14** $(3x+2)^2$ **15** $(x+y)^2$

16 $(1-2x)(1+2x)$ **17** $(2x-y)^2$ **18** $(3-2x)(3+2x)$

19 $(6+x)^2$ **20** $(5x-4)(8x+3)$ **21** $(7x+30)(x-5)$

22 $(6-5x)(6+5x)$ **23** $(x-y)(x+y)$ **24** $(9x-2y)^2$

25 $(7-6x)^2$ **26** $(5x-2y)(5x+2y)$

27 $(6x+5y)^2$ **28** $(2x-3y)(2x+y)$

29 $(3x+4y)(2x+y)$ **30** $(7pq-2)^2$ **31** not possible

32 $2(x+1)^2$ **33** $(x+2)(x+1)$ **34** $3(x+5)(x-1)$

35 not possible **36** not possible **37** not possible

38 $2(x-2)^2$ **39** $3(x-2)(x+1)$ **40** $2(x^2-3x+4)$

41 $3(x-4)(x+2)$ **42** $(x-6)(x+2)$ **43** not possible

44 $4(x-5)(x+5)$ **45** $5(x^2-5)$ **46** not possible

47 not possible **48** not possible

Exercise 3

1 $x=-2$ or $x=-3$ **2** $x=2$ or $x=-3$ **3** $x=3$ or $x=-2$

4 $x=-2$ or $x=-4$ **5** $x=1$ or $x=3$ **6** $x=1$ or $x=-3$

7 $x=-1$ or $x=-\dfrac{1}{2}$ **8** $x=2$ or $x=\dfrac{1}{4}$ **9** $x=1$ or $x=-5$

10 $x=8$ or $x=-9$ **11** $-1, 3$ **12** $1, 4$

13 $1, 5$ **14** $2, -5$ **15** $-2, 7$

16 $2, 7$ **17** $x=2$ or $x=5$ **18** $x=3$ or $x=-5$

19 $x=4$ or $x=-1$ **20** $x=3$ or $x=4$ **21** $x=\dfrac{1}{3}$ or $x=-1$

22 $x=-1$ or $x=-6$ **23** $x=0$ or $x=2$ **24** $x=-1$ or $x=-\dfrac{1}{4}$

25 $x=\dfrac{2}{3}$ or $x=-1$ **26** $x=0$ or $x=-\dfrac{1}{2}$ **27** $x=0$ or $x=-6$

28 $x=0$ or $x=10$ **29** $x=0$ or $x=\dfrac{1}{2}$ **30** $x=5$ or $x=-4$

31 $x=2$ or $x=-\dfrac{4}{3}$ **32** $x=2$ or $x=-1$ **33** $x=0$ or $x=1$

34 $x=0$ or $x=2$ **35** $x=3$ or $x=-1$ **36** $x=-1$ or $x=\dfrac{1}{2}$

Exercise 4

1 4 **2** 1 **3** 9

4 25 **5** 2 **6** $\dfrac{25}{4}$

7 192 **8** 81 **9** 200

10 $\dfrac{1}{4}$ **11** $\dfrac{1}{3}$ **12** $\dfrac{9}{8}$

13 $x=-4\pm\sqrt{17}$ **14** $x=1\pm\sqrt{3}$ **15** $x=-\dfrac{1}{2}\left(1\pm\sqrt{5}\right)$

16 $x=-\dfrac{1}{2}\left(1\pm\sqrt{3}\right)$ **17** $x=-\dfrac{1}{2}\left(3\pm\sqrt{5}\right)$ **18** $x=\dfrac{1}{4}\left(1\pm\sqrt{17}\right)$

19 $x=-2\pm\sqrt{6}$ **20** $x=-\dfrac{1}{6}\left(1\pm\sqrt{13}\right)$ **21** $x=\dfrac{1}{2}\left(-2\pm3\sqrt{2}\right)$

22 $x=\dfrac{1}{2}\left(1\pm\sqrt{13}\right)$ **23** $x=-\dfrac{1}{8}\left(1\pm\sqrt{17}\right)$ **24** $x=\dfrac{1}{4}\left(3\pm\sqrt{41}\right)$

Exercise 5

1 $x=-2\pm\sqrt{2}$ **2** $x=\dfrac{1}{4}\left(1\pm\sqrt{17}\right)$ **3** $x=\dfrac{1}{2}\left(-5\pm\sqrt{21}\right)$

4 $x=\dfrac{1}{4}\left(1\pm\sqrt{33}\right)$ **5** $x=2\pm\sqrt{3}$ **6** $x=\dfrac{1}{4}\left(1\pm\sqrt{41}\right)$

7 $x=\dfrac{1}{6}\left(1\pm\sqrt{13}\right)$ **8** $x=-\dfrac{1}{6}\left(1\pm\sqrt{13}\right)$ **9** $x=-0.260$ or -1.540

10 $x=2.781$ or 0.719 **11** $x=1.883$ or -0.133

12 $x=0.804$ or -1.554 **13** $x=0.804$ or -1.554

14 $x=0.724$ or 0.276 **15** $x=7.873$ or 0.127

16 $x=3.303$ or -0.303

Exercise 6

1 4 **2** $-\dfrac{5}{3}$ **3** 1

4 -3 **5** $\dfrac{4}{3}$ **6** $\dfrac{2}{5}$

7 real and different **8** not real

9 real and different **10** real and equal **11** real and different

12 real and equal **13** real and different

14 not real **15** real and different

16 real and equal **17** $k=\pm12$

18 $a=\dfrac{9}{4}$ **19** $p=2$ **22** $q^2=4p$

Exercise 7

1

x	-2	1
y	-1	2

2 $x=-1, y=3$ **3** $x=2, y=3$

4

x	-1	$\dfrac{1}{2}$
y	4	1

5

x	2	$-\dfrac{1}{2}$
y	-3	2

6

x	$\dfrac{7}{2}$	-2
y	$-\dfrac{1}{2}$	5

7

x	1	2
y	2	1

8

x	-1	3
y	-4	4

9 $x=1, y=5$

10

x	6	-6
y	2	-4

11 $x=\dfrac{1}{2}, y=-1$

12

x	1	0
y	$\dfrac{1}{3}$	$\dfrac{2}{3}$

13 $x=-1, y=-\dfrac{1}{2}$ **14** $x=1, y=-\dfrac{1}{3}$

15

x	$-\dfrac{1}{3}$	$\dfrac{2}{3}$
y	$-\dfrac{1}{2}$	$\dfrac{1}{4}$

16 $x=1, y=\dfrac{1}{2}$

17

x	-3	6
y	-3	$\dfrac{3}{2}$

18

x	1	$-\dfrac{1}{4}$
y	-2	3

19

x	-1	2
y	$\dfrac{1}{3}$	$-\dfrac{1}{6}$

20

x	$\dfrac{1}{2}$	0
y	$\dfrac{1}{2}$	1

21

x	1	$3\dfrac{1}{2}$
y	1	-4

22

x	-1	$7\dfrac{1}{2}$
y	2	$-\dfrac{7}{5}$

Review

1 $3(x-1)(x-2)$ **2** $4(x-3)(x+3)$ **3** $-1, 6$

4 $3\pm\sqrt{14}$ **5** $\dfrac{1}{4}\left(-3\pm\sqrt{17}\right)$ **6** $\dfrac{1}{3}\left(-2\pm\sqrt{19}\right)$

7 $x=4, y=5$ or $x=-22, y=31$

8 $x=2, y=5$ or $x=4, y=9$

9 $5, \dfrac{2}{3}$ **a** rational   **b** it factorises

10 a not real   **b** real and different
   c real and equal   **d** real and different

11 $4, -1$

Assessment

1 b $p=6$ or 2 **c** $x^2-4x+4=0; x^2=0$

2 a $\left(x-\dfrac{3}{2}\right)^2-\dfrac{33}{4}; a=-\dfrac{3}{2}, b=-\dfrac{33}{4}$

   b $x=\dfrac{3\pm\sqrt{33}}{2}$

3 a $\left(x-\dfrac{5}{2}\right)^2+\dfrac{3}{4}; a=-\dfrac{5}{2}, b=\dfrac{3}{4}$

   b $\sqrt{-\dfrac{3}{4}}$ is not a real number.

4 $x=-6, y=17; x=2, y=1$ **5 a** 9   **7 i** $(x-3)^2+2$

3 Algebraic Division

Exercise 1

1 quotient $2x+1$, remainder -5
2 quotient $x-2$, remainder 6
3 quotient $2x^2+x+1$, remainder 0
4 quotient $2x^2+3x+6$, remainder 14

7 $(2x-1)(x^2+x+5); a=2, b=-1, p=1, q=1, r=5$
8 $b=3, c=-6$ **9** 5
10 $p=11, q=-5$ **12** $a=1, b=2, c=-1$

Review

1 $2x+7, 18$ **2** $3x^2-5x+4; -7$
3 $3x-14-\dfrac{43}{x+3}$ **4** $x^2-3x-3, 2; x^2-3x-3+\dfrac{2}{x-1}$
5 -7 **6** yes, when $x=1$ then $x^3-2x^2+1=0$
7 $(x-3)(x-1)^2$ **8** $(x-1)(x+1)(2x-1)$ **9** -6

Exercise 2

1 3   **2** 18   **3** 47
4 $\dfrac{35}{16}$ **5** $-\dfrac{16}{27}$ **6** a^3-2a^2+6
7 c^2-ac+b **8** $\dfrac{1}{a^4}-\dfrac{2}{a}+1$ **9** -7

Exercise 3

1 yes   **2** no   **3** neither are
5 a $(x-1)(x+1)(x+2)$ **b** $(x-2)(x^2+x+1)$
   c $(2x-1)(x^2+1)$
6 20

Assessment

1 $x^2-3x+2+\dfrac{1}{x-1}$ **2 b** $(x+1)^2(x-1)$
   $a=1, b=-3, c=2, d=1$
3 a -4 **c** $(x-3)(x^2+1)$
4 a $a=4, b=1$ **b** $(x-1)(x+2)(x+3)$
5 a 5   **b** $x=2$ or 3
6 a ii $(x+3)(x-1)(x+5)$ **b** 35
7 a i 36   **iii** $(x+2)(x^2-5x+10)$ **iv** -2

4 Functions and Graphs

Exercise 1

1 a $\dfrac{11}{4}$ **b** 3   **c** 4

2 a   **b**   **c**   **d**   **e**

3 a

b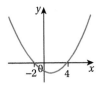

c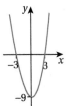

d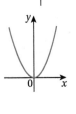

e

f

4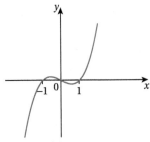

5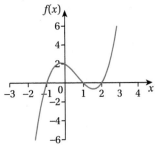

6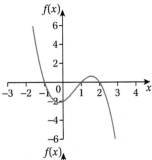

7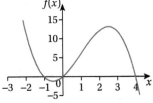

2 a $9-(x+2)^2$ **b** $(-2, 9)$

c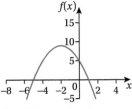

3 a $(x-1)(x^2+x+1)$ **b**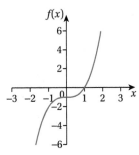

4 a $(x+1)^2(x-2)$ **b**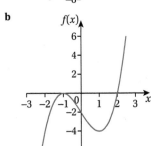

6 a 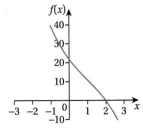   **b** 1

7 a $(3, -6)$ **b**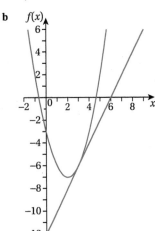

9 a $x^2+(12-p)x+25=0$ **b** 2, 22

Exercise 3

1 $x < \dfrac{7}{2}$ **2** $x > 4$ **3** $x > -3$

4 $x > -2$ **5** $x < -\dfrac{1}{2}$ **6** $x < -3$

7 $x < -\dfrac{1}{4}$ **8** $x > \dfrac{8}{3}$ **9** $x > \dfrac{3}{8}$

10 $x > 2$ and $x < 1$ **11** $x \geq 5$ and $x \leq -3$ **12** $-4 < x < 2$

13 $x \geq \dfrac{1}{2}$ and $x \leq -1$ **14** $x > 2 + \sqrt{7}$ and $x < 2 - \sqrt{7}$

15 $-\dfrac{1}{2} < x < \dfrac{1}{2}$ **16** $-4 \leq x \leq 2$ **17** $x > 1$ and $x < -\dfrac{2}{5}$

18 $x \geq \dfrac{3}{2}$ and $x \leq -5$ **19** $x > 4$ and $x < -2$

20 $\dfrac{1}{2}\left(-3 - \sqrt{17}\right) \leq x \leq \dfrac{1}{2}\left(-3 + \sqrt{17}\right)$ **21** $x > 7$ and $x < -1$

22 $k < \dfrac{49}{12}$ **23** $-4 < b < 4$ **24** $k < 0$ and $k > 2$

25 true for all real values of a

Exercise 4

1

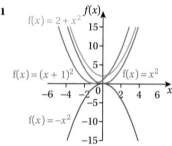

$f(x) = 2 + x^2$
$f(x) = (x + 1)^2$
$f(x) = x^2$
$f(x) = -x^2$

2 a one way stretch by scale factor $\dfrac{1}{2}$ parallel to the y-axis

 b one way stretch by scale factor 3 parallel to the y-axis

3 $y = 3^{2-x} - 5$ **4** $y = 2(3^{x+2} - 5)$ **5** $y = 5 + \sqrt{2x - 7}$

6 a one way stretch by scale factor $\dfrac{1}{2}$ parallel to the y-axis OR one way stretch scale factor 2 parallel to x-axis

 b one way stretch by scale factor 2 parallel to the y-axis OR one way stretch scale factor $\dfrac{1}{2}$ parallel to x-axis

 c translation $\begin{bmatrix} 0 \\ -2 \end{bmatrix}$ **d** translation $\begin{bmatrix} 2 \\ 0 \end{bmatrix}$

7 $\dfrac{1}{2}\sqrt{16 - x^2}$ **8** $\dfrac{2}{1+x}$

9 one way stretch by scale factor 2 parallel to the x-axis

10 $\dfrac{\sqrt{x^2 + 2x + 6}}{x + 1} + 5$

Review

1 a least value $\dfrac{11}{4}$ where $x = \dfrac{3}{2}$ **b** least value $\dfrac{41}{8}$ where $x = \dfrac{7}{4}$

 c least value -9 where $x = -2$

---

2 a $(x + 1)(x^2 - x + 1)$ **b**

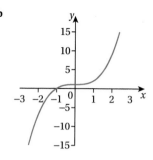

3 b $(-4, 3)$ and $\left(\dfrac{1}{2}, \dfrac{15}{2}\right)$

4 a $x^3 - x^2 + x - 1 = 0$ **b** $x = 1$

6 a $x < 1$ **b** $x > \dfrac{3}{2}$ **c** $x < -2$ and $x > 3$

 d $-\dfrac{2}{3} < x < \dfrac{3}{2}$ **e** $-\sqrt{13} < x < \sqrt{13}$ **f** $x < -1$ and $x > 7$

7 a one way stretch parallel to the y-axis by scale factor 2

 b one way stretch parallel to the x-axis by scale factor 2

 c translation $\begin{bmatrix} 0 \\ -3 \end{bmatrix}$ **d** reflection in the x-axis

8 a $-\dfrac{x}{1+x}$ **b** $\dfrac{x}{3-x}$ **c** $\dfrac{2x+1}{x+1}$

Assessment

1 a $-\dfrac{41}{4} + \left(x + \dfrac{7}{2}\right)$ **b** $x = -\dfrac{7}{2}$

 c $\left(-\dfrac{7}{2}, -\dfrac{41}{4}\right)$ **d**

2 b $(0, 0)$

 c translation by $\begin{bmatrix} 0 \\ 1 \end{bmatrix}$

3 c $(0, 2)$ when $k = 2$, and $(1, 3)$ when $k = 3$

4 a $x(x + 1)(x - 1)$ **e** $k \neq -\dfrac{1}{2}$

5 a $2\sqrt{2x^2 - 1}$ **b** $-\sqrt{2x^2 - 1}$

6 a $\left(x + \dfrac{3}{2}\right)^2 - \dfrac{1}{4}$

 b **i** $(-1.5, -0.25)$ **ii** $x = -1.5$

 c $y = x^2 - x + 4$

7 a i $2\left(x + \dfrac{3}{2}\right)^2 + \dfrac{1}{2}$ **ii** 0.5

 b ii $\dfrac{1}{2}\sqrt{10}$

5 Coordinate Geometry

Exercise 1

1 a 5 **b** $\sqrt{2}$ **c** $\sqrt{13}$

2 a $\left(\dfrac{5}{2}, 4\right)$ **b** $\left(\dfrac{5}{2}, \dfrac{1}{2}\right)$ **c** $\left(3, \dfrac{7}{2}\right)$

3 a $\sqrt{109}, \left(\dfrac{1}{2}, 1\right)$ **b** $\sqrt{5}, \left(-\dfrac{1}{2}, -1\right)$ **c** $2\sqrt{2}, (-2, -3)$

4 $\sqrt{65}$ **5** $\sqrt{13}$ **6** $(2, -4)$ **8 b** $\left(-\dfrac{7}{2}, -\dfrac{1}{2}\right)$

9 a $\sqrt{5}\left(2 + \sqrt{2}\right)$ **b** $\left(0, \dfrac{9}{2}\right)$ **c** $2\dfrac{1}{2}$ **11** $(-5, -3)$

Exercise 2

1 a 3   **b** $\dfrac{3}{2}$ **c** $\dfrac{1}{3}$

   d $\dfrac{3}{4}$ **e** -4 **f** 6

   g $-\dfrac{7}{3}$ **h** $-\dfrac{3}{2}$ **i** $\dfrac{k}{h}$

2 a yes   **b** no   **c** yes   **d** yes

3 a parallel   **b** perpendicular   **c** perpendicular
   d neither   **e** parallel

Exercise 3

1 a $y = 2x$ **b** $x + y = 0$ **c** $3y = x$
   d $4y + x = 0$ **e** $y = 0$
2 a $2y = x + 2$ **b** $3y + 2x = 0$ **c** $y = 4x$

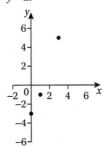

3 a $y = 1$ **b** $y + 2x - 6 = 0$ **c** $y = 2x + 5$
4 a $2y + x = 0$ **b** $2x - 3y = 0$ **c** $2x + y = 0$
5 a $x - 3y + 1 = 0$ **b** $2x + y - 5 = 0$
6 a $5x - y - 17 = 0$ **b** $x + 7y + 11 = 0$
7 $3x + 4y - 48 = 0, 5$

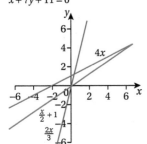

Exercise 4

1 a $y = 3x - 3$ **b** $5x + y - 6 = 0$ **c** $x - 4y - 4 = 0$
   d $y = 5$ **e** $2x + 5y - 21 = 0$ **f** $30x + 80y - 277 = 0$

2 a $3x - 2y + 2 = 0$ **b** $3x - 2y + 7 = 0$ **c** $x = 3$
3 a, c and **d**   **4** $x + y - 7 = 0$
5 a $x + 2y - 5 = 0$ **b** $16x - 6y + 19 = 0$ **c** $10x - 16y + 23 = 0$
6 $2x + y = 0$ **7** $4x + 5y = 0$ **8** $5x - 4y = 0$
9 $x + 2y - 11 = 0$ **10** $3x - 4y + 19 = 0$

Exercise 5

2 $\dfrac{1}{4}$ sq. units   **3** $\left(-\dfrac{4}{3}, \dfrac{11}{3}\right), (-5, 0), (6, 0)$ **4** $10x - 26y - 1 = 0$

5 $x + 3y - 11 = 0, \left(\dfrac{13}{5}, \dfrac{14}{5}\right)$ **6** $\left[\dfrac{2}{5}(2 + 2a - b), \dfrac{1}{5}(8 - 2a + b)\right]$

7 $y = 2x - 3$ **8** $8by - 2ax + 8b^2 + 3a^2 = 0$

9 $\left(\dfrac{9}{10}, \dfrac{17}{10}\right)$ and $\left(-\dfrac{18}{10}, \dfrac{26}{10}\right), \left(-\dfrac{27}{10}, -\dfrac{1}{10}\right)$ or $\left(\dfrac{36}{10}, \dfrac{8}{10}\right), \left(\dfrac{27}{10}, -\dfrac{19}{10}\right)$

10 $(1, 2), (5, 2), (3, 6)$

Review

1 c   **2 d**   **3 b**   **4 a**

5 a $-\dfrac{2}{3}$ **b** $2y - 3x = 8$ **c** $\left(-\dfrac{14}{13}, \dfrac{31}{13}\right)$

6 a $9x + 13y + 14 = 0$ **7** $\dfrac{1}{6}$

Assessment

1 a $-\dfrac{2}{3}$ **b** $2x + 3y - 3 = 0$ **c** $3x - 2y - 11 = 0$

2 a $9x + 13y + 14 = 0$ **b** $\dfrac{3}{4}, -\dfrac{4}{3}$ **c** $\left(\dfrac{7}{2}, -\dfrac{7}{2}\right)$

3 a $\left(-\dfrac{2}{3}, 0\right), (0, -2)$ **b** $\dfrac{2}{3}$ square units   **c** $y = 3x$

4 a $p = 3$ **b** 3   **c** $\left(-\dfrac{1}{2}, \dfrac{15}{2}\right)$ **d** $x + 3y - 4 = 0$

5 a $p = 13$ **b** $\left(\dfrac{7}{13}, \dfrac{9}{13}\right)$ **c** $\dfrac{32\sqrt{26}}{13}$

6 a $-\dfrac{7}{3}$ **b i** $7x + 3y = 2$ **ii** $(4, -5)$ **c** $(-2, 9)$

7 a i $-\dfrac{7}{4}$ **ii** $7x + 4y = 1$
   b i $(1, -1.5)$ **ii** $y + \dfrac{3}{2} = \dfrac{4}{7}(x - 1)$ **c** $1, -\dfrac{11}{5}$

6 Differentiation

Exercise 1

1 $5x^4$ **2** $-3x^{-4}$ **3** $\dfrac{4}{3}x^{\frac{1}{3}}$ **4** $-\dfrac{1}{x^2}$

5 $10x^9$ **6** $-\dfrac{2}{x^3}$ **7** $\dfrac{3}{2}\sqrt{x}$ **8** $-\dfrac{1}{2}x^{\frac{3}{2}}$

9 $-\dfrac{4}{x^5}$ **10** $\dfrac{1}{3}x^{-\frac{2}{3}}$ **11** $-\dfrac{1}{4}x^{-\frac{5}{4}}$ **12** 1

13 $\dfrac{7}{2}\sqrt{x^5}$ **14** $\dfrac{-7}{x^8}$ **15** $\dfrac{1}{7}x^{-\frac{6}{7}}$ **16** $3x^2$

7 $-\dfrac{3}{4}x^{-\frac{7}{4}} - \dfrac{3}{4}x^{-\frac{1}{4}} + 1$ **8** $9x^2 - 8x + 9$ **9** $\dfrac{3}{2}x^{\frac{1}{2}} - \dfrac{1}{2}x^{-\frac{1}{2}} - \dfrac{1}{2}x^{-\frac{3}{2}}$

10 $\dfrac{1}{2\sqrt{x}} + \dfrac{3\sqrt{x}}{2}$ **11** $\dfrac{-2}{x^3} + \dfrac{3}{x^4}$ **12** $\dfrac{-1}{2\sqrt{x^3}} + \dfrac{2}{x^2}$

13 $\dfrac{-1}{2}x^{-\frac{3}{2}} + \dfrac{9}{2}x^{\frac{1}{2}}$ **14** $\dfrac{1}{4}x^{-\frac{3}{4}} - \dfrac{1}{5}x^{-\frac{4}{5}}$ **15** $-\dfrac{12}{x^4} + \dfrac{3x^2}{4}$

16 $-\dfrac{4}{x^2} - \dfrac{10}{x^3} + \dfrac{18}{x^4}$ **17** $\dfrac{3}{2\sqrt{x}} - 3$ **18** $1 + 2x^{-2} - 9x^{-4}$

19 $\dfrac{3}{2}\sqrt{x} - \dfrac{5}{2}x\sqrt{x}$ **20** $\dfrac{-3\sqrt{x}}{2x^3} + \dfrac{3\sqrt{x}}{2}$

Exercise 2

1 $3x^2 - 2x + 5$ **2** $6x + \dfrac{4}{x^2}$ **3** $\dfrac{1}{2\sqrt{x}} - \dfrac{1}{2x\sqrt{x}}$

4 $8x^3 - 8x$ **5** $3x^2 - 4x - 8$ **6** $2x + \dfrac{5}{2\sqrt{x}}$

Exercise 3

1 $\dfrac{dy}{dx} = 2x + 2$ **2** $\dfrac{dz}{dx} = -4x^{-3} + x^{-2}$ **3** $\dfrac{dy}{dx} = 6x + 11$

4 $\dfrac{dy}{dz} = 2z - 8$ **5** $\dfrac{ds}{dt} = -\dfrac{3}{2t^4}$ **6** $\dfrac{ds}{dt} = \dfrac{1}{2}$

7 $\dfrac{dy}{dx} = 1 - \dfrac{1}{x^2}$ **8** $\dfrac{dy}{dx} = \dfrac{5z^2 - 1}{2\sqrt{z}}$ **9** $\dfrac{dy}{dx} = 18x^2 - 8$

10 $\dfrac{ds}{dt} = 2t$ **11** $\dfrac{ds}{dt} = 1 - \dfrac{7}{t^2}$ **12** $\dfrac{dy}{dx} = -\dfrac{3\sqrt{x} + 28}{2x^3}$

Exercise 4

1 $2; -\dfrac{1}{2}$ **2** $-\dfrac{1}{3}; 3$ **3** $\dfrac{1}{4}; -4$ **4** $6; -\dfrac{1}{6}$

5 $1; -1$ **6** $5; -\dfrac{1}{5}$ **7** $11; -\dfrac{1}{11}$ **8** $-11; \dfrac{1}{11}$

9 $4; -\dfrac{1}{4}$ **10** $\dfrac{4}{27}; -\dfrac{27}{4}$ **11** $\dfrac{5}{4}; -\dfrac{4}{5}$ **12** $2; -\dfrac{1}{2}$

13 $(2, 2)$ and $(-2, 4)$ **14** $(1, 0)$ and $\left(-\dfrac{1}{3}, \dfrac{4}{27}\right)$

15 $(3, 0)$ and $(-3, 18)$ **16** $(-1, -2)$ and $(1, 2)$

17 $(1, -16)$ **18** $\left(-2, \dfrac{1}{4}\right)$ **19** $(0, -5)$ **20** $(1, -2)$ and $(-1, 2)$

Exercise 5

1 a $y = 2x - 5$ **b** $2y + x + 5 = 0$
2 a $y = 4x - 2$ **b** $4y + x + 8 = 0$
3 a $y + x + 2 = 0$ **b** $y = x$
4 a $y = 5$ **b** $x = 0$
5 a $y + x = 3$ **b** $y = x - 1$
6 a $y = 19x + 26$ **b** $19y + x + 230 = 0$
7 $4y + x + 12 = 0$ **8** $y = 7x - 29$

9 $y + x = 1, 2y = 2x - 3; \left(\dfrac{5}{4}, -\dfrac{1}{4}\right)$ **10** $4y - x + 1 = 0, 4y + x - 5 = 0$

11 $y = 5x - 1, 3y + 9x + 19 = 0$ **12** $y = 7x - 4, y + 5x + 28 = 0$

13 $(2, 8), y = 8x - 8$ **14** $\left(\dfrac{1}{2}, -\dfrac{1}{4}\right)$ **15** $y + x + 1 = 0$

16 $2y = x + 2$ **17** $k = -\dfrac{7}{2}$ **18** $8y + 121 = 0$

19 $(1, -1)$ **20** $p = 12, q = 8; (-2, 4)$

Exercise 6

1 $x = 0$ **2** $x = \dfrac{3}{4}$ **3** $x = 0, x = \dfrac{8}{3}$

4 $x = \pm\dfrac{1}{2}$ **5** $x = 0, x = \dfrac{4}{3}$ **6** $x = \pm 1$

7 $x = 4$ **8** $x = \pm 3$ **9** $x = 1, x = -\dfrac{4}{3}$

10 $x = \pm\dfrac{5}{9}\sqrt{3}$ **11** $x = 1, x = -4$ **12** $x = \pm\dfrac{2}{3}\sqrt{3}$

13 $(3, 3), (-3, -3)$ **14** $(1, -7), \left(\dfrac{1}{3}, -\dfrac{185}{27}\right)$ **15** $\left(\dfrac{1}{2}, -\dfrac{25}{4}\right)$

16 $\left(\dfrac{1}{3}, \dfrac{-2}{9}\sqrt{3}\right)$ **17** $(1, 2)$ **18** $(4, 10), (-4, 6)$

Exercise 7

1 $(1, 1)$ max   **2** $(-1, -2)$ min; $(1, 2)$ max

3 $(3, 6)$ min; $(-3, -6)$ max   **4** $(0, 0)$ max; $\left(\dfrac{10}{3}, -\dfrac{500}{27}\right)$ min

5 $(0, 0)$ min   **6** $\left(1, \dfrac{3}{2}\right)$ min

7 $(-1, 1)$ max; $(0, 0)$ min; $(1, 1)$ max

8 $(0, 0)$ min   **9** $\left(\dfrac{5}{4}, -\dfrac{49}{8}\right)$ min

10 $(-1, 4)$ max; $(1, -4)$ min   **11** $(-2, -16)$ min; $(0, 0)$ max; $(2, -16)$ min
12 $(-2, 8)$ min; $(2, 8)$ min   **13** $(-1, -2)$ max; $(1, 2)$ min

14 $\left(\dfrac{1}{2}, \dfrac{11}{4}\right)$ min   **15** $\left(-\dfrac{1}{2}, -\dfrac{5}{16}\right)$ min, $(0, 0)$ max, $(1, -2)$ min

16 $5x^4 + 3x^2 + 4 = 0$ has no real roots so there is no stationary point

Exercise 8

1 a $80 - 2x$ **c** $800 \text{ m}^2; 20 \text{ m} \times 40 \text{ m}$

2 a $\dfrac{4}{x^2}$ **c** $2 \text{ m} \times 2 \text{ m} \times 1 \text{ m}$

3 a $r = \sqrt{9 - h^2}$ **c** $12\pi\sqrt{3} \text{ cm}^3$

4 a $10 - x$ **b** $x(10 - x), 5 \text{ cm}$

5 $\sqrt{35} \text{ cm}$ (a square)   **6** $a = 1, b = -2, c = 3$

Review

1 a $-3x - 4 - 3x^2$ **b** $\dfrac{1}{2}x^{-\frac{1}{2}} + \dfrac{1}{2}x^{-\frac{3}{2}}$ **c** $-\dfrac{2}{x^3} - \dfrac{6}{x^4}$

2 a $\dfrac{dy}{dx} = \dfrac{3}{2}x^{\frac{1}{2}} - \dfrac{2}{3}x^{-\frac{1}{3}} - \dfrac{1}{3}x^{-\frac{4}{3}}$ **b** $\dfrac{dy}{dx} = \dfrac{1}{2\sqrt{x}} + \dfrac{1}{x^2} - \dfrac{3}{x^4}$

    c $-\dfrac{3}{4x^{\frac{7}{4}}} + \dfrac{1}{4x^{\frac{5}{4}}}$

3 a 5 **b** 5 **c** 17
4 a 5 **b** 6
5 a 1 **b** 7 and -7

6 a $(1, 10)$ and $(-1, 6)$ **b** $\left(\dfrac{1}{3}, 7\dfrac{7}{9}\right)$ and $\left(-\dfrac{1}{3}, 8\dfrac{2}{9}\right)$

7 $2y = x - 1; \left(-\dfrac{3}{2}, -\dfrac{5}{4}\right)$ **8** $y = 3x + 6, y = 3x + 2$

9 $x + 2y + 9 = 0$ **10** $(-2, 16)$ max, $(2, -16)$ min

11 min $(1, 2)$, max $(-1, -2)$ **12 a** $5 - x$

13 a $h = \dfrac{1}{2}(7 - 2r - \pi r)$

Assessment

1 b 4 **c** maximum, since $\dfrac{d^2 y}{dx^2} < 0$

2 a $4x^3 - 6x^2 + 3$ **b** $12x^2 - 12x$ **c** $y = x - 6$

3 a $-\dfrac{3}{2}$ **4 a** $\dfrac{32}{x^2}$ **c** $4 \text{ m} \times 4 \text{ m} \times 2 \text{ m}$

5 a $10 - x$ **c** 12.5 cm^2

6 a i $5x^4 - 6x + 1$ **ii** $20x^3 - 6$ **b** $y = 12(x + 1)$

7 a $\dfrac{t^3}{2} - 2t$ **b i** $-1\dfrac{1}{2}$ **ii** $\dfrac{dy}{dt} < 0$ so decreasing
    c i 4 **ii** minimum

7 Integration

Exercise 1

1 $\dfrac{1}{6}x^6 + K$ **2** $-\dfrac{1}{4}x^{-4} + K$ **3** $\dfrac{4}{5}x^{\frac{5}{4}} + K$

4 $-\dfrac{1}{2}x^{-2} + K$ **5** $-\dfrac{2}{3}x^{-\frac{3}{2}} + K$ **6** $2x^{\frac{1}{2}} + K$

7 $\dfrac{1}{2}x^2 + K$ **8** $\dfrac{3}{2}x^{\frac{2}{3}} + K$ **9** $x + \dfrac{1}{3}x^3 + K$

10 $x^2 - \dfrac{2}{3}x^{\frac{3}{2}} + K$ **11** $x - \dfrac{1}{x} + K$ **12** $\dfrac{1}{2}x^2 + \dfrac{1}{3}x^3 + K$

13 $2x + \left(\dfrac{7}{2}\right)x^2 - 5x^3 + K$ **14** $2\sqrt{x} + \dfrac{2}{3}x^{\frac{3}{2}} + K$ **15** $-\dfrac{1}{2x^2} + \dfrac{2}{x} + K$

16 $2\sqrt{x} + \dfrac{2}{3}x^{\frac{3}{2}} + \left(\dfrac{2}{7}\right)x^{\frac{7}{2}} + K$ **17** $x - x^2 + \dfrac{1}{3}x^3 + K$

18 $\dfrac{1}{2}x^2 - \dfrac{1}{4}x^4 + K$ **19** $-\dfrac{1}{x} + \dfrac{2}{\sqrt{x}} + K$ **20** $y = x^3 - x^2 + 1$

21 $y = \dfrac{2}{3}x^{\frac{3}{2}} - \dfrac{13}{3}$ **22** $y = \dfrac{2}{3}x^{\frac{3}{2}} - 6x^{\frac{1}{2}} + 2$ **23** $y = x^3 - x^2 + 3x + 2$

Exercise 2

1 4 **2** $\dfrac{2}{7}(8\sqrt{2} - 1)$ **3** $26\dfrac{2}{3}$ **4** $12\dfrac{2}{3}$ **5** 15

6 −2 **7** $2\dfrac{1}{3}$ **8** $6\sqrt{2} - 4$ **9** 1

Exercise 3

Answers are in square units.

1 $5\dfrac{1}{3}$ **2** $12\dfrac{2}{3}$ **3** $2\dfrac{2}{3}$ **4** $13\dfrac{1}{2}$

5 $5\dfrac{1}{3}$ **6** 60 **7** $5\dfrac{1}{3}$ **8** $4\dfrac{7}{8}$

9 24 **10** $\dfrac{4}{3}$ **11** $15\dfrac{1}{4}$ **12** $\dfrac{1}{3}$

13
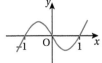
a $\dfrac{1}{4}$ **b** $\dfrac{1}{4}$

14
a 4 **b** 4

15

a −2 **b** 2 **c** 0

16 a $\dfrac{8}{3}$ **b** $\dfrac{16}{3}$ **17** $\dfrac{1}{6}$ **18** $\dfrac{4}{3}$ **19** $\dfrac{1}{3}$ **20** $4\sqrt{3}$

Exercise 4

1 22 **2** 1.39 **3** 18 **4** 10.5 **5** 25.2 **6** 0.512

Review

1 $\dfrac{1}{3}x^3 + \dfrac{1}{x} + K$ **2** $\dfrac{3}{4}x^{\frac{4}{3}} + K$

3 $\dfrac{2}{3}x^{\frac{3}{2}} + 2x^{\frac{1}{2}} + K = \dfrac{2}{3}\sqrt{x}(x + 3) + K$ **4** $\dfrac{1}{2}x^2 - 2x^{\frac{1}{2}} + K$

5 $\dfrac{1}{2}x^2 + \dfrac{1}{x} + K$ **6** $\dfrac{2}{5}x^{\frac{5}{2}} - 2x^{\frac{1}{2}} + K$ **7** 9

8 $\dfrac{297}{10}$ **9** $\dfrac{135}{4}$ **10** $\dfrac{3}{4}$

11 $10\dfrac{2}{3}$ **12** $\dfrac{1}{6}$

13 a 4 square units **b** $-4\dfrac{1}{2}$

Assessment

1 b $x + \dfrac{6}{x} - \dfrac{3}{x^3} + K$ **c** $\dfrac{5}{8}$

2 a $\dfrac{34}{3}$ **b** $\dfrac{22}{3}$

3 a $2x + 5y - 13 = 0$ **b** $y = \dfrac{2}{3}x^{\frac{3}{2}} - 2x^{\frac{1}{2}} + x - \dfrac{13}{3}$

4 a $27\dfrac{1}{2}$ **b** $27\dfrac{1}{2}$

 c a is exact because the area under the line is that of a trapezium.

5 a stretch by a factor of 2 parallel to the x-axis

 b $\dfrac{1}{2}(5\sqrt{2} + 8) = 7.54$ (3 significant figures)

6 i 8 **ii** 2
7 a 6.43 **b** increase the number of strips

8 Sequences and Series

Exercise 1

1 **a** $1, \frac{1}{4}, \frac{1}{9}, \frac{1}{16}, \frac{1}{25}, \frac{1}{36}$ convergent

b $-2, 4, -8, 16, -32, 64$ divergent

c $-\frac{1}{2}, \frac{1}{4}, -\frac{1}{8}, \frac{1}{16}, -\frac{1}{32}, \frac{1}{64}$ convergent

2 **a** $3, 0.57, 0.73, 0.70, 0.71, 0.71$

b $-4, -10, -88, -7654, -5.9 \times 10^7, -3.4 \times 10^{15}$

c $0.5, 1.2, 0.88, 1.1, 0.97, 1.0$

d $1, 0.8, 0.87, 0.85, 0.86, 0.85$

3 **a** Undefined after u_2, divergent **b** divergent

c divergent

4 **a** converges (to 0.4) **b** converges (to 0.4)

c converges (to 0.4)

Exercise 2

1 **a** $\displaystyle\sum_{n=1}^{5} n^3$ **b** $\displaystyle\sum_{n=1}^{10} 2n$

c $\displaystyle\sum_{n=1}^{49} \frac{1}{n+1}$ **d** $\displaystyle\sum_{n=1}^{\infty} \frac{1}{3^{n-1}}$

e $\displaystyle\sum_{n=1}^{8} (-7+3n)$ **f** $\displaystyle\sum_{n=1}^{\infty}\left(\frac{8}{2^{n-1}}\right) = \sum_{n=1}^{\infty} \frac{1}{2^{n-4}}$

2 **a** $1 + \frac{1}{2} + \frac{1}{3} + \cdots$ **b** $0 + 2 + 6 + \cdots + 20$

c $\frac{1}{2} + \frac{4}{15} + \frac{5}{28} \cdots + \frac{22}{861}$ **d** $\frac{1}{2} + \frac{1}{5} + \frac{1}{10} \cdots$

e $6 + 24 + 72 \cdots + 720$ **f** $-a - a^2 + a^3 \cdots$

3 **a** $8; 9$ **b** $13; 8$

c $\frac{1}{462}; \infty$ **d** $\left(\frac{1}{2}\right)^n; \infty$

e $-48; 23$ **f** $4; 10$

Exercise 3

1 **a** $9, 2n-1$ **b** $16, 4(n-1)$ **c** $15, 3n$

d $17, 3n+2$ **e** $-2, 8-2n$ **f** $p+4q, p+(n-1)q$

g $18, 8+2n$ **h** $17, 4n-3$ **i** $0, \frac{1}{2}(5-n)$

j $8, 3n-7$

2 **a** 100 **b** 180 **c** 165

d 185 **e** -30 **f** $5(2p+9q)$

g 190 **h** 190 **i** $-\frac{5}{2}$

j 95

3 $a = 27.2, d = -2.4$ **4** $d = 3; 30$ **5** $1, \frac{1}{2}, 0; -\frac{17}{2}$

6 **a** $28\frac{1}{2}$ **b** 80 **c** 400

d 80 **e** 108 **f** $3n(1-6n)$

g 40 **h** $2m(m+3)$

7 $4, 2n-4$ **9** $2, 364$ **10** 39

11 64 **12** **a** $1, 4$ **b** 270

13 **a** $a = 21, d = -3$ **b** less than 4 or more than 11

Exercise 4

1 **a** $32, 2n$ **b** $\frac{1}{8}, \frac{1}{2^{n-2}}$ **c** $48, 3(-2)^{n-1}$

d $\frac{1}{2}, (-1)^{n-1}\left(\frac{1}{2}\right)^{n-4}$ **e** $\frac{1}{27}, \left(\frac{1}{3}\right)^{n-2}$

2 **a** 189 **b** -255 **c** $2-\left(\frac{1}{2}\right)^{19}$

d $\frac{781}{125}$ **e** $\frac{341}{1024}$ **f** 1

3 $\frac{1}{2}, 2$ **4** $-\frac{1}{2}$ **5** $-\frac{1}{2}, \frac{1}{1024}$ **6** 13.21 to 4 s.f.

7 **a** $\frac{x-x^{n+1}}{1-x}$ **b** $\frac{x\left(1-\frac{1}{x^n}\right)}{1-\frac{1}{x}}$ **c** $\frac{1+(-1)^{n+1}y^n}{1+y}$

d $\frac{x(2^n-x^n)}{2^{n-1}(2-x)}$ **e** $\frac{x(2^n-x^n)}{2^{n-1}(2-x)}$

8 62 or 122 **9** 8.493 to 4 s.f.

Exercise 5

1 **a** yes **b** no **c** yes

d yes **e** no **f** yes

2 **a** $-1 < x < 1$ **b** $x < -1, x > 1$ **c** $-\frac{1}{2} < x < \frac{1}{2}$

d $0 < x < 2$ **e** $-1-a < x < 1-a$ **f** $x < -(1+a), x > 1-a$

3 **a** 6 **c** $13\frac{1}{3}$ **d** $\frac{5}{9}$ **f** $\frac{9}{4}$

4 $\frac{1}{2}$ **5** $8, 4, 2, 1$ or $24, -12, 6, -3$

Exercise 6

1 **a** $1 + 36x + 594x^2 + 5940x^3$

b $1 - 18x + 144x^2 - 672x^3$

c $1024 + 5120x + 11520x^2 + 15360x^3$

d $1 - \frac{20}{3}x + \frac{190}{9}x^2 - \frac{380}{9}x^3$

e $128 - 672x + 1512x^2 - 1890x^3$

f $\frac{19683}{512} + \frac{59049}{128}x + \frac{19683}{8}x^2 + \frac{15309}{2}x^3$

2 **a** $336x^2$ **b** $-10x$ **c** $-21840x^{11}$

d $3360p^6q^4$ **e** $34992a^7b$ **f** $7920x^4$

g $63x^5$ **h** $56a^3b^5$

3 **a** $1 - 8x + 27x^2$ **b** $1 + 19x + 160x^2$ **c** $2 - 19x + 85x^2$

d $1 - 68x + 2136x^2$

4 **a** $x^5 + 5x^4y + 10x^3y^2 + 10x^2y^3 + 5xy^4 + y^5$

b $32 + 80(0.01) + 80(0.01)^2 + 40(0.01)^3 + 10(0.01)^4 + (0.01)^5$

c 32.8080401001

Review

1 $\frac{1}{2}, \frac{2}{5}, \frac{3}{10}, \frac{4}{17}, \frac{5}{26}, \frac{6}{37}$ converges to 0

2 **a** $1, 0, 0, 0, 0, 0$ converges to 0

b $0.5, -0.25, 0.3125, -0.215, 0.261, -0.193$ converges (slowly to 0)

3 **a** cycles $2, -1, \frac{1}{2}$ **b** undefined for $r > 1$

4 $\dfrac{2}{3}$

5 $\dfrac{1}{2}(1+3^{11})=88574$

6 $\dfrac{ab^4(1-b^2)^n}{1-b^2}$

7 $2(n+5)(n-4)$

8 1

9 2.5 or −1

10 $1, 7, 19, 37; 3n^2-3n+1$

11 $16, \dfrac{8}{3}$

12 1

13 $a=2\pm\sqrt{2}, r=\dfrac{1}{4}\left(2\pm\sqrt{2}\right)$

14 $1+18x+144x^3, 512x^9$

15 $512-6912x+41472x^2-145152x^3$

Assessment

1 a $1+15x+90x^2+270x^3+405x^4+243x^5$
　b $1-15x+90x^2-270x^3+405x^4-243x^5$
2 a $r=\pm2$ 　　　　　　　　**b** $a=\pm3$
3 a $A=-0.5, B=5.5$ 　　　　**b** 11
4 a $4, 2, 6, -2$
5 a $a=1, b=-2n, c=2n(n-1), d=\dfrac{4n}{3}(n-1)(n-2)$ 　**b** 0.88584
6 a 3 　　　　　　　　　　**b** 39
7 a i $8+12y+6y^2+y^3$ 　　**ii** $16+12y^{-4}$
　b i $16x-4x^{-3}+c$ 　　　**ii** 19.5

9 Coordinate Geometry and Circles

Exercise 1

1 a $x^2+y^2-2x-4y=4$
　b $x^2+y^2-8y+15=0$
　c $x^2+y^2+6x+14y+54=0$
　d $x^2+y^2-8x-10y+32=0$

2 a $(-4, 1); 5$
　b $\left(-\dfrac{1}{2}, -\dfrac{3}{2}\right); 3\dfrac{\sqrt{2}}{2}$
　c $(-3, 0); \sqrt{14}$
　d $\left(\dfrac{3}{4}, -\dfrac{1}{2}\right); \dfrac{\sqrt{5}}{4}$
　e $(0, 0); 2$
　f $(2, -3); 3$
　g $(1, 3); 3$
　h $\left(-1, \dfrac{1}{2}\right); \dfrac{\sqrt{69}}{6}$

3 a and **f**

4 $(-2, 0)$ and $(6, 0)$

5 a $(0, 2)$
　b $\left(\dfrac{24}{5}, \dfrac{22}{5}\right)$

Exercise 2

1 $(x-1)^2+(y-4)^2=100$ 　**2** $x^2+y^2-12x-4y+4=0$ 　**3** $\begin{pmatrix}-3\\0\end{pmatrix}$

Exercise 3

1 a $(x-5)^2+(y-3)^2=25$ 　**2** $(4, 4)$
3 $a^2+b^2-9a-b+14=0$
4 a $\left(-\dfrac{115}{18}, -\dfrac{29}{6}\right)$ 　　**b** 10.8 units to 3 s.f.

Exercise 4

1 a $2x+y=0$ 　　**b** $\left(-\dfrac{4}{5}, \dfrac{8}{5}\right)$ 　**2** $x+y-14=0$
3 a $3y+4x=23$ 　**b** $4y=x+22$ 　**c** $3y=2x+25$
4 $x+2y=0$ 　　**5** $3y+x=11$ 　　**6** $y=2x-7$

Review

1 $a=b=1, f=g=0, c=-9$
2 a not a circle, the coefficients of x^2 and y^2 are not equal
　b not a circle, the coefficients of x^2 and y^2 are not equal
　c a circle, centre $(-1, 1)$ radius 1
　d not a circle, the radius is not a real number
3 a $(1, 2)$ 　　　　　**b** $(x-1)^2+(y-2)^2=100$
4 a $(2, 4), 2\sqrt{5}$ 　　**5 a** $\sqrt{13}, (-1, 3)$
6 a $(x-4)^2+(y+3)^2=25$ 　**b** $(x-4)^2+(y-3)^2=8$
　c $(x-2)^2+(y-7)^2=29$ 　**d** $x^2+(y-3)^2=9$
7 a $(4-\sqrt{21}, 0)$ and $(4+\sqrt{21}, 0)$ 　**b** $2y=\sqrt{21}(x-4)+21$
8 $3x+2y-3=0$

Assessment

1 a $(x-3)^2+(y+1)^2=100$ 　　**b** $\sqrt{91}-1, -\sqrt{91}-1$
　c $3x+2y-7=0$
2 a $(1, 3), \sqrt{17}$ 　　　　　　**b** $7, -1$
3 a $(x-5)^2+(y-3)^2=25$ 　　**c** $3x-4y+35=0$
4 a $(2, -3), 5$ 　　　　　　　**b** $\dfrac{7}{6}$ 　　**c** $-\dfrac{6}{7}$
5 a $\left(3, \dfrac{17}{2}\right)$ 　　　　　　　**b** $\dfrac{1}{2}\sqrt{53}$
　c $(x-3)^2+\left(y-\dfrac{17}{2}\right)^2=\dfrac{53}{4}$ 　**d** $2x+7y-92=0$
6 a $(x+1)^2+(y-3)^2=50$ 　**b i** C(−1, 3) 　**ii** $5\sqrt{2}$
　c −2, 8 　　　　　　　　　　**d** 7
7 a $(x-3)^2+(y+8)^2=100$ 　**b** $(-3, 9)$
　c $2x+2y+34=0$ 　　　　　**d ii** $-3\pm\sqrt{11}$

10 Trigonometry

Exercise 1

1 $\sin A=\dfrac{12}{13}, \cos A=\dfrac{5}{13}$ 　**2** $\tan X=\dfrac{3}{4}, \sin X=\dfrac{3}{5}$
3 $\cos P=\dfrac{9}{41}, \tan P=\dfrac{40}{9}$ 　**4** $\sin A=\dfrac{1}{\sqrt{2}}=\cos A$
5 $\sin Y=\dfrac{1}{3}\sqrt{5}, \tan Y=\dfrac{1}{2}\sqrt{5}$ 　**6** $\cos A=\dfrac{1}{2}\sqrt{3}; 30°$
7 $\cos X=\dfrac{24}{25}$ 　**8** 80° or 100° 　**9** 105°

10 52° 　　　　**11** 150° 　　　**12** 99°
13 57° 　　　　**14** 90° 　　　　**15** 89°
16 $\dfrac{5}{13}$ 　　　**17** 53° or 127° 　**18** 150°
19 a yes, 90° 　　**b** yes, 0 　　　**c** no
20 $A+B=180°$ 　**21** $\dfrac{\sqrt{2}}{2}$ 　　　**22** 45°, 135°

Exercise 2

1 11.1 cm	**2** 10.2 cm	**3** 156 cm	
4 113 cm	**5** 7.01 cm	**6** 141 cm	
7 16.3 cm	**8** no; you do not know any angles in the triangle		
9 18°	**10** 58° or 122°	**11** 17°	
12 35°	**13** 57° or 123°	**14** 30°	

Exercise 3

1 5.29 cm	**2** 12.9 cm	**3** 53.9 cm	**4** 4.04 cm
5 101 cm	**6** 12.0 cm	**7** 64.0 cm	**8** 31.8 cm
9 38°	**10** 55°	**11** 45°	**12** 94°
13 a 18°	**b** 126°	**14** 29°	
15 a 11.4 cm	**b** 68°		

Exercise 4

1 $b = 87.4$ cm   **2** $B = 30.4°; c = 23.8$ cm
3 $c = 17.5$ cm   **4** $B = 81.0°; a = 112$ cm
5 $a = 164$ cm; $c = 272$ cm   **6** $B = 34°; a = 37.0$ cm
7 40.5°, 53.0°, 86.5°

Exercise 5

1 12 300 cm²	**2** 2 190 cm²	**3** 1 680 cm²
4 453 square units	**5** 42.9 square units	**6** 51.0°, 21.0 cm²
7 10.6 cm, 59.8 cm²	**8** 52.4°, 151 cm	**9** 5.25 cm
10 $h = c \sin A$		

Review

1 a 116°   **b** 86°   **2** $-\dfrac{24}{25}$

3 a $\dfrac{5\sqrt{39}}{39}$ **b** $-\dfrac{5\sqrt{39}}{39}$

4 a $\dfrac{\sqrt{2}}{2}, -\dfrac{\sqrt{2}}{2}$ **b** $\dfrac{3\sqrt{13}}{13}, -\dfrac{2\sqrt{13}}{13}$ **5** $-\dfrac{5}{13}$

6 9.05 cm   **7** 4.82   **8** 83.3°

9 54.1°, 125.9°   **10** 22°, 50°, 108°   **12** 75.8 cm²

13 a $\dfrac{1}{2}$

     b yes, the triangles where two adjacent sides are each 6 cm and the included angle is either 30° or 150° both have the same area of 9 cm²

Assessment

1 a 98°	**b** 19.8 cm²	**c** 3.96 cm
2 a 6.4 cm	**b** 15.9 cm	**c** 11.6°

3 a $\left(\dfrac{5}{3}, \dfrac{5}{3}\right), \left(\dfrac{5}{2}, 0\right)$ **b** 45°   **c** $\dfrac{25}{12}$ square units

4 a 364 m	**b** 201 m	**c** 477 m
5 a 889 m	**b** 196 000 m²	**c** 785 m
6 a 15.6 cm	**b** 13.7 cm	
7 b 25.16 m	**c** 11.5°	

11 Trigonometric Functions and Equations

Exercise 1

1 a $\dfrac{\pi}{4}$ **b** $\dfrac{5\pi}{6}$ **c** $\dfrac{\pi}{6}$ **d** $\dfrac{\pi}{2}$
   e $\dfrac{3\pi}{2}$ **f** $\dfrac{2\pi}{3}$ **g** $\dfrac{\pi}{3}$ **h** $\dfrac{\pi}{8}$
   i $\dfrac{4\pi}{3}$ **j** $\dfrac{5\pi}{3}$ **k** $\dfrac{7\pi}{4}$ **l** $\dfrac{3\pi}{4}$
   m $\dfrac{7\pi}{6}$ **n** $\dfrac{5\pi}{4}$

2 a 30°	**b** 180°	**c** 18°	**d** 60°
e 150°	**f** 15°	**g** 210°	**h** 135°
i 20°	**j** 270°	**k** 80°	**l** 45°
m 108°	**n** 22.5°		
3 a 0.61	**b** 0.82	**c** 1.62	**d** 4.07
e 0.25	**f** 2.04	**g** 6.46	
4 a 97.4°	**b** 190.2°	**c** 57.3°	**d** 119.7°
e 286.5°	**f** 360.0°		
5 a 0.8660	**b** 0.5	**c** 0	**d** 0.5
e 1	**f** 0	**g** 1	**h** −1
i −1	**j** −0.5		

6 a 0, 2π **b** $\dfrac{\pi}{4}, \dfrac{5\pi}{4}$ **c** $\dfrac{\pi}{6}, \dfrac{5\pi}{6}$ **d** $\dfrac{\pi}{3}, \dfrac{5\pi}{3}$
   e $\dfrac{3\pi}{2}$ **f** π **g** $\dfrac{\pi}{2}$ **h** $\dfrac{3\pi}{4}, \dfrac{7\pi}{4}$
   i $\dfrac{7\pi}{6}, \dfrac{11\pi}{6}$ **j** 0, π, 2π **k** $\dfrac{\pi}{2}, \dfrac{3\pi}{2}$ **l** $\dfrac{5\pi}{4}, \dfrac{7\pi}{4}$

7 a 0.932	**b** 0.939	**c** 9.89	**d** −0.801
8 a 0.284	**b** 0.929	**c** 0.644	**d** 0.0226

Exercise 2

1 $\dfrac{2\pi}{3}$ cm   **2** $\dfrac{25\pi}{2}$ cm   **3** 2.4 rad   **4** 0.692 rad   **5** $\dfrac{15}{\pi}$ cm

6 $\dfrac{25}{\pi}$ cm   **7** 4π cm   **8** $\dfrac{60}{\pi}$ cm   **9** $\dfrac{360°}{\pi}$ **10** 146.4°

Exercise 3

1 4.19 cm²   **2** 75.4 cm²   **3** $\dfrac{\pi}{2}$ **4** 0.96 rad

5 $\dfrac{125\pi}{3}$ cm²   **6** $\dfrac{15}{\pi}$ cm, $\dfrac{225}{2\pi}$ cm²   **7** 6 cm   **8** 23.1 cm

9 8 cm   **10** 0.288 rad   **11 a** 12 cm²   **b** 23.2 cm²
12 14.5 mm², 139 mm²   **13 a** 15.2 cm   **b** 32.5 cm²

Exercise 4

1 $\dfrac{1}{2}\sqrt{3}$ **2** 0   **3** $-\dfrac{1}{2}\sqrt{3}$ **4** $\dfrac{1}{2}$

5 $\dfrac{\pi}{2}, \dfrac{5\pi}{2}, \dfrac{9\pi}{2}$ **6** $-\dfrac{1}{2}\pi, -\dfrac{5}{2}\pi$ **7** $\sin 55°$ **8** $-\sin 70°$

9 $-\sin 60°$ **10** $-\sin\dfrac{\pi}{6}$

11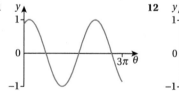

12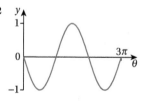

13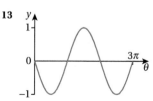

14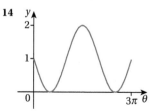

4 $\dfrac{2}{3}$

5 $\dfrac{1}{2}(1+3^{11})=88574$

6 $\dfrac{ab^4(1-b^2)^n}{1-b^2}$

7 $2(n+5)(n-4)$

8 1

9 2.5 or -1

10 $1, 7, 19, 37; 3n^2-3n+1$

11 $16, \dfrac{8}{3}$

12 1

13 $a=2\pm\sqrt{2}, r=\dfrac{1}{4}\left(2\pm\sqrt{2}\right)$

14 $1+18x+144x^3, 512x^9$

15 $512-6912x+41\,472x^2-145\,152x^3$

1 a $1+15x+90x^2+270x^3+405x^4+243x^5$
   b $1-15x+90x^2-270x^3+405x^4-243x^5$

2 a $r=\pm2$ **b** $a=\pm3$

3 a $A=-0.5, B=5.5$ **b** 11

4 a $4, 2, 6, -2$

5 a $a=1, b=-2n, c=2n(n-1), d=\dfrac{4n}{3}(n-1)(n-2)$ **b** 0.88584

6 a 3   **b** 39

7 a i $8+12y+6y^2+y^3$ **ii** $16+12y^{-4}$
   b i $16x-4x^{-3}+c$ **ii** 19.5

9 Coordinate Geometry and Circles

Exercise 1

1 a $x^2+y^2-2x-4y=4$ **b** $x^2+y^2-8y+15=0$
   c $x^2+y^2+6x+14y+54=0$ **d** $x^2+y^2-8x-10y+32=0$

2 a $(-4,1); 5$
   b $\left(-\dfrac{1}{2}, -\dfrac{3}{2}\right); 3\dfrac{\sqrt{2}}{2}$

   c $(-3,0); \sqrt{14}$
   d $\left(\dfrac{3}{4}, -\dfrac{1}{2}\right); \dfrac{\sqrt{5}}{4}$

   e $(0,0); 2$
   f $(2,-3); 3$

   g $(1,3); 3$
   h $\left(-1, \dfrac{1}{2}\right); \dfrac{\sqrt{69}}{6}$

3 **a** and **f**

4 $(-2,0)$ and $(6,0)$

5 a $(0,2)$
   b $\left(\dfrac{24}{5}, \dfrac{22}{5}\right)$

Exercise 2

1 $(x-1)^2+(y-4)^2=100$ **2** $x^2+y^2-12x-4y+4=0$ **3** $\begin{pmatrix}-3\\0\end{pmatrix}$

Exercise 3

1 a $(x-5)^2+(y-3)^2=25$ **2** $(4,4)$

3 $a^2+b^2-9a-b+14=0$

4 a $\left(-\dfrac{115}{18}, -\dfrac{29}{6}\right)$ **b** 10.8 units to 3 s.f.

Exercise 4

1 a $2x+y=0$
   b $\left(-\dfrac{4}{5}, \dfrac{8}{5}\right)$ **2** $x+y-14=0$

3 a $3y+4x=23$
   b $4y=x+22$ **c** $3y=2x+25$

4 $x+2y=0$ **5** $3y+x=11$ **6** $y=2x-7$

Review

1 $a=b=1, f=g=0, c=-9$

2 a not a circle, the coefficients of x^2 and y^2 are not equal
   b not a circle, the coefficients of x^2 and y^2 are not equal
   c a circle, centre $(-1, 1)$ radius 1
   d not a circle, the radius is not a real number

3 a $(1,2)$ **b** $(x-1)^2+(y-2)^2=100$

4 a $(2,4), 2\sqrt{5}$ **5 a** $\sqrt{13}, (-1,3)$

6 a $(x-4)^2+(y+3)^2=25$ **b** $(x-4)^2+(y-3)^2=8$
   c $(x-2)^2+(y-7)^2=29$ **d** $x^2+(y-3)^2=9$

7 a $(4-\sqrt{21}, 0)$ and $(4+\sqrt{21}, 0)$ **b** $2y=\sqrt{21}(x-4)+21$

8 $3x+2y-3=0$

1 a $(x-3)^2+(y+1)^2=100$ **b** $\sqrt{91}-1, -\sqrt{91}-1$
   c $3x+2y-7=0$

2 a $(1,3), \sqrt{17}$ **b** $7, -1$

3 a $(x-5)^2+(y-3)^2=25$ **c** $3x-4y+35=0$

4 a $(2,-3), 5$ **b** $\dfrac{7}{6}$ **c** $-\dfrac{6}{7}$

5 a $\left(3, \dfrac{17}{2}\right)$ **b** $\dfrac{1}{2}\sqrt{53}$

   c $(x-3)^2+\left(y-\dfrac{17}{2}\right)^2=\dfrac{53}{4}$ **d** $2x+7y-92=0$

6 a $(x+1)^2+(y-3)^2=50$ **b i** C$(-1,3)$ **ii** $5\sqrt{2}$
   c $-2, 8$ **d** 7

7 a $(x-3)^2+(y+8)^2=100$ **b** $(-3,9)$
   c $2x+2y+34=0$ **d ii** $-3\pm\sqrt{11}$

10 Trigonometry

Exercise 1

1 $\sin A=\dfrac{12}{13}, \cos A=\dfrac{5}{13}$ **2** $\tan X=\dfrac{3}{4}, \sin X=\dfrac{3}{5}$

3 $\cos P=\dfrac{9}{41}, \tan P=\dfrac{40}{9}$ **4** $\sin A=\dfrac{1}{\sqrt{2}}=\cos A$

5 $\sin Y=\dfrac{1}{3}\sqrt{5}, \tan Y=\dfrac{1}{2}\sqrt{5}$ **6** $\cos A=\dfrac{1}{2}\sqrt{3}; 30°$

7 $\cos X=\dfrac{24}{25}$ **8** $80°$ or $100°$ **9** $105°$

10 $52°$ **11** $150°$ **12** $99°$
13 $57°$ **14** $90°$ **15** $89°$

16 $\dfrac{5}{13}$ **17** $53°$ or $127°$ **18** $150°$

19 a yes, $90°$ **b** yes, 0   **c** no

20 $A+B=180°$ **21** $\dfrac{\sqrt{2}}{2}$ **22** $45°, 135°$

Exercise 2

1 11.1 cm   **2** 10.2 cm   **3** 156 cm
4 113 cm   **5** 7.01 cm   **6** 141 cm
7 16.3 cm   **8** no; you do not know any angles in the triangle
9 18°   **10** 58° or 122°   **11** 17°
12 35°   **13** 57° or 123°   **14** 30°

Exercise 3

1 5.29 cm   **2** 12.9 cm   **3** 53.9 cm   **4** 4.04 cm
5 101 cm   **6** 12.0 cm   **7** 64.0 cm   **8** 31.8 cm
9 38°   **10** 55°   **11** 45°   **12** 94°
13 a 18°   **b** 126°   **14** 29°
15 a 11.4 cm   **b** 68°

Exercise 4

1 $b = 87.4$ cm   **2** $B = 30.4°$; $c = 23.8$ cm
3 $c = 17.5$ cm   **4** $B = 81.0°$; $a = 112$ cm
5 $a = 164$ cm; $c = 272$ cm   **6** $B = 34°$; $a = 37.0$ cm
7 40.5°, 53.0°, 86.5°

Exercise 5

1 12 300 cm²   **2** 2 190 cm²   **3** 1680 cm²
4 453 square units   **5** 42.9 square units   **6** 51.0°, 21.0 cm²
7 10.6 cm, 59.8 cm²   **8** 52.4°, 151 cm   **9** 5.25 cm
10 $h = c \sin A$

Review

1 a 116°   **b** 86°   **2** $-\dfrac{24}{25}$
3 a $\dfrac{5\sqrt{39}}{39}$ **b** $-\dfrac{5\sqrt{39}}{39}$
4 a $\dfrac{\sqrt{2}}{2}, -\dfrac{\sqrt{2}}{2}$ **b** $\dfrac{3\sqrt{13}}{13}, -\dfrac{2\sqrt{13}}{13}$ **5** $-\dfrac{5}{13}$
6 9.05 cm   **7** 4.82   **8** 83.3°
9 54.1°, 125.9°   **10** 22°, 50°, 108°   **12** 75.8 cm²
13 a $\dfrac{1}{2}$
   b yes, the triangles where two adjacent sides are each 6 cm and the included angle is either 30° or 150° both have the same area of 9 cm²

Assessment

1 a 98°   **b** 19.8 cm²   **c** 3.96 cm
2 a 6.4 cm   **b** 15.9 cm   **c** 11.6°
3 a $\left(\dfrac{5}{3}, \dfrac{5}{3}\right), \left(\dfrac{5}{2}, 0\right)$ **b** 45°   **c** $\dfrac{25}{12}$ square units
4 a 364 m   **b** 201 m   **c** 477 m
5 a 889 m   **b** 196 000 m²   **c** 785 m
6 a 15.6 cm   **b** 13.7 cm
7 b 25.16 m   **c** 11.5°

11 Trigonometric Functions and Equations

Exercise 1

1 a $\dfrac{\pi}{4}$ **b** $\dfrac{5\pi}{6}$ **c** $\dfrac{\pi}{6}$ **d** $\dfrac{\pi}{2}$
   e $\dfrac{3\pi}{2}$ **f** $\dfrac{2\pi}{3}$ **g** $\dfrac{\pi}{3}$ **h** $\dfrac{\pi}{8}$
   i $\dfrac{4\pi}{3}$ **j** $\dfrac{5\pi}{3}$ **k** $\dfrac{7\pi}{4}$ **l** $\dfrac{3\pi}{4}$
   m $\dfrac{7\pi}{6}$ **n** $\dfrac{5\pi}{4}$
2 a 30°   **b** 180°   **c** 18°   **d** 60°
   e 150°   **f** 15°   **g** 210°   **h** 135°
   i 20°   **j** 270°   **k** 80°   **l** 45°
   m 108°   **n** 22.5°
3 a 0.61   **b** 0.82   **c** 1.62   **d** 4.07
   e 0.25   **f** 2.04   **g** 6.46
4 a 97.4°   **b** 190.2°   **c** 57.3°   **d** 119.7°
   e 286.5°   **f** 360.0°
5 a 0.8660   **b** 0.5   **c** 0   **d** 0.5
   e 1   **f** 0   **g** 1   **h** −1
   i −1   **j** −0.5
6 a $0, 2\pi$ **b** $\dfrac{\pi}{4}, \dfrac{5\pi}{4}$ **c** $\dfrac{\pi}{6}, \dfrac{5\pi}{6}$ **d** $\dfrac{\pi}{3}, \dfrac{5\pi}{3}$
   e $\dfrac{3\pi}{2}$ **f** π **g** $\dfrac{\pi}{2}$ **h** $\dfrac{3\pi}{4}, \dfrac{7\pi}{4}$
   i $\dfrac{7\pi}{6}, \dfrac{11\pi}{6}$ **j** $0, \pi, 2\pi$ **k** $\dfrac{\pi}{2}, \dfrac{3\pi}{2}$ **l** $\dfrac{5\pi}{4}, \dfrac{7\pi}{4}$
7 a 0.932   **b** 0.939   **c** 9.89   **d** −0.801
8 a 0.284   **b** 0.929   **c** 0.644   **d** 0.0226

Exercise 2

1 $\dfrac{2\pi}{3}$ cm   **2** $\dfrac{25\pi}{2}$ cm   **3** 2.4 rad   **4** 0.692 rad   **5** $\dfrac{15}{\pi}$ cm

6 $\dfrac{25}{\pi}$ cm   **7** 4π cm   **8** $\dfrac{60}{\pi}$ cm   **9** $\dfrac{360°}{\pi}$ **10** 146.4°

Exercise 3

1 4.19 cm²   **2** 75.4 cm²   **3** $\dfrac{\pi}{2}$ **4** 0.96 rad
5 $\dfrac{125\pi}{3}$ cm²   **6** $\dfrac{15}{\pi}$ cm, $\dfrac{225}{2\pi}$ cm²   **7** 6 cm   **8** 23.1 cm
9 8 cm   **10** 0.288 rad   **11 a** 12 cm²   **b** 23.2 cm²
12 14.5 mm², 139 mm²   **13 a** 15.2 cm   **b** 32.5 cm²

Exercise 4

1 $\dfrac{1}{2}\sqrt{3}$ **2** 0   **3** $-\dfrac{1}{2}\sqrt{3}$ **4** $\dfrac{1}{2}$
5 $\dfrac{\pi}{2}, \dfrac{5\pi}{2}, \dfrac{9\pi}{2}$ **6** $-\dfrac{1}{2}\pi, -\dfrac{5}{2}\pi$ **7** $\sin 55°$ **8** $-\sin 70°$
9 $-\sin 60°$ **10** $-\sin\dfrac{\pi}{6}$

11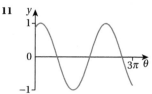

12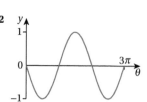

13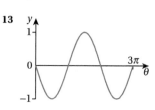

14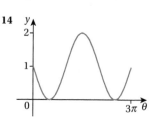

15

16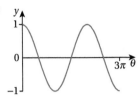

17 The curve $y = \sin 3\theta$ is a one-way stretch of the curve $y = \sin \theta$ by a factor $\frac{1}{3}$ parallel to the x-axis.

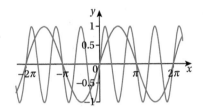

18 a 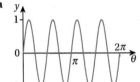 **b**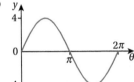

Exercise 5

1 a $-\cos 57°$ **b** $-\cos 70°$ **c** $\cos 20°$ **d** $-\cos 26°$

2 a $-\dfrac{\sqrt{3}}{2}$ **b** 0 **c** $-\dfrac{1}{\sqrt{2}}$ **d** 1

3 a 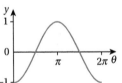 **b**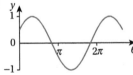

c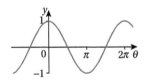

4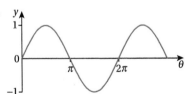

$$\sin\theta = \cos\left(\theta - \frac{1}{2}\pi\right)$$

$$\cos\theta = -\sin\left(\theta - \frac{1}{2}\pi\right)$$

5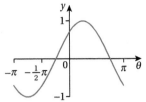

a $\theta = \dfrac{1}{4}\pi$ **b** $\theta = -\dfrac{3}{4}\pi$ **c** $\theta = -\dfrac{1}{4}\pi$ and $\dfrac{3}{4}\pi$

6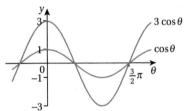

7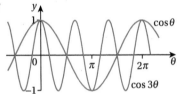

8 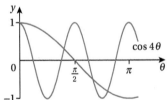 $\dfrac{\pi}{8}, \dfrac{3\pi}{8}, \dfrac{5\pi}{8}, \dfrac{7\pi}{8}$

Exercise 6

1 a 1 **b** $-\sqrt{3}$ **c** $\sqrt{3}$ **d** -1

2 a $\tan 40°$ **b** $-\tan\dfrac{2}{7}\pi$ **c** $-\tan 50°$ **d** $-\tan\dfrac{2}{5}\pi$

3 a $\dfrac{1}{4}\pi, \dfrac{5}{4}\pi$ **b** $\dfrac{3}{4}\pi, \dfrac{7}{4}\pi$ **c** $0, \pi, 2\pi$

d $\dfrac{1}{2}\pi, \dfrac{3}{2}\pi$

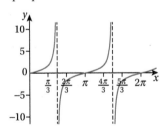

Exercise 7

1 a 0.412 rad, 2.73 rad, −5.87 rad, −3.55 rad

b $-\dfrac{4}{3}\pi, -\dfrac{2}{3}\pi, \dfrac{2}{3}\pi, \dfrac{4}{3}\pi$

c 0.876 rad, 4.02 rad, −2.27 rad, −5.41 rad

2 a 141.3°, 321.3°, 501.3°, 681.3°

b 191.5°, 348.5°, 551.5°, 708.5°

c 84.3°, 275.7°, 444.3°, 635.7°

3 a 0.644 **b** −0.644° **c** 0.464 rad

4 $0, \pi, 2\pi$ **5** 60°, 120° **6** 90°, 270°

7 120°, 300° **8** 194.5°, 345.5° **9** 120°, 240°

10 45°, 225° **11** 30°, 150°, 210°, 330°

12 $-\pi, \dfrac{-2\pi}{3}, 0, \dfrac{2\pi}{3}, \pi$ **13** ±0.723 rad   **14** $-\pi, -\dfrac{1}{3}\pi, \dfrac{1}{3}\pi, \pi$

15 $-\pi, -\dfrac{2}{3}\pi, 0, \dfrac{2}{3}\pi, \pi$ **16** $-\pi, -\dfrac{1}{6}\pi, 0, \dfrac{1}{6}\pi, \pi$

Exercise 8

1 $75°$ **2** $165°$ **3** none

4 $135°$ **5** none   **6** $60°$

7 $\dfrac{\pi}{3}$ **8** $\dfrac{2\pi}{3}$ **9** $\dfrac{\pi}{4}$

10 $\dfrac{\pi}{12}$ **11** $\dfrac{\pi}{12}$

Review

1 e   **2** d   **3** d

4 $\sin\beta = \pm\dfrac{\sqrt{3}}{2}, \tan\beta = \pm\sqrt{3}$ **5** $\dfrac{7\pi}{12}, -\dfrac{5\pi}{12}$

6 $\pm70.5°, \pm180°, 0°$ **7** 30 **8** $-90°, 30°, 150°$

Assessment

1 a $30°, 150°$ **c** $76°, 108°, 256°, 288°$

2 a 8.38 cm   **b** 16.8 cm² **c** 8 cm **d** 11.0 cm²

3 b $\dfrac{1}{2}r(20-2r)$ **4 a** $120°$ **b** $0, \pi$

5 a $\sin x = 1, -\dfrac{4}{5}$ **b** $90°$ **6 a** 0.8 **b** 10 cm²

7 b $25°, 65°, 115°, 155°$

12 Exponentials and Logarithms

Exercise 1

1 a $16, 8, 4, 2, 1, \dfrac{1}{2}, \dfrac{1}{4}, \dfrac{1}{8}, \dfrac{1}{16}$

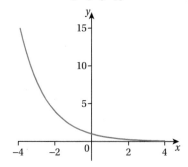

b $y = \left(\dfrac{1}{2}\right)^x$ is the reflection of $y = 2^x$ in the y-axis.

2 a
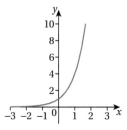

c
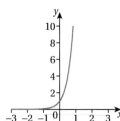

Exercise 2

1 $\log_{10} 1000 = 3$ **2** $\log_2 16 = 4$ **3** $\log_{10} 10\,000 = 4$

4 $\log_3 9 = 2$ **5** $\log_4 16 = 2$ **6** $\log_5 25 = 2$

7 $\log_{10} 0.01 = -2$ **8** $\log_9 3 = \dfrac{1}{2}$ **9** $\log_5 1 = 0$

10 $\log_4 2 = \dfrac{1}{2}$ **11** $\log_{12} 1 = 0$ **12** $\log_8 2 = \dfrac{1}{3}$

13 $\log_q p = 2$ **14** $\log_x 2 = y$ **15** $\log_p r = q$

16 $10^5 = 100\,000$ **17** $4^3 = 64$ **18** $10^1 = 10$

19 $2^2 = 4$ **20** $2^5 = 32$ **21** $10^3 = 1000$

22 $5^0 = 1$ **23** $3^2 = 9$ **24** $4^2 = 16$

25 $3^3 = 27$ **26** $36^{\frac{1}{2}} = 6$ **27** $a^0 = 1$

28 $x^z = y$ **29** $a^b = 5$ **30** $p^r = q$

31 2 **32** 6 **33** 6

34 4 **35** 2 **36** 3

37 $\dfrac{1}{2}$ **38** -2 **39** -1

40 $\dfrac{1}{2}$ **41** 0 **42** 1

43 $\dfrac{1}{3}$ **44** 0 **45** $\dfrac{1}{3}$ **46** 3

47 a 0.477 **b** 0.380 **c** -0.697 **d** 1.24

e 0.748 **f** 2.40

Exercise 3

1 $\log p + \log q$ **2** $\log p + \log q + \log r$ **3** $\log p - \log q$

4 $\log p + \log q - \log r$ **5** $\log p - \log q - \log r$ **6** $2\log p + \log q$

7 $\log q - 2\log r$ **8** $\log p + \dfrac{1}{2}\log q$

9 $2\log p + 3\log q - \log r$ **10** $\dfrac{1}{2}\log q - \dfrac{1}{2}\log r$

11 $n\log q$ **12** $n\log p + m\log q$

13 $\log pq$ **14** $\log p^2 q$ **15** $\log \dfrac{q}{r}$

16 $\log q^3 p^4$ **17** $\log \dfrac{p^n}{q}$ **18** $\log \dfrac{pq^2}{r^3}$

19 $\log 5 + \log x$ **20** $\log 5 + 2\log x$

21 $\log 3 + \log(x+1)$ **22** $\log x - \log(x+1)$

23 $\log 2 + \log x - \log(x-1)$ **24** $\log x + 2\log y$

25 $\log x + \log(x+4)$ **26** $\log(x+1) + \log(x-1)$

27 $2\log x + \log(x+y)$ **28** $2\log a + \log x + \log(x-b)$

29 $\log 2x$ **30** $\log\left(\dfrac{3}{x}\right)$ **31** $\log\left(\dfrac{x^2}{4}\right)$

32 $\log\left(\dfrac{x}{(1-x)^2}\right)$ **33** $\log\left(\dfrac{x^2}{\sqrt{x-1}}\right)$ **34** $\log\left(\dfrac{x^2}{y^3}\right)$

Exercise 4

1 2 **2** -2 **3** $\dfrac{3}{2}$ **4** 1.63

5 1.16 **6** 0.861 **7** 2.77 **8** $\dfrac{1}{4}$

9 1   10 16   11 1, 4   12 $\dfrac{3}{2}$

13 $\log_x\left(\dfrac{5}{9}\right), \dfrac{1}{3}\sqrt{5}$ 14 1   15 $\log_3\left(\dfrac{y}{x^2}\right), y=3x^2$ 16 $y^2, \dfrac{1}{2}$

Review

1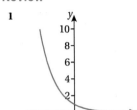

2 **a** 7   **b** $\dfrac{1}{2}$
   c 0   **d** 5
3 **a** $3\log a - \log b - 2\log c$
   b $n\log a - \log b$
   c $\log a + \log b - \log c$
   d $\log a + \dfrac{1}{2}\log(1+b)$
4 **a** $\log\dfrac{a^3}{b}$ **b** $-\log a$

5 **a** $\log\dfrac{x}{y}$ **b** $\log 100\,(x+1)$ **c** $\log Ax$ **d** $\log y$
6 3   7 2   8 2.10   9 2

Assessment

1 **a** 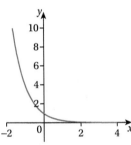   **c** 0.132

2 **a** $\log\left(\dfrac{5\sqrt{x}}{4}\right)$ **b** 6.4×10^9 3 **a** $\log_3\dfrac{3x}{2}$ **b** $y=\dfrac{x}{6}$

4 **a** $\log\dfrac{x}{y^3}$ **b** $x=y^3$ 5 **b** 1

6 **a** 1.39   **b** 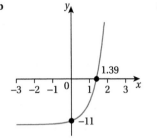   **c** $-\dfrac{5}{12}$

7 **a** 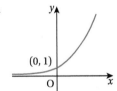   **b** translation by $\begin{bmatrix} 0 \\ -5 \end{bmatrix}$
   (0, 1)   **c ii** 2.332

8 **a** $b=a^c$

13 Probability

Exercise 1

1 **a** $\dfrac{1}{2}$ **b** 1   **c** $\dfrac{2}{3}$ **d** $\dfrac{1}{2}$ **e** $\dfrac{5}{6}$
2 **a** $\dfrac{3}{8}$ **b** $\dfrac{5}{8}$ **c** 0   **d** $\dfrac{4}{5}$
3 **a** 0.3   **b** 0.75
4 **a i** $\dfrac{1}{13}$ **ii** $\dfrac{1}{2}$ **iii** $\dfrac{3}{52}$ **b** $\dfrac{7}{25}$
5 **a i** 0.4   **ii** 0.5   **b** 0.25
6 **a** $\dfrac{2}{7}$ **b** $\dfrac{3}{7}$ 7 **a** 0.64   **b** 0.68
8 **a** $\dfrac{5}{7}$ **b** $\dfrac{41}{140}$ **c** $\dfrac{41}{140}$ **d** $\dfrac{3}{5}$
9 **a** $\dfrac{1}{2}$ **b** $\dfrac{3}{4}$ 10 **a** $\dfrac{1}{4}$ **b** $\dfrac{3}{4}$ **c** $\dfrac{3}{8}$
11 **a i** $\dfrac{1}{36}$ **ii** $\dfrac{1}{12}$ **iii** 0   **iv** 0   **b** 6, 12

Exercise 2

1 **a** $\dfrac{1}{2}$ **b** $\dfrac{1}{2}$ **c** $\dfrac{5}{6}$ **d** $\dfrac{1}{6}$
2 **a** $\dfrac{11}{30}$ **b** $\dfrac{9}{30}$ **c** $\dfrac{19}{30}$
3 **a** $\dfrac{4}{17}$ **b** $\dfrac{4}{51}$ **c** $\dfrac{5}{17}$ **d** $\dfrac{5}{17}$ **e** 0

4 **a** 0.41   **b** 0.005   **c** 0.98
5 **a** 0.35   **b** 0.55   **c** 0.15   **d** 0.75
6 **a** $\dfrac{7}{36}$ **b** $\dfrac{1}{6}$ **c** $\dfrac{5}{18}$ **d** $\dfrac{1}{12}$
7 **a** 0.2   **b** 0.7   **c** 0.3   **d** 0.5
8 **a** 0.8   **b** 0.2   9 **a** 0.6   **b** 0.3
10 **a** 0.4   **b** 0.9   **c** 0.5   11 **a** 0.75   **b** 0
12 **a** 0.4   **b** 0.35   **c** 0.4   **d** 0
13 **a** No, $P(A \text{ and } B)\neq 0$ **b** Yes; $P(A \text{ and } C)=0$
    c No, $P(B \text{ and } C)\neq 0$

Exercise 3

1 **a** $\dfrac{1}{3}$ **b** 0   2 **a** 0.05   **b** 0.5   **c** 0.0625
3 **a** $\dfrac{5}{8}$ **b** $\dfrac{3}{13}$ **c** $\dfrac{1}{20}$ 4 **a** $\dfrac{3}{7}$ **b** $\dfrac{3}{8}$
5 **a** $\dfrac{7}{41}$ **b** $\dfrac{43}{82}$ **c** $\dfrac{13}{25}$ **d** $\dfrac{7}{20}$ **e** $\dfrac{118}{241}$
6 **a** 0.09   **b** 0.3   **c** 0.66
7 **a** 0.28   **b** 0.56   **c** 0.98   **d** 0.02
8 **a** $\dfrac{9}{38}$ **b** $\dfrac{21}{380}$ **c** $\dfrac{10}{19}$ **d** $\dfrac{39}{95}$
9 **a** $\dfrac{1}{2704}$ **b** $\dfrac{1}{16}$ **c** $\dfrac{25}{169}$
10 **a i** 0.821   **ii** 0.480   **b** 0.282

Exercise 4

1 **a** **i** $\dfrac{1}{36}$ **ii** $\dfrac{5}{18}$ **iii** $\dfrac{11}{36}$ **iv** $\dfrac{1}{9}$ **b** $\dfrac{1}{8}$

2 **a** 0.000625   **b** 0.04875   **3** **a** $\dfrac{3}{4}$ **b** $\dfrac{15}{16}$ **c** $\dfrac{37}{64}$

4 **a** 0.06   **b** 0.09

5 **a** **i** 0.36   **ii** 0.06   **iii** 0.81
   b **i** 0.162   **ii** 0.108   **c** 0.07776

Exercise 5

1 **a** $\dfrac{5}{14}$ **b** $\dfrac{1}{3}$ **c** $\dfrac{17}{42}$

2 **a** $\dfrac{15}{38}$ **b** $\dfrac{1}{2}$ **3** **a** 0.65   **b** $\dfrac{56}{65}$

4 **a** **i** $\dfrac{8}{27}$ **ii** $\dfrac{19}{27}$ **b** **i** $\dfrac{5}{21}$ **ii** $\dfrac{16}{21}$

5 **a** $\dfrac{1}{3}$ **b** $\dfrac{3}{11}$ **6** **a** $\dfrac{1}{4}$ **b** **i** $\dfrac{1}{16}$ **ii** $\dfrac{3}{8}$

   c **i** $\dfrac{27}{64}$ **ii** $\dfrac{9}{64}$ **iii** $\dfrac{5}{32}$ **iv** $\dfrac{27}{32}$ **d** $\dfrac{1}{256}$

7 **a** $\dfrac{3}{10}$ **b** $\dfrac{1}{3}$ **8** **a** **i** 0.36   **ii** 0.48   **b** 0.01024

9 **a** 0.364   **b** 0.086   **c** $\dfrac{18}{43}$

10 **a** **i** 0.28   **ii** 0.54   **b** $\dfrac{47}{110}$

Exercise 6

1 **a** No, $P(B) \neq P(B \mid A)$ **b** 0.66

2 **a** 0.1   **b** $P(A) \times P(B) \neq P(A \text{ and } B)$ **c** $\dfrac{2}{7}$

3 **a**

	A	A'	
B	0.21	0.09	0.3
B'	0.49	0.21	0.7
	0.7	0.3	1

   b **i** Yes, $P(A) \times P(B) = P(A \text{ and } B)$
     ii No, $P(A \text{ and } B) \neq 0$
     iii Yes, $P(A') \times P(B) = P(A' \text{ and } B)$
4 No, $P(X' \text{ and } Y') \neq 0$

5 **a**

	C	C'	Total
Full-time teacher	45	25	70
Part-time teacher	12	18	30
Total	57	43	100

   b **i** 0.12   **ii** 0.25   **iii** 0.82   **iv** $\dfrac{12}{57}$

   c No, $P(C) \times P(F) \neq P(C \text{ and } F)$ **d** F and P; C and C'

Review

1 **a** $\dfrac{7}{30}$ **b** $\dfrac{1}{6}$ **c** $\dfrac{17}{30}$

2 **a** 0.85   **b** 0.37   **c** 0.12
3 **a** no heads   **b** fewer than 2 heads
4 **a** 0.375   **b** 0.16
5 **a** **i** 0.375   **ii** 0.5   **b** $\dfrac{37}{64}$

6 **a** 0.75   **b** 0.35   **c** $\dfrac{3}{7}$

7 0.591   **8** **a** $\dfrac{14}{23}$ **b** 0.226

9 **a** 0.976   **b** 0.385   **10** 0.710
11 **a** **i** $\dfrac{5}{8}$ **ii** $\dfrac{4}{5}$ **iii** $\dfrac{37}{80}$ **iv** $\dfrac{3}{10}$ **b** 0.0642
12 **a** **i** 0.4   **ii** 0.35   **iii** 0.65   **b** 0.61
13 **a** **i** 0.45   **ii** 0.47   **iii** 0.77   **b** 0.16875
14   **i** is more likely since $p = 0.518$; for **ii** $p = 0.491$

Assessment

1 **a**

	J	J'	Total
W	0.55	0.1	0.65
W'	0.15	0.2	0.35
Total	0.7	0.30	1

   b 0.25

   c **i** Reasons include $P(W \text{ and } J) = 0.55 \neq 0$,
     ii Reasons include $P(W) \times P(J) = 0.455 \neq P(W \text{ and } J)$

2 **a** $p - 0.3$ **b** $\dfrac{0.2}{p - 0.3}$ **c** 0.7

3 **a** **i** 0.024   **ii** 0.336   **iii** 0.084   **b** **i** 0.18   **ii** 0.22
4 **a** **i** 0.065   **ii** 0.17   **iii** 0.298   **b** **i** 0.655   **ii** 0.345
5 **a** 0.096   **b** 0.188   **c** 0.976
6 **a** **i** 0.72   **ii** 0.12   **b** **i** 0.504   **ii** 0.328

14 Discrete Random Variables

Exercise 1

1 **a** 61   **b** 52   **c** 73   **d** 21
2 **i** **a** 28   **b** 207   **c** 22
   ii **a** 2562   **b** 2104.5   **c** 765.5
3 **a** 3.05, 3.45   **b** 3.12   **c** 3.321   **d** 50%
4 **a** **i** 5, 2   **ii** 8.5, 1.80   **iii** 18.8, 6.46
   b 16.8, 6.46   **5** 11.52, 0.827
6 **a** 69   **b** 69.3   **c** 1.7
7 **a** 0.6   **b** 0.24
8 **a** **i** 63.87   **ii** 29.47   **b** 133, 144
9 1.014, 0.0102   **10** 29, 2.429   **11** 5.84, 203.7
12 5.099   **13** **a** 2.236   **b** 4.33
14 **a** 49.85   **b** 0.5275   **15** 3.838
16 **a** 1850   **c** 74.72, 8.234
17 **a** 61, 73, 83.6   **b** 181, 21, 42.0
   c **i** median; not affected by outlier
     ii IQR; not unduly affected by outlier
18 **a** 39.35, 12.36   **b** 42, 24
   c Reasons include

   i If mode exists it must be 58 or greater than 60, so it would not be representative
   ii The maximum value is unknown

Exercise 2

1 **a** 0.1   **b** 0.85   **c** 0.55   **d** 0.5
   e 0.45   **f** 0.15
2 **a** $\dfrac{1}{12}$ **b** $\dfrac{1}{2}$ **c** $\dfrac{1}{16}$

3 **a**

r	12	13	14
$P(R = r)$	$2k$	$3k$	$4k$

   b $\dfrac{1}{9}$ **c** $\dfrac{2}{3}$

4 **a** $\dfrac{1}{40}$ **b** $\dfrac{17}{40}$ **5** **a** 12   **b** $\dfrac{3}{11}$

6 **a**

x	−1	0	1	3	4	5
$P(X = x)$	0.1	0.1	a	a	0.3	0.1

   b 0.2   **c** 0.6

7 b

b	0	1	2
P($B=b$)	$\frac{1}{11}$	$\frac{16}{33}$	$\frac{14}{33}$

8 a ii

x	0	1	2
P($X=x$)	0.36	0.48	0.16

b ii

x	0	1	2	3
P($X=x$)	0.216	0.432	0.288	0.064

iii 0.352

Exercise 3

1 a 2.9   **b** 0.6   **2 a** 6.9   **b** 35.5
3 a 1.4   **b** 5.6   **c** 1.8

4 a

x	0	1	2
P($X=x$)	$\frac{2}{11}$	$\frac{6}{11}$	$\frac{3}{11}$

b $\frac{12}{11}$

5 a $a+b=0.5$ **b** $10a+30b=7$
   c $a=0.4, b=0.1$ **d** 16

6

x	10	20
P($X=x$)	0.4	0.6

7 a $\frac{1}{32}$ **b** 1   **c** $\frac{63}{32}$

8 a 4.6   **b** $\frac{13}{30}$ **c** 31   **d** 2
9 a 3   **b** 9.5   **c** 27.5   **d** 0.5
10 $\frac{1}{4}$ **11 a** 2.25, 1.1875   **b ii** 33

Exercise 4

1 a 5.8   **b** 3.36   **2 a** 4.2   **b** $7\frac{1}{3}$
3 a 0.1   **b** 0.34   **c** 0.982
4 a $3, 5, 7, 9, 11, 13; \frac{1}{9}, \frac{1}{9}, \frac{2}{9}, \frac{2}{9}, \frac{2}{9}, \frac{1}{9}$ **b** $8\frac{1}{3}$ **c** 2.98
5 a 0.2   **b** 8   **c** 11.6
6 b $\frac{49}{99}$ **c** $3\frac{58}{99}$ **d** 1.23   **7** 5.85

8 a $1\frac{5}{6}$ **b i** $4\frac{5}{36}$ **ii** $66\frac{2}{9}$ **iii** $16\frac{5}{9}$
9 a 121, 144   **b** 58, 36   **c** 6, 1   **d** $-80, 64$
10 2.5   **11 a** $a=1.5, b=-10$ **b** 65
12 a $\frac{1}{2}$ **c** $\frac{1}{30}$

Exercise 5

1 a 64   **b** 155   **c** 7.68
2 a i 2.9, 0.89   **ii** 0.6, 0.44   **b** 8.8, 3.82
   c i 3   **ii** 2.2   **3 a** 6.5, 4.5   **b** 57
4 11, 2.5   **5** 100, 4.47   **6 a** 80, 12   **b** 80, 6
7 a 72, 24   **b** 72, 9.80

Review

1 a 26, 16
   b i No unique value
     ii Two values are unknown, so s.d. cannot be calculated
   c 26, 9.42

2 a

Coin		Die			
		1	2	3	6
H		2	4	6	12
T		1	2	3	6

c

s	1	2	3	4	6	12
P($S=s$)	$\frac{1}{8}$	$\frac{1}{4}$	$\frac{1}{8}$	$\frac{1}{8}$	$\frac{1}{4}$	$\frac{1}{8}$

e 11

3 a 0.1   **b** 1   **c** 1.75   **d** 0.5   **e** 1.2875
4 a 1.47, 1.14   **b** 0.19   **5** 2.56
6 a 0.1   **b** 2, 1   **c** 0.4
7 a $3a+b=0.9, 5a+2b=1.6, a=0.2, b=0.3$ **b** 1.61
   c 15.2   **d** 40.25
8 a 1.3275   **b i** $T=c+nX$ **ii** $c+1.85n, 1.3275n^2$
9 a 0.25   **b** 10   **10 a** 8   **b** 32

Assessment

1 a 70.4, 2.03   **b** 10.4, 2.03   **2 a** 50, 10.13   **b** 750, 101
3 a 0.09   **b i** $a=0.3, b=0.1$ **iii** 5.2, 2.27
4 a ii 2.35   **iv** 0.417   **b** 185, 124
5 a 0.5   **c** 1.875, 0.324
6 a i 25, 10   **ii** 255   **b** 0.8844
7 a 60   **b** 148   **c** 100, 9.49   **8** 9, 3

15 Bernoulli and Binomial Distributions

Exercise 1

1 0.36, 0.48   **2** $\frac{7}{12}$, 0.493   **3 a** 0.2   **b** 0.499
4 a $\frac{1}{3}$ **b** $\frac{2}{9}$ **5** 0.2, 0.8

Exercise 2

1 a 0.233   **b** 0.0368   **c** 0.00000590   **d** 0.0282
2 a 0.0231   **b** 0.208   **c** 0.886   **d** 0.000381
3 a 0.102   **b** 0.143   **c** 0.000965
4 a 0.583   **b** 0.157   **5 a** 0.1827   **b** 0.414
6 a 0.819   **b** 0.997   **7 a** 0.919   **b** 0.319
8 a 0.193   **b** 0.00432   **c** 0.064
9 a 0.290   **b** 0.0188   **c** 0.159   **d** 0.745
10 a 0.0425   **b** 0.167
11 a 0.3125   **b** 0.5   **c** 0.1875   **d** 0.78125
12 a 0.0638   **b** 0.465   **c** 0.267

13

x	0	1	2	3	4
P($X=x$)	q^4	$4p^1q^3$	$6p^2q^2$	$4p^3q^1$	p^4

Exercise 3

1 a 0.8499   **b** 0.2793   **c** 0.0175   **d** 0.5141   **e** 0.0918
2 a 0.9493   **b** 0.8730   **c** 0.8747   **d** 0.7984
3 a 0.3668   **b** 0.0650   **c** 0.1734   **d** 0.5332   **e** 0.9425
4 a 0.9845   **b** 0.7757   **c** 0.2063
5 Random selection, 0.1827
6 a 0.3222   **b** 0.3020   **c** 0.2682   **7 a** 0.1172   **b** 0.0003
8 a 0.0081   **b** 0.9470   **c** 0.2669

Exercise 4

1 a 0.183   **b** 0.002   **c** 0.2001
2 a 0.0789   **b** 0.8411   **c** 0.0251   **d** 0.1224
3 0.1121   **4** p is constant for each shot; 0.766
5 a 0.914   **b** 0.974   **c** 0.0048
6 a 0.6678   **b i** 0.1287   **ii** 0.7608   **iii** 0.8162   **iv** 0.1723
7 a i 0.2173   **ii** 0.0950   **b** 0.0405
8 a 0.1926   **b** 0.8041   **c** 0.9937

1 a 5.04   **b** 3.2256   **c** 1.80
2 0.180   **3 a** 23   **b** 1.92
4 a 8   **b** 0.0467   **c** 1.30
5 a 5, 1.58   **b i** 0.0107   **ii** 0.0107   **c** 0.0215
6 a 0.25   **b** 5, 1.94   **c** 0.561
7 a 0.0081   **b** 1.6   **c** 0.84
8 a 0.3   **b** 8   **c** 0.0100
9 a 9, 0.4   **b** 0.232
10 a 0.1   **b** 0.354   **c** 6   **d** 25
11 a 0.13   **b** 2.6   **c** 2.1
12 a 0.216, 0.288   **b** 1.2, 0.72   **c** 1.2, 0.72

Review

1 a 0.0173   **b** 0.118   **2** 0.737
3 a 2.4   **b** 0.439   **c** 2.04
4 a 0.0749   **b** 9, 6.75   **5** 0.133

6 a i 0.925   **ii** 0.0011   **iii** 0.3535
   b 0.0593
   c i 5, 1.43
     ii means same, variances similar, claim appears valid
7 a 0.086   **b** 4.2   **c** 25
8 a 0.624   **b** 2.5   **c** 1.275
9 a 0.0008   **b i** 0.720   **ii** 0.186
10 a i 0.1742   **ii** 0.4325   **iii** 0.5003   **b** 4.8, 2.02

Assessment

1 a 0.247   **b** 0.144   **2 a** 4.05   **b** 1.49
3 a 0.930   **b** 0.0109   **c** 0.000729
4 a 0.309   **b** 6   **c** 0.977
5 a 0.086   **b** 6   **6 a** 0.207   **b** 0.499   **c** 22, 4.14
7 a i 0.969   **ii** 0.140   **iii** 0.891
   b $\mu = 10$, $\sigma^2 = 8$; means similar, variance much larger than expected, doubt validity of claim
8 a i 0.8725   **ii** 0.0940   **iii** 0.338   **b** 60, 52.8

16 Displacement, Speed, Velocity and Acceleration

Exercise 1

1 a, d and **f** are vectors; **b, c, e, g** and **h** are scalars
2 a i 2.3 m   **ii** 0   **iii** −1.2 m
   b i 3.4 m   **ii** 14 m   **c** 0.08 ms⁻¹

Exercise 2

1 a Correct
   b Incorrect; the direction is changing all the time so velocity not constant
   c Incorrect; do not know whether the speed is constant
2 a 1 ms⁻¹   **b** −2 ms⁻¹   **c** 2 ms⁻¹   **d** −0.5 ms⁻¹
3 −4 ms⁻²   **4 a** −2 ms⁻²   **b** −3 ms⁻²
5 a 23 ms⁻¹   **b** −7 ms⁻¹   **c** 5 ms⁻¹

Exercise 3

1 a i 24 ms⁻¹   **ii** 0 ms⁻¹   **iii** −1.6 ms⁻¹
   b i 0 ms⁻¹
     ii Ball hits wall and its direction of motion reverses
   c After 3 s

2

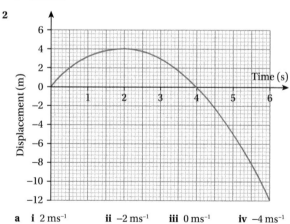

   a i 2 ms⁻¹   **ii** −2 ms⁻¹   **iii** 0 ms⁻¹   **iv** −4 ms⁻¹
   b i 2 ms⁻¹   **ii** 2 ms⁻¹   **iii** 2 ms⁻¹   **iv** 4 ms⁻¹
   c 4 ms⁻¹   **d** 2 s

3 a

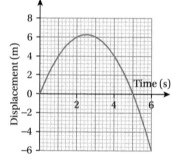

   b 2.5 s   **c** −5 ms⁻¹   **d i** −1 ms⁻¹   **ii** 3.08 ms⁻¹
4 a 1 s, 1.75 s, 2.5 s, 3 s, 3.5 s, 3.8 s
   b i 1 ms⁻¹   **ii** 0.8 ms⁻¹   **iii** 0.7 ms⁻¹
   c i −1 ms⁻¹   **ii** 0 ms⁻¹   **iii** 0 ms⁻¹   **d** −0.22 ms⁻¹

Exercise 4

1 a 3 ms⁻²   **b** 1 ms⁻²   **c i** 152 m   **ii** 400 m   **iii** 200 m; the car moves in the same direction all the time.

2

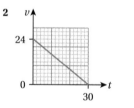

60 m

3

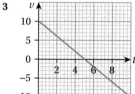

   a $t = 5$ **b** 16 m   **c** 34 m

4 a **i** Accelerates at a reducing rate to zero acceleration

    ii Constant velocity   **iii** Accelerates at varying rate

  b 185 m; less, because trapeziums have smaller area than that under the curve

5 a **i** after $8\frac{1}{2}$ s   **ii** after 20 s   **b** 6 s, 15 s after starting

  c The girl stops accelerating and begins to slow down.

6 a 3 ms^{-2} **b** v(m/s)

  c 1350 m

  d 42 m

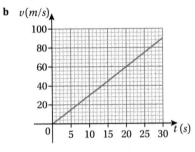

5 a

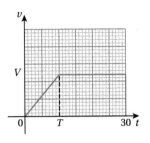

7 a

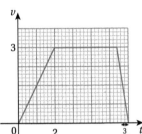

    b $\frac{3}{4}$ s   **c** 8.04 s

Assessment

1 a v ms^{-1} **b** 13.3 seconds   **c** 7.75 s

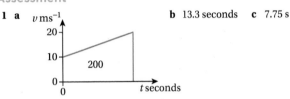

2 a 547.5 m   **b** 16.6 ms^{-1} **3 a** 400 s   **b** 1215 m

4 a 36 m   **b** 1.8 ms^{-1} **c** -0.5 ms^{-1}

5 a y **b** $t=2$ **c** 6 ms^{-1}

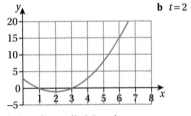

  d **i** 2 ms^{-1} **ii** 3.5 ms^{-1}

6 a 16 m, B   **b** 18 s   **7 a** 120 m   **b** 5.71 ms^{-1}

17 Motion in a Straight Line

Exercise 1

1 5   **2** -1.2 **3** 6   **4** 60

5 2   **6** -2 **7** 6   **8** 1.5

9 -54 **10** -8 **11 a** 16 m   **b** 7 m

12 b 8.22 s   **13** 97.2 m   **14** $-\frac{1}{30}$ ms^{-2}

15 a 6.5 ms^{-1} **b** -1.5 ms^{-2} **c** 8.7 s   **16** 612 m

Exercise 2

1 a 120 m   **b** 11 m/s   **2 a** 31 m   **b** 5 s

3 a 4.9 m   **b** 20 m   **c** 25 m

4 a 11.0 m   **b** 15 m/s   **5 a** 3.2 s   **b** 31 m/s

6 50 m   **7** 28 m/s

8 a 1.1 s   **b** 2.6 s; 18 m/s in each case

9 18 m   **10 a** 11 m   **b** 2.6 s

11 a 5.6 ms^{-1} **b** 0.4 m   **12** 440 m

Exercise 3

1 85 ms^{-1} **2** 50 ms^{-2} **3** 26 ms^{-2}

4 $82\frac{2}{3}$ m   **5** 17 ms^{-1}; 12 m   **6** $t=2$ and 4

7 $t=\frac{2}{3}$ and $\frac{1}{3}$ **8** $v=t^2; s=\frac{1}{3}t^3$ **9** $29\frac{1}{3}$ ms^{-1}; $47\frac{1}{3}$ m

10 a $t=\frac{2}{3}$ s, $t=2$ s   **b** $-\frac{32}{27}$ m, 0 m

11 $v=3\left(t+\frac{1}{3}\right)^2+\frac{35}{3}$ which is always positive

12 a $31\frac{1}{2}$ ms^{-1} **b** 27 m   **13 a** 2.53 ms^{-1} **b** 2.5 ms^{-1}

14 a 8 ms^{-1} **b** $5\frac{1}{3}$ m   **15** $2\left(1+\frac{1}{t^3}\right)$

Assessment

1 a 14.3 ms^{-1} **b** 1.22 s

2 a 7.67 ms^{-1} **b** 5.42 ms^{-1} **c** 1.11 s

3 a 9 ms^{-2} **b** -3.17 m   **4 a** 50 s   **b** 0.16 ms^{-2}

5 a $v=\frac{1}{2}t^2-6t+10$ **b** $t=2$ and $t=10$ **c** 42.7 m

6 a **i** 40 s   **ii** 0.2 ms^{-2}

  b **i** 0.4 ms^{-2} **ii** 40 ms^{-1} **c** 22.4 ms^{-1}

7 a **i** -2 ms^{-2} **ii** 10 seconds

18 Forces and Newton's Laws

Exercise 1

1

2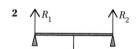

3

4

5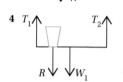

6 a   **b**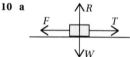

Exercise 2

1 $P = 10$

2 $P = 20, Q = 90$

3 $P = 18$

4 $P = 12, Q = 4$

5 $P = 12, Q = 26$

6 Yes, horizontal

7 Yes, vertical

8 No

9 Yes, vertical

10 a   **b i** $T > F$ **ii** $T = F$

Exercise 3

1 $1.5\,\text{ms}^{-2}$

2 28 N

3 40 kg

4 a 38 N   **b** $P = 10\,\text{N}, Q = 8\,\text{N}$
   c $P = 34\,\text{N}, Q = 30\,\text{N}$

5 a $6\,\text{ms}^{-2}, \rightarrow$ **b** $16\,\text{ms}^{-2}, \rightarrow$ **c** $\dfrac{20}{3}\,\text{ms}^{-2}, \rightarrow$

6 a $m = 4\,\text{kg}, P = 23\,\text{N}$ **b** $m = 5\,\text{kg}, P = 30\,\text{N}$
   c $m = 10\,\text{kg}, P = 40\,\text{N}$

7 $P = 8\,\text{N}, Q = 31\,\text{N}$

8 8 N

9 31.25 m

10 a $5\,\text{ms}^{-2}$ **b** 30 N

11 a $5\,\text{ms}^{-2}$ **b** 16 kg

12 100 m

13 $14\,\text{ms}^{-1}$

Exercise 4

1 a 49 N   **b** 15 kg   **c** 0.59 N

2 a 6 N   **b** 122.5 kg   **c** 0.072 N   **3** 31.2 N

4 a 28.8 N   **b** 101 N   **c** 58.8 N

5 a 5.47 kg   **b** 87.5 kg   **c** 7.14 kg   **d** 7.14 kg

6 a i 18 000 N   **ii** 6790 N   **b** 750 kg

7 10 920 N

8 a 59.2 N   **b** 19.2 N

Exercise 5

1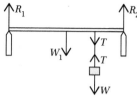

2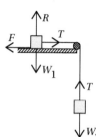

3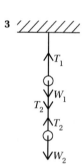

4

Exercise 6

1 a $a = \dfrac{g}{2}, T = 3g$ **b** $a = \dfrac{g}{5}, T = 12g$

   c $a = \left(\dfrac{M-m}{M+m}\right)g, T = \dfrac{2Mmg}{M+m}$

2 a $a = \dfrac{g}{3}, T = \dfrac{20g}{3}$ **b** $a = \dfrac{g}{5}, T = \dfrac{48g}{5}$

3 a $\dfrac{g}{3}$ **b** $\dfrac{25g}{6}\,\text{m}$

4 a $\dfrac{3g}{8}$ **b** $\dfrac{15g}{8}$

5 a $\dfrac{g}{2}\,\text{ms}^{-2}$ **b** $g\,\text{m}$ **c** $g\,\text{ms}^{-1}$

6 a $\dfrac{g}{9}$ **b** $\dfrac{\sqrt{2g}}{3}$

7 a $\dfrac{g}{2}$ **b** $g\,\text{N}$ **c** $\sqrt{\dfrac{g}{2}}$

8 a 130 N   **b** 600 N   **9** 1400 N

Exercise 7

1 $\dfrac{1}{3}$ **2** 6.4 N   **3** 98.7 N

4 a 40 N, no   **b** 40 N, no (just on the point of moving)
   c 40 N, yes

5 a $\dfrac{30g}{10}\,\text{ms}^{-2}$ **b** $28g\,\text{N}$

6 a $\left(\dfrac{4-g}{6}\right)\text{ms}^{-2}$ **b** $40\left(\dfrac{4-g}{6}\right)\text{N}$

Review

1 c, d, f

2 a $1\,\text{ms}^{-2}$ **b** 2 N   **3** 6.5 N

4 $0.2g\,\text{ms}^{-2}, 1.2g\,\text{N}$ **5 a** $\dfrac{g}{3}\,\text{ms}^{-2}$ **b** $\sqrt{\dfrac{g}{3}}\,\text{ms}^{-1}$

1 a 50 N   **b**   1100 N   **2 a** 3.92 ms^{-2} **b** 41.2 N

3 a

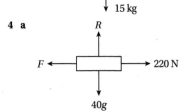

b 69.2 N

4 a

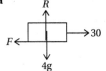

b 220 N   **c** 0.561

5 a 3.27 ms^{-2} **b** 1.96 ms^{-2}

   c decreased: before $T = \dfrac{8g}{3}$, after $T = \dfrac{12g}{5}$

6 a 3.92 ms^{-2} **b** 3.07 ms^{-1}
   c −0.785 ms^{-2} **d** 2.83 ms^{-1}

   e If the size of the block is not negligible there will be mixed friction on the block as it passes from the smooth to rough sections of the surface.

7 a

b 39.2 N

   c 11.8 N   **d** 4.56 ms^{-2}

19 Momentum and Impulse

Exercise 1

1 a 120 Ns   **b** 24 000 Ns   **c** 11 040 Ns
   d 1177×10^4 Ns   **e** 4 Ns

2 a 84 Ns   **b** 72×10^4 Ns   **c** 88 Ns

3 2 N   **4** −2 N   **5** 25 ms^{-1}

6 −2 ms^{-1} **7** $\dfrac{8}{3}$ s   **8** 3 s

9 a 26 ms^{-1} **b** −6 ms^{-1} **10** 7 N

11 32 N   **12 a** $\dfrac{236}{15}$ ms^{-1} **b** 12 ms^{-1}

13 a 31 200 Ns   **b** 31 200 Ns   **c** 15 600 N

14 12.1 Ns   **15** 10.2 Ns   **16** 12.5 Ns

17 0.72 Ns   **18** 110 Ns

Exercise 2

1 3 ms^{-1} **2** 0 ms^{-1} **3** −1.5 ms^{-1}

4 0 ms^{-1} **5** 6 kg   **6** 1.5 kg

7 3 ms^{-1} **8** 6 ms^{-1} **9** $u = 2$, $v = 7$

10 a 0.57 ms^{-1} **b** 0.73 ms^{-1} **c** 36 Ns

Review

1 b

4 a 3.8 ms^{-1}

5 a 14 ms^{-1}

6 5.4 ms^{-1}

2 a

   b 2280 Ns

   b 49 Ns

7 6.75 Ns

3 c

Assessment

1 a 8.85 ms^{-1} **b** 19.3 Ns   **c** 19.3 Ns

2 1.67 ms^{-1} **3** 0.143 kg

4 a 1.6 Ns   **b** 2.89 ms^{-1}

5 $v = \dfrac{6 - 3m}{m}$ **6 a** 6 Ns   **b** 40 kg

7 0.5 kg   **8** 7.5

Index

A

acceleration 301–2, 310
  equations of motion with constant
    acceleration 314–19, 326
  motion in a straight line variable
    acceleration 323–6, 326
addition law of probability 183–5, 213
algebraic expressions 2
  coefficients 3
angle units 148
angles 154
  angle units 148–50
  trigonometric ratios of acute angles
    134–5, 145, 154
  trigonometric ratios of obtuse angles 135–7
arcs 150–2, 162
arithmetic series 107, 120
  sum of an arithmetic series 108
  sum of the first n natural numbers 108–10
asymptotes 167
averages 222

B

base 10
Bernoulli distribution 268–9, 292–3
  Bernoulli distribution with parameter p 268
  Bernoulli trial 268
  mean and variance 269–70
binomial coefficients 117
binomial distribution 271–2, 293
  binomial probabilities and the binomial
    expansion of $(q + p) n$ 274–7
  binomial theorem 274
  conditions for a binomial distribution 272
  cumulative binomial distribution
    function 278–85
  deciding whether a binomial distribution
    is appropriate 272–4
  formula for calculating binomial
    probabilities 274–8, 293
  further applications 289–92
  mean, variance and standard deviation of
    a binomial distribution 285–9, 293
binomial theorem 115–19, 120, 274

C

calculators 169
  using a calculator in statistical mode 227–8
Cartesian coordinates 56
chords 70, 85
circles 124, 131
  area of a sector 152–3, 162
  effect of a translation on equation of a
    circle 126–7
  geometric properties of circles 127–8
  length of an arc 150–2, 162
  radians 148–50, 162
  recognising the equation of a circle 124–6
  tangents to circles 128–30
coalescing 357
coefficients 3
  binomial coefficients 117
  coefficient of friction 346–8

collinear points 60–1
combined events 182–3
common difference 107
common factors 18–19
common ratio 111
complements 178, 213
completing the square 22–3
conditional events 191–2
  conditional probability 191, 213
connected particles 342–5
conservation of linear momentum
  357–60, 360
constant of integration 90
  finding the constant of integration 91–2
constants 40
  constant of integration 90–2
  differentiating constants and multiples
    of x 74
continuous variables 222
convergent sequences 103–4
convergent series 113
coordinate geometry 56
  effect of a translation on equation of a
    circle 126–7
  equation of a circle 124–6, 131
  equation of a straight line 61–6, 67
  geometric properties of circles 127–8
  gradient of straight lines 59–61, 67
  lines joining two points 56–9, 67 56
  tangents to circles 128–30
cosine rule 140–1, 145
  general triangle calculations 143
  using the cosine rule to find an angle 141–3
cosines 134, 145
  cosine function 156–8, 162
  cosine of an obtuse angle 135
cube roots 6
cubic equations 44
cubic polynomials 34
cumulative binomial distribution function
  278–85
curved lines 70–2
  differentiation 72–5
  gradients of tangents and normals 75–8
  maximum and minimum points 80–2
  stationary points 79–80, 85
cyclic functions 155

D

definite integration 93–4
  finding area by definite integration 94–5
  finding compound areas 96–7
  meaning of a negative result 95–6
denominators 8–10
derivatives 72
differentiation 72, 85
  applications 82–5
  differentiating constants and multiples
    of x 74
  differentiating products and fractions
    74–5
  differentiating x^n 73–4
  notation 73

discrete random variables 231, 261
  comparing the distributions of $2X$ and $X_1 +$
    X_2 255–7
  $E(X)$, expectation of X 237–45, 261
  further applications 257–60
  notation 231
  probability distribution 232
  probability functions 234–7
  sum of independent observations of a
    discrete random variable 253–4
  sum of probabilities 232–4
  sum or difference of two independent
    random variables 251–2, 262
  variation and standard deviation
    of X 245–51, 262
discrete variables 222, 260
  measures of central tendency 223–4
  measures of spread 224–5, 261
discriminant 25–7
displacement 298–300
  displacement-time graphs 302–5
dividends 32
division
  division of a polynomial by $x - a$ 32–3
  factor theorem 34–5, 36
  factors of $a^3 - b^3$ and $a^3 + b^3$ 35–6
  remainder theorem 33–4, 36
divisors 32
dynamic friction 345–8
  coefficient of friction 346–8

E

equally likely outcomes 178, 212
equations
  cubic equations 44
  equation of a circle 124–7, 131
  equation of a straight line 61–6, 67
  equations containing logarithms or x as a
    power 172–3
  equations of motion with constant
    acceleration 314–19, 326
  equations of tangents and normals 77–8
  graphical interpretation of equations
    43–6
  quadratic equations 19–29
  simultaneous equations 27–9
  trigonometric equations 159–62
equilibrium 331
events 178
  combined events 182–3
  conditional events 191–2, 213
  independent events 196–200, 213
  mutually exclusive events 187–90, 213
  showing whether events are
    mutually exclusive or independent
    206–9
expansion of two brackets 3, 14
  binomial probabilities and the binomial
    expansion of $(q + p)^n$ 274–7
  difference of two squares 4
  harder expansions 5–6
  important expansions 5
  squares 4

expectation of X 237, 261
  calculating $E(X)$ 238–40
  $E(g(X))$, expectation of $g(X)$ 240–1, 261
  $E(X)$ 237
  expected mean 237
  expected value 237
  mean of a simple function of X 242–5
  mean of X 237
  practical approach 237–8
  theoretical approach 238
exponential functions 166, 173
  shape of the graph of $f(x) = a^x$ 166–7

F

factorials 117
factors
  common factors 18–19
  factor theorem 34–5, 36
  factorising quadratic equations 19–20
  factorising quadratic polynomials 16–19
  factors of $a^3 - b^3$ and $a^3 + b^3$ 35–6
finite series 105–6
forces 330
  connected particles 342–5
  drawing diagrams 331–3
  dynamic friction 345–8
  friction 331
  Newton's first law of motion 333–5, 348
  Newton's second law of motion 335–40, 348
  Newton's third law of motion 341–2, 348
  normal reaction 331, 348
  resultant force 333–5
  tension 331, 348
  weight 330
fractions 32, 64–5
free fall motion under gravity 319–23
friction 331, 348
  dynamic friction 345–8
functions 40, 52
  cubic functions 42–3
  derived function 72
  exponential functions 166–7, 173
  gradient function 72
  increasing and decreasing functions 78–9
  linear functions 40
  probability functions 234–7
  quadratic functions 41–2
  trigonometric functions 154–9

G

geometric series 110–11, 120
  convergence of series 113
  sum of a geometric series 110–13
  sum to infinity of geometric series 113–15
gradient of curved lines 70–2, 85
  gradient function 72
  gradients of tangents and normals 75–8
gradient of straight lines 59–61, 67
  collinear points 60–1
  parallel lines 60
  perpendicular lines 60
graphical interpretation of equations
  cubic equations 44
  intersections 44–6
  quadratic equations 43

graphs 40–3
  displacement-time graphs 302–5
  graphical interpretation of equations 43–6
  shape of the graph of $f(x) = a^x$ 166–7
  transformations of graphs 48–51, 52
  velocity-time graphs 305–10
  vertical line graphs 232
gravity 319–23

I

improper fractions 32
impulse 352, 353, 360
  impulsive forces 355–7
  units of impulse 353–5
indefinite integration 90–2
independent events 196–200, 213
  showing whether events are mutually
    exclusive or independent 206–9
indices (index) 10–13, 14
  base and index 10
  Law 1 10
  Law 2 10–11
  Law 3 13
  Law 4 11–12
inequalities 46, 52
  solving linear inequalities 46–7
  solving quadratic inequalities 47–8
infinite series 105–6
integration 90, 99
  definite integration 93–4
  finding area by definite integration 94–7
  finding area by indefinite integration 92–3
  finding the constant of integration 91–2
  integrating a sum or difference of
    functions 91
  trapezium rule 97–9, 99
intersections 44–6, 52, 66–7
irrational numbers 7

L

like terms 2
limiting value 178
linear combinations 251
linear equations 27–9
linear functions 40
linear inequalities 46–7
linear momentum 357–60, 360
lines joining two points 56–9, 67
  Cartesian coordinates 56
  distance between two points 56–7
  midpoint of the line joining two given
    points 57–9
logarithms 168, 173
  equations containing logarithms or x as a
    power 172–3
  evaluating logarithms 169
  laws of logarithms 170–2, 173
  using a calculator 169
long-term relative frequency 178, 212

M

maximum points 80–2
  maximum value 80
mean 222, 260
  Bernoulli distribution 269–70

binomial distribution 285–9
  mean of a simple function of X 242–5
  mean of X 237
measures of central tendency 222
  mean 222, 260
  median 223, 261
  mode 222, 260
measures of spread 224
  interquartile range 224–5
  range 224, 261
median 223, 261
minimum points 80–2
  minimum value 80
mode 222, 260
momentum 352–3, 360
  conservation of linear momentum
    357–60, 360
  units of momentum 353–5
motion in a straight line 314
  equations of motion with constant
    acceleration 314–19, 326
  free fall motion under gravity 319–23
  Newton's first law of motion 333–5, 348
  Newton's second law of motion 335–40, 348
  Newton's third law of motion 341–2, 348
  variable acceleration 323–6, 326
multiples 256
  differentiating constants and multiples
    of x 74
multiplication law of probability 192–6
  independent events 196–200, 213
  two events 193–4, 213
mutually exclusive events 187–90, 213
  addition law 187
  showing whether events are mutually
    exclusive or independent 206–9

N

Newton's first law of motion 333–5, 348
  resultant force 333–5
Newton's second law of motion 335–40, 348
  weight 338–40
Newton's third law of motion 341–2, 348
normal reaction 331, 348
normals 71, 85
  equations of normals 77–8
  gradients of normals 75–6

O

one-way stretches 48, 50
ordinates 92
outcomes 178
  equally likely outcomes 178, 212

P

parabolas 41
parallel lines 60
Pascal's Triangle 116, 274
period 155
periodic functions 155
perpendicular lines 60
polynomials 16–19
  cubic polynomials 34
  division of a polynomial by $x - a$ 32–3
  factorising quadratic polynomials 16–19

possibility space 178
powers 10
    equations containing logarithms or x as a
        power 172–3
probability 176–7, 212
    addition law of probability 183–5, 213
    combined events 182–3
    conditional events 191–2
    definitions and notation 178–82
    equally likely outcomes 178, 212
    further applications 209–12
    multiplication law of probability
        192–200
    mutually exclusive events 187–90, 213
    practical probability 177–8
    probability distribution 232
    probability functions 234–7
    probability tables 185–7, 213
    showing whether events are mutually
        exclusive or independent 206–9
    sum of probabilities 232–4
    theoretical probability 178
    tree diagrams 200–6, 214
products 74–5
proper fractions 32

Q

quadratic equations 19, 29
    discriminant and the nature of roots
        25–7
    formula for solving a quadratic equation
        23–4
    graphical interpretation 43
    losing a solution 21
    properties of the roots of a quadratic
        equation 24–7, 29
    rearranging the equation 20–1
    solution by completing the square 22–3
    solution by factorising 20
    solution of one linear and one quadratic
        equation 27–9
quadratic functions 41–2
quadratic inequalities 47–8
quadratic polynomials 16
    common factors 18–19
    factorising quadratic polynomials 16–18
    harder factorising 18
quartiles 224–5, 261
quotients 32, 33

R

radians 148–50, 162
range 224, 261
    interquartile range 224–5
rational numbers 7
real numbers 25
recurrence relations 103
reflections 48, 49–51
relative frequency tables 185, 213
remainders 32, 33
    remainder theorem 33–4, 36
resultant force 333–5
roots 6
    properties of the roots of a quadratic
        equation 24–7, 29

S

sampling without replacement 193
sectors 152–3, 162
segments 152
sequences 102, 120
    behaviour of u_n as $n \to \infty$ 103–4
    convergent sequences 103–4
    defining a sequence 102–3
series 102, 120
    arithmetic series 107–10, 120
    finite series 105–6
    geometric series 110–15, 120
    infinite series 105–6
simultaneous equations 27
    solution of one linear and one quadratic
        equation 27–9
sine rule 137–8, 145
    ambiguous case 138–40
sine waves 155
sines 134, 145
    sine function 154–6, 162
    sine of an obtuse angle 135–6
smooth contact 331
speed 300–1, 310
square roots 6
    cube roots 6
    other roots 6
squares 4
    completing the square 22–3
    difference of two squares 4
standard deviation 225–31, 261
    mean, variance and standard deviation of
        a binomial distribution 285–9, 293
    using a calculator in statistical mode 227–8
standard deviation of X 245, 262
    practical approach 245
    theoretical approach 245–7
    variance of a simple function of
        X 247–51, 262
stationary points 79–80, 85
    investigating the nature of stationary
        points 80–2
    stationary value 79
straight lines 59–61, 67
    equation of a line passing through $(x_1, y_1$
        and x_2, y_2) 64–6
    equation of a line with gradient m and
        passing through the point (x_1, y_1) 64
    finding the equation of a straight line 64–6
    general form of the equation of a line 62–3
    intersections 44–6, 52, 66–7
summary data 226
sums 256
    integrating a sum or difference of
        functions 91
    sum of a geometric series 110–13
    sum of an arithmetic series 108–10
    sum of independent observations of a
        discrete random variable 253–7
    sum of probabilities 232–4
    sum or difference of two independent
        random variables 251–2, 262
    sum to infinity of geometric series 113–15
surds 7, 14
    multiplying surds 7–8

rationalising a denominator 8–10
    simplifying surds 7
symmetry 239

T

tangents 70, 85
    equations of tangents 77–8
    gradients of tangents 75–6
    tangent function 158–9, 162
    tangents of an angle 134, 145
    tangents to circles 128–30
tension 331, 348
terms 2
transformations of graphs 48–51, 52
    reflections 49–51
    translations 48–9
translations 48–9
    effect on equation of a circle 126–7
trapezium rule 97–9, 99
tree diagrams 200–6, 214
triangles 134, 145
    area of a triangle 144–5
    general triangle calculations 143
    sine rule and cosine rule 137–43
trigonometric equations 159–61
    equations involving compound angles
        161–2
trigonometric functions 154–9
    cosine function 156–8, 162
    definition of trigonometric ratios 154
    general definition of an angle 154
    sine function 154–6, 162
    tangent function 158–9, 162
trigonometric ratios 154
trigonometric ratios of acute angles
    134–5, 145
    exact values 135
trigonometric ratios of obtuse angles
    135–7, 154
turning points 80

U

unlike terms 2

V

variables 222
variance 261
    Bernoulli distribution 269–70
    mean, variance and standard deviation of
        a binomial distribution 285–9, 293
    variation and standard deviation of
        X 245, 262
    practical approach 245
    theoretical approach 245–7
    variance of a simple function of
        X 247–51, 262
velocity 300–1, 310
    acceleration 301–2
    velocity-time graphs 305–10
vertical line graphs 232
vertices (vertex) 41

W

weight 330
    Newton's second law of motion 338–40